ROUTLEDGE LIBRARY EDITIONS:
ETHNOSCAPES

Volume 8

ENVIRONMENTAL PSYCHOLOGY IN EUROPE

ENVIRONMENTAL PSYCHOLOGY IN EUROPE

From Architectural Psychology to Green Psychology

ENRIC POL

LONDON AND NEW YORK

First published in 1993 by Avebury (Ashgate Publishing Limited)

This edition first published in 2025
by Routledge
4 Park Square, Milton Park, Abingdon, Oxon OX14 4RN

and by Routledge
605 Third Avenue, New York, NY 10158

Routledge is an imprint of the Taylor & Francis Group, an informa business

© 1993 E. Pol

All rights reserved. No part of this book may be reprinted or reproduced or utilised in any form or by any electronic, mechanical, or other means, now known or hereafter invented, including photocopying and recording, or in any information storage or retrieval system, without permission in writing from the publishers.

Trademark notice: Product or corporate names may be trademarks or registered trademarks, and are used only for identification and explanation without intent to infringe.

British Library Cataloguing in Publication Data
A catalogue record for this book is available from the British Library

ISBN: 978-1-032-86590-4 (Set)
ISBN: 978-1-032-83324-8 (Volume 8) (hbk)
ISBN: 978-1-032-83329-3 (Volume 8) (pbk)
ISBN: 978-1-003-50880-9 (Volume 8) (ebk)

DOI: 10.4324/9781003508809

Publisher's Note
The publisher has gone to great lengths to ensure the quality of this reprint but points out that some imperfections in the original copies may be apparent.

Disclaimer
The publisher has made every effort to trace copyright holders and would welcome correspondence from those they have been unable to trace.

New Series Introduction to
RLE: Ethnoscapes

The neologism *Ethnoscapes*[1] was created by David Canter and David Stea in 1987 when they happened both to be in Yogjakarta at the same time. They wanted a term to cover the rapidly emerging multidisciplinary field of research into many aspects of how individuals, groups and cultures interact and transact with their surroundings. It was derived as follows:

Ethno (combining form) indicating race, people or culture.

Scape (suffix-forming nouns) indicating a scene or view of something.

Ethnoscapes (plural noun) Scholarly and/or scientific explorations of the relationships people and their activities, have with the places they create and/or inhabit; historical, psychological, anthropological, sociological, and related disciplines that study the experiences of places, attitudes towards them, or the processes of shaping, managing, or designing them. The term was subsequently used to provide an umbrella for a series of books. These cover topics that are so multidisciplinary that they do not sit comfortably in any of the constrained silos of academic and scholarly research. As indicated on the opening page of the first book in the series, many disciplines "have developed marauding sub-groups who move freely across each others' borders, carrying ideas almost like contraband, without declaring that they have crossed any disciplinary boundaries."

They include domains labelled as Behavioural or Perceptual Geography, Environmental/Architectural Psychology, Urban History, Social Ecology, Behavioural Archaeology, Urban Planning, Behavioural Architecture, and Landscape Architecture. There are also many other areas of research and practice that, whilst not being overtly psychological, social, or cultural, do explore and act on the built and natural environment in a way that recognises the importance of the human transactions with those settings. These professions include interior and product design, comparative linguistics, and even aspects of criminology and mental health providers.

Like all such implicit and explicit transactions between different domains, a community of interest and support has emerged in which those who cross the boundaries often find they have more in common with other transgressors than with their mother disciplines. This has given rise to common means and forms of communication, with a shared understanding of the issues and approaches that are of value. Although, of course, these are not always understood in the same way by all those involved,

The *Ethnoscapes* series of books provides a forum for these multifarious, cross-disciplinary, determinedly international, studies and practices. Each of the books takes on board one or more of the environmental challenges that that individuals, societies and cultures are facing. Emphasising a social perspective, rather than the dominant 'hard' science viewpoints embedded in physical, geological and climate changes.

It may now be regarded as rather prescient that it was over three decades ago that the need and importance was recognised of bringing together the many strands of environmental social research and practice. But there is no doubt that there were academics and professionals exploring Ethnoscape topics, going back to the 1960s, often in isolation and with little recognition, that are today front-page, and podcast, news. The challenges in the environmental social sciences that Ethnoscapes explores are just as pertinent now as they were when initially identified.

The series, in essence, deals with four challenges the environmental social sciences embrace.

1. Addressing "the awareness of governments and public alike of the problems of environmental degradation and pollution."

This includes the challenge of providing acceptable housing and related environmental conditions that also encompassed the support for environmental and related cultural heritage. It also requires detailed consideration of the assessment and evaluation of designs and design proposals as well as background research on policy related issues.

2. Developing ways of conceptualising human interactions with the physical surroundings.

This may seem somewhat abstract but has practical implications. The dominant view that people are passively controlled by their surroundings supports a paternalistic, management of what it is assumed people need. That ignores the active way in which people make sense of their environment, drawing on cultural and historical influences. This recognises the importance of user participation in decisions about built and natural settings. That, in turn, requires a much richer understanding of how people interact with where they are or want to be.

3. A much wider range of ways of exploring people's transactions with the environment is needed to contribute to policy and practice as well as developing richer insights into human experiences.

The stock in trade of surveys, or the inevitably artificial laboratory-based experiments, whilst of value for some explorations, need to be augmented by methodologies that enrich an understanding of what the experiences are of being in, acting on, and developing places. They need to connect not just with the endeavours of individuals but also with how cultures and societies express these transactions.

4. Finding ways to enable practitioners and researchers to express their own encounters with the contexts they are influencing or studying.

Much of the research that is carried out in what are curiously called 'Ivory Towers', even when it is studying the big wide world, allows the pretence of distancing from the direct experiences of the issues being studied. Yet the challenges of moving across disciplinary boundaries are as much personal challenges of finding new ways of thinking, communicating, and acting, as an academic demand to develop more effective intellectual systems. The Ethnoscapes series recognises the value of exploring these challenges by hosting a variety of formats. Many of these go beyond the staid and limited formulations that academic discourse assumes to be the norms.

The Ethnoscapes series brings together a vibrant mix of cutting-edge explorations, from all over the world, of human transactions with the built and natural environments. This includes, for example, consideration of vernacular architecture that contrasts with the architecture and urbanism of the colonial enterprise, the meaning of home, aesthetics, well-being and health, and consideration of how environmental psychology has become 'green'. All of these topics, and more, provide an exciting basis for dealing with current challenges in the environmental social sciences.

Note

[1] Not to be confused by the term *Ethnoscape* later concocted by Arun Appadurai in 1990, to refer to **human migration**, the flow of people across boundaries. This includes migrants, refugees, exiles, and tourists, among other moving individuals and groups, all of whom appear to affect the politics of (and between) nations to a considerable degree. Ignorant of the lexicographical origins of the term 'scape' he rather confusingly added it to many ideas of flow, such as the flow of technology – technoscapes and the flow of ideas ideoscapes. Appadurai, A. (1990). "Disjuncture and difference in the global cultural economy." *Theory, Culture and Society* 7(2–3): 295–310.

Routledge Library Editions: Ethnoscapes

1. *Environmental Perspectives* David Canter, Martin Krampen & David Stea (Eds) (1988)
ISBN 978-1-032-81616-6

2. *Environmental Policy, Assessment, and Communication* David Canter, Martin Krampen &
David Stea (Eds) (1988) ISBN 978-1-032-81635-7

3. *New Directions in Environmental Participation* David Canter, Martin Krampen & David
Stea (Eds) (1988) ISBN 978-1-032-81646-3

4. *Vernacular Architecture: Paradigms of Environmental Response* Mete Turan (Ed.) (1990)
ISBN 978-1-032-82023-1

5. *Forms of Dominance: On the Architecture and Urbanism of the Colonial Enterprise* Nezar
AlSayyad (Ed.) (1992) ISBN 978-1-032-84164-9

6. *The Meaning and Use of Housing: International Perspectives, Approaches and Their
Applications* Ernesto G. Arias (Ed.) (1993) ISBN 978-1-032-84781-8

7. *Placemaking: Production of Built Environment in Two Cultures* David Stea & Mete Turan
(1993) ISBN 978-1-032-86434-1

8. *Environmental Psychology in Europe: From Architectural Psychology to Green
Psychology* Enric Pol (1993) ISBN 978-1-032-83324-8

9. *Housing: Design, Research, Education* Marjorie Bulos & Necdet Teymur (Eds) (1993)
ISBN 978-1-032-86388-7

10. *Architecture, Ritual Practice and Co-determination in the Swedish Office* Dennis Dox-
tater (1994) ISBN 978-1-032-81774-3

11. *On the Aesthetics of Architecture: A Psychological Approach to the Structure
and the Order of Perceived Architectural Space* Ralf Weber (1995)
ISBN 978-1-032-82034-7

12. *The Home: Words, Interpretations, Meanings and Environments*
by David N. Benjamin (Ed.) (1995) ISBN 978-1-032-86411-2

13. *Tradition, Location and Community: Place-making and Development* Adenrele Awotona
& Necdet Teymur (Eds) (1997) ISBN 978-1-032-84608-8

14. *Aesthetics, Well-being and Health: Essays within Architecture and Environmental Aesthetics* Birgit Cold (Ed.) (2001) ISBN 978-1-032-86577-5

Other Ethnoscapes series titles also available:

Integrating Programming, Evaluation and Participation in Design: A Theory Z Approach
Henry Sanoff (1992) HBK 978-1-138-20338-9; EBK 978-1-315-47173-0;
PBK 978-1-138-20339-6

Directions in Person-Environment Research and Practice Jack Nasar & Wolfgang
F. E. Preiser (Eds) (1999) HBK 978-1-138-68674-8; EBK 978-1-315-54255-3;
PBK 978-1-138-68677-9

Psychological Theories for Environmental Issues Mirilia Bonnes, Terence Lee & Marino
Bonaiuto (Eds) (2003) HBK 978-0-75461-888-1; EBK 978-1-315-24572-0;
PBK 978-1-138-27742-7

Housing Space and Quality of Life David L. Uzzell, Ricardo Garcia Mira, J. Eulogio Real &
Joe Romay (Eds) (2005) HBK 978-0-81538-952-1; EBK 978-1-351-15636-3; PBK 978-1-
138-35596-5

Doing Things with Things: The Design and Use of Everyday Objects Alan Costall & Ole
Dreier (Eds) (2006) HBK 978-0-75464-656-3; EBK 978-1-315-57792-0;
PBK 978-1-138-25314-8

Rethinking the Meaning of Place: Conceiving Place in Architecture-Urbanism Lineu Castello
(2010) HBK 978-0-75467-814-4; EBK 978-1-315-60616-3; PBK 978-1-138-25745-0

ENVIRONMENTAL PSYCHOLOGY IN EUROPE

"The serpent replied:
You will not die,
... you will be as the angels,
who know good and evil.
... seeing that the tree was good to eat
... that it was tempting to acquire knowledge,
he took its fruit and he ate it".

(Genesis 3:4,5,6)

This was to be the first ecological crime...

Environmental Psychology in Europe

From architectural psychology to green psychology

ENRIC POL

Department of Social Psychology
University of Barcelona

Avebury

Aldershot · Brookfield USA · Hong Kong · Singapore · Sydney

© E. Pol 1993

All rights reserved. No part of this publication may be reproduced, stored in a retrieval system, or transmitted in any form or by any means, electronic, mechanical, photocopying, recording or otherwise without the prior permission of the publisher.

Published by
Avebury
Ashgate Publishing Limited
Gower House
Croft Road
Aldershot
Hants GU11 3HR
England

Ashgate Publishing Company
Old Post Road
Brookfield
Vermont 05036
USA

British Library Cataloguing in Publication Data

Pol, Enric
    Environmental Psychology in Europe: From
    Architectural Psychology to Green
    Psychology. — 2Rev.ed. — (Ethhnoscapes:
    Current Challenges in the Environmental
    Social Sciences)
    I. Title II. Robson, Colin III. Series
    155.9

Revised and updated version of
Psicologia Ambiental en Europa. Analisis Sociohistórico
published by Editorial Anthropos©, Barcelona.

Translated by:
Colin Robson
Tau Traduccions
Vilamarí 58
08015 Barcelona

ISBN 1 85628 528 6

Printed and Bound in Great Britain by
Athenaeum Press Ltd, Newcastle upon Tyne.

Contents

Figures and tables		vii
Once upon a time...		ix
Preface of the English edition		xi
Preface of the Spanish edition		xiii
Part 1	GENERAL APPROACHES	
1	The Science as a social organisation: objectives and processes	1
2	The definition of a space for Environmental Psychology	14
PART 2	THE ORIGIN AND DEVELOPMENT OF ENVIRONMENTAL PSYCHOLOGY IN THE PRINCIPAL EUROPEAN CULTURAL AREAS	23
3	Environmental Psychology in Great Britain	26
4	Environmental Psychology in Sweden and the Scandinavian Countries	41
5	Environmental Psychology in France and the Francophone Area	48
6	Environmental Psychology in the Germanic Area	60
7	Environmental Psychology in the ex-USSR	68
8	Environmental Psychology in Italy	74
9	Environmental Psychology in Spain	80
Epilogue to part two: The Netherlands, Portugal, Greece and Turkey		91

| PART 3 | EUROPE AS A WHOLE | 95 |

| 10 | Environmental Psychology:
Pluridisciplinary Science versus Applied Psychology | 99 |

| 11 | Who is who: the most notable authors and nuclei | 116 |

| 12 | The Invisible Colleges | 139 |

| 13 | Prospective and general conclusion | 156 |

POSTSCRIPT: 1984-1992   161

| 14 | From Architectural Psychology to Green Environmental Psychology | 163 |

Epilogue   189

APPENDIX

Statistical appendix   191

Bibliographical appendix: Historical and Functional Classics   223

Bibliographical references   235

Thirty-one Years Later...
Preface of the Reissue

Thirty-one years later... it's really an honor and a pleasure to have received the invitation to reprint my book on the history of Environmental Psychology. An honor, accompanied by a touch of sadness...

An honor, in terms of the recognition of the work done and the small contribution that all this documentary and personal research meant, during the 80s of the last century... ("last century" sounds very strong... !!!).

But also a sadness since, despite the time that has passed, I have no evidence of significant contributions that allow us to expand the understanding of the origin and evolution of a "disciplinary" area (or "sub-disciplinary", depending on who you listen to). In addition ("apparently", rather than "substantively") main object has been changing the focus, for not to mention the object of study.

Apparently, if we look at the dominant name in Europe during the 70s of the 20th century, that of "Architectural Psychology", it has little to do with the dominant names (or adjectives) in the first quarter of the 21st century, such as those of "Psychology of (or for) Climate Change", "Psychology of Sustainability," among many others. The question then is obvious: are we talking about the same thing? Or do we push ourselves to create our 'own' study spaces, as a defense of particular (and low-level) interests?

Re-reading the 1993 text, I realize that in a very high percentage, it is still in full force. That in fact, in the documentation analyzed to build the history of our discipline (where the texts of Lenelis Kruse and Carl Graumann played - and continue to play – a fundamental role, in addition to those of Gerhard Kaminski and some more recent ones such as those of Hartmut Günther and collaborators), already points to the common ground between what seemed to be the main object of the Architectural Psychology and the object of the Psychology of Climate Change.

The Architectural Psychology wanted to create comfortable spaces to facilitate people's well-being. Psychology oriented to Climate Change – dominant at present –, wants to face/adapt to the climate emergency, changing habits, individual and social behaviors, ways of doing things, ways of consuming, production technologies... etc. In other words, everything that facilitates, harms or simply affects the well-being of people, in their *'behavior setting'*, that is, their *'ecosystem'*.

And here we find one of the several oldest links that, in an apparently paradoxical way, put us in front of the most modern interpretation of the object of study. In the transition from the 19th century to the 20th century, the Jakob von Uexküll's peculiar way of understanding the *Umwelt* laid the foundations of the modern concept of *"ecosystem"*, including the symbolic dimension of the place, which is different for each species. The *Behavior Setting* of Roger Barker (author who calls his research *Ecological Psychology*, but he only works in architectural environments), we can consider it to be a consequence, although possibly without awareness of this relationship.

Here we are faced with the need to look in a different way at the dominant globalization, the overconsumption, the overproduction of emissions that it entails. We have also to question the difficulty (for some uselessness or even inadvisability due to its negative effects) of generating universal laws of environmental behavior in people and social groups, when the characteristics of each ecosystem are different, and require behaviors adjusted to the needs and possibilities of these ecosystems.

In fact, if we dare to look further back, we should see how the cultures, the cultural differentiations of the world and humanity, have to do with the effects of the environmental peculiarities of each place, of each ecosystem. But we forget this too often, and we end up with a simplified, reductionist (but apparently rational) idea of our history, our evolution, and why we are here, with the problems of our present and future.

As the Catalan writer, Josep Maria Espinàs ("only" literate, neither scientific nor pseudoscientific), said coincidentally also in 1993, *The Ecologism is an Egoism*. It is the fear – he said – of losing what has brought us to the ways of life of the present, the ones we are used to and the ones we don't want to give up.

But all this is waiting to finish showing, to finish arguing, and to see if we still have time – altruistically or selfishly – to survive the present threats with some dignity, but always bearing in mind that it is not only an environmental resources problem. It is, inevitably, a problem of the model of society, of economic interests (which dominate and manipulate ideologies and political interests). And we have to be careful: emphasizing only the "scientific" dimension implies – or sometimes hides – strategies of control and power... But this would take us to dimensions that exceed what I wanted – and I still want – to contribute with this text on the history of Environmental Psychology.

Enric Pol
Barcelona, April 2024

Figures and tables

Figures and tables in the text

Table	1.1	Organisational elements of the science and their indicators	8
Figure	1.1	Synopsis of work procedure	12
Table	10.1	Authors with the function of classics and to whom the European 'eminents' recognize their dept	108
Table	10.2	Classic historical authors to cultural areas	109
Table	10.3	Classic functional authors to cultural areas	111
Table	10.4	Functional Classics	113
Table	11.1	The most productive authors: number of works and conferences where they were presented	118
Table	11.2	The most productive authors in each cultural area	121
Figure	11.0	The most productive nuclei of collaboration. Code	123
Figure	11.1	The most productive nuclei of collaboration in G.B.-1	123
Figure	11.2	The most productive nuclei of collaboration in G.B.-2	124
Figure	11.3	The most productive nuclei of collaboration in Sweden -1	125
Figure	11.4	The most productive nuclei of collaboration in Sweden -2	126
Figure	11.5	The most productive nuclei of collaboration. France & Switzerland	127
Figure	11.6	The most productive nuclei of collaboration. France & Italy	128
Figure	11.7	The most productive nuclei of collaboration. New Zealand	129
Figure	11.8	The most productive nuclei of collaboration. USA	130
Figure	11.9	The most productive nuclei of collaboration in Spain	131
Table	11.3	The most visible authors with & without self-quotation. Percentage of self-quotation and country of origin	133
Table	11.4	Impact of the 10 most productive authors in the conferences	135
Table	11.5	The most productive authors arranged in order of quotes received	136
Table	11.6	Outstanding authors according to the criteria of eminence	137
Figure	12.1	Institutional collaboration in the IAPS Board, Conference Organization and invited lecturers	142
Figure	12.2	Individual relationship network through institutional participation in the IAPS activities	144
Table	12.1	Intellectual links of outstanding authors in the different cultural areas	147

vii

Figure 12.3	Synthesis of quotations, co-quotations and mutual influences among outstanding authors in the Conferences	150
Figure 12.4	Invisible Colleges in European Environmental Psychology that express themselves through the IAPS Conferences	152
Figure 14.1	Conceptualisation of environment and social roles in environmental intervention	179

Tables and figures in the statistical appendix

Table 1	Commitees composition, organization department, invited lecturers, sponsors and publishers.	193
Table 2	Synthesis of data and index of active participation and collaboration in the nine conferences.	195
Figure 1	Growth of number of authors	196
Figure 2	Growth of number of signatures	196
Figure 3	Growth of number of works	196
Figure 4	Growth of number of attendents	196
Table 3	Frecuency of the professions (globally)	197
Figure 5	Evolution of active participation of professions in percentages	198
Figure 6	Evolution of active participation of professions in direct value	198
Table 4	Evolution of the professional composition in Great Britain	199
Table 5	Professional evolution of the participation in USA	199
Table 6	Professional evolution in Germany	200
Table 7	Professional evolution in the Francophone area	200
Table 8	Professional evolution in Sweden	201
Table 9	Evolution of the institutional links	201
Figure 7	Evolution of institutional links in direct value	202
Figure 8	Evolution of institutional links in percentages	202
Table 10	The institutional link of professions	203
Table 11	Author's distribution and signatures according to production	203
Table 12	Author's distribution and signature-works (globally)	204
Table 13	Works by countries	205
Table 14	Résumé of the most dealt-with themes per conference	206
Table 15a	Evolution of the perspectives of analysis	207
Table 15b	Evolution of the types of environment analysed	208
Table 16	Thematic interest in the most productive countries (approaches)	209
Table 17	Analysed environment in the most productive countries	210
Table 18	Frecuency of subjects in each conference	211
Figure 9	Bradford Areas applied to the quotations	216
Table 19	Emited references from each conference to the other	217
Table 20	Received quotations for each conference from the other	217
Table 21	The most visible authors in the SSCI	217
Table 22	Summary of the most visible authors in each cultural area (MVI)	218
Table 23	Summary of authors who receive three quotations or more in two or more cultural areas	219
Table 24	Received quotations from the 13 journals with more impact per conference	220
Table 25	Year of publication of articles in the most quoted journals	221

Once upon a time...

It is a cold night on the last day of February 1969. The fire is alight and warms a typical room of a British club, which is filling with people by the moment. The compasses of a suite by Britten can be heard in the background.

A skinny young man with a chestnut beard, visibly nervous, attentively greets the recent arrivals, calling them by their full name and finally convincing them of the importance of their presence. The conversation flows, warm and polite. There are very few introductions. Almost everyone knows each other, in spite of the time that has passed since they last met, perhaps not since '63 or '67.

Outside it is raining, as it almost always does. In Gareloch, by the slopes of the mountains of Scotland, there is a swell as if it were learning to be the sea. In Dalandhui, the University of Strathclyde's Country club, a conversation is about to begin that has not yet concluded.

Over fifteen years, against a different backdrop, with Debussy, Mozart, Ravel or Mompou in the background, in shorts or dinner jackets, between Esperanto and Babel, the scene has repeated itself twelve times in Europe, marking the rhythm of the development of that branch of science which was baptised with the waters of Gareloch as the Psychology of Architecture, and in its successive marriages, following the northern European tradition, has changed its surname, calling itself "Environmental", "Ecological", or taking on airs of sophistication as it has become polygamous, adopting the suggestive and rakish name of "The study of Man and his Physical surroundings".

Preface of the English edition

In this book I try to explain the nature of Environmental Psychology, how it was born, how it has developed, what are its dominant subjects, its principal actors and what is its present state in Europe.

I start from the consideration that science, and therefore environmental psychology as well, is a social construction. Sociology of Science has been demonstrating this for many years, as I explain in the first chapter. So science, as a social construction, requires a clearly defined object of study, with an internal logic. But this is not enough. To understand its development it is essential to understand the contexts that have favoured its emergence and the personalities that drive it.

Without the authors who characterise it -people with all the virtues and miseries of the human race-, with scientific motivations but also with the desire for personal fame, for individual leadership in society or in more or less defined scientific circles, we could not understand some of the changes, orientations, reorientations, theoretical discrepancies justified by logic and coherence, or differentiations more related to personal notoriety and leadership than to the internal logic of ideas.

In the same way, the groups, alliances, the scientific circles or even the "Invisible colleges" do not only answer to a logical coherence of science, but they are sensitive to social and institutional conditioners, to historic contexts, but also to relational, intellectual and affective factors, as is shown throughout the whole book, especially in the third part and the postscript.

Europe is a complex, plurilingual and pluricultural reality. It has a common history which is more a product of the construction from the interests of economic unity and the (recent) unifying effects of the mass media, of the forms of modern production and consumption, than of a pre-existent social reality. The regional, cultural, social and political wealth and diversity obliges us to analyse the large cultural and/or socio-political areas in a differentiated way, with not always similar philosophical backgrounds, which give slightly different products in environmental psychology, albeit with a strong unifying tendency in the last periods, as is shown in the second part.

This study was published for the first time in Spanish in 1988. When we considered the possibility of translating and publishing the book in English, I thought of the need to bring it up to date or at least to complete some of its parts. Some years had passed since the conclusion of the ground work which gave shape to the text. In fact, the quantitative analysis was finished in 1984. Therefore, to consider publishing it in 1993 required some kind of revision. Immediately, however, I realised the difficulty of the project and even the unsuitability of going through with it. The complexity of the process of historicisation undertaken would require me to completely reconsider all the statistical analyses of the bibliometrical part and to carry out an important reconstruction of the most classical sources. On the other side, the study in itself was still valid. On the other hand, for the turns that things had been taking within Environmental Psychology, it could be that it was not particularly appropriate nor representative to centre the analysis of the period 84-92 principally on the IAPS conferences, as I had done for the previous period. The thematic diversification, of nuclei of study and platforms of exchange or debate, and above all the new perspectives that seemed to be opening out with the changes in the dominant social philosophy in the world politics and geopolitics, required a more qualitative than quantitative appreciation, more evaluative than expositive, more prospective than historiographic.

The form of the postscript seemed to be the most adequate to me, for what it represented in respecting the analysis made at the time, and for the freedom it gave me to pickup another type of discourse and reflection. However, in the analysis of each cultural area I have incorporated a brief descriptive note of the changing tendencies which have come about in this period. I was able to converse again with almost all of the personalities that I had interviewed for the first version, which guaranteed a certain nexus of continuity and of collective endorsement of the evolutions detected. Evolutions of a theoretical and methodological nature, but also of the power groups, areas of influence and above all, of the object of study. We are witnessing an important step from "psychology of architecture" to what some call "green environmental psychology". This formula has given me the freedom to go into some themes and aspects that I had not considered in the first draft in Catalan (1986) nor in the first edition in Spanish (1988).

I do not want to finish this presentation of the English edition without expressing my gratitude to David Canter for making the publication of the book in English possible. Also to Perla Korosec-Serfaty, Lenelis Kruse, Rikart Küller, Mirilia Bonnes, Thomas Niit, Cleopatra Karaletsou, Sue Ann Lee, Jonathan Sime and David Canter for agreeing to converse on the subject during the 12th IAPS Conference in the summer of 1992, as well as the authors who have facilitated their own documentation and on their cultural areas like Tommy Gärling, Manuel Carreiras, Ybonne Bernard, Marion Segaud and specially Erminelda Mainardi Peron.

I would also like to express my thanks to my colleagues Lupicinio Iñiguez, Juan Muñoz, Pep Garcia-Bores, Esther Busch and Antonia Ferrer for their help in the preparation of this edition; to Colin Robson for his laborious, patient and polished translation; to Frederic Munné, Javier Serrano, Sergi Valera, Louis Sauer and Perla Korosec for reading the postscript and to give me interesting suggestions. To Joan Guardia, who as well as encouraging me in the work, lent me his house in the Empordà, a fantastic setting of Hellenic resonances, perfect for the necessary concentration for the reconsideration and rewriting of the study. To Emi (Emilia Moreno), as well as a colleague, my wife, who has not been sparing in her encouragement nor supportive efforts and work, nor in patience to endure my encloisterings. To my children, Gerard who is already grown and patient, and the one who will be born just when this book is published, but who has already suffered it.

ENRIC POL
Albons, Costa Brava.
December 1992

Preface of the Spanish edition

For two long decades, in a complex context that will be analysed throughout this book, a field of study has been defined, hinged between various disciplines, which is concerned with the relationships and mutual influences between the human being and his surroundings or environment (M-E). With strongly interdisciplinary origins, the emphasis placed on man as an individual and as a social animal has lead this issue to heel towards psychology. Different terms have been used to denominate it: Psychology of Architecture, Ecological Psychology, Ecopsychology, Psychology of the State, etc., the most widely accepted being that of Environmental Psychology, in spite of its polysemy on referring both to a specific centre of interest for psychology, and to the multidisciplinary field previously mentioned. This has brought about notable inexactitudes and confusions that question even its own existence.

In spite of this, in the period mentioned we can find a group of public manifestations that give credit to the reality of a community with certain levels of social and scientific organisation that confer a substantivity onto this field that goes beyond its immediately relevant definition. One of these most obvious manifestations is the celebration of a goodly number of meetings, conferences and congresses in the whole ambit of western culture, many of which have taken place in the different European cultural areas. In the beginning, with a low level of academic and institutional development and a relatively small volume of publications, the direct and little formalised communication that facilitates the meetings with scientific pretensions has marked a milestone in its development.

This field of study has concentrated our interest both on the applied aspects and in the epistemological and historical aspects in the last years. Via this line of sustained work the impact, -and the orientation of this impact-, of social, institutional, cultural and personal factors (especially the ambition for power) has attracted our attention, which in principle outside the internal logic of its scientific approaches, would strongly mark the direction of its development. This has lead us to enter the field known as sociology of science, adapting the so-called bibliometry as a methodological instrument for a part of this study, apart from other historiographic procedures.

In this study we propose to expose a series of qualitative and quantitative data on environmental psychology, concentrating our efforts on the analysis of its antecedents, its origins, the context of its evolution and development, and its social organisation as a scientific community, with its directive groups and its invisible colleges. All this enclosed in a well defined geographic framework, which at the same time is complex and plural from the point of view of its wealth in cultural traditions. This leads us to reflect on and question ourselves about whether it is possible to speak of a European Environmental Psychology, both for the possibility that an internally homogeneous product should emerge from such cultural diversity and for the possibility that it should have sufficient entity and identity within the global framework of reference of the so-called western world.

The book is arranged in three parts:

In the first, the theoretical objectives and frameworks from where they will be tackled are defined. From how we can understand the social organisation of the science and how some of its manifestations can be adapted as objective indicators of its reality, among them the specialised congresses. We end this section with a reference to the methodological processes used.

In the second part the origin and development of Environmental Psychology is revised within the various most productive European cultural areas in this field of discipline. From the information collected in the documents analysed, but also in the interviews with the most important active authors and the visits to the centres where they develop their activity, together with the data that springs from the analysis of the international conferences that have been celebrated in Europe, we have tried to draw up the panorama of each one of the cultural areas. Although the structure of the script for each of them seeks to approach a common standard for them all, in practice it has been marked by the quantity and quality of the information that we have been able to obtain on the zone.

In the third part we analyse the European reality as a whole, giving weight to the data of the analysis of the conferences. Through this we raise the question of the substantive or subsidiary nature of Environmental Psychology. We analyse the evolution of the names it has taken, the incidence of the general historic context and what are its recognised predecessors, its level of consolidation and impact in society. Through the analysis of the scientific production we will evaluate the weight of each cultural area, as well as who are the most outstanding authors and nuclei. Through the analysis of the quotations and co-quotations and collaboration in the joint signing of studies, we will endeavour to outline the intellectual connections and dependencies. Finally, the social organisation that it adopts and how the nuclei of power and Invisible colleges are constituted will be analysed. To these three sections a statistical appendix and a bibliography are added.

The original version took up more than twice the volume that we now present and was defended as a doctoral thesis. In this edition we have tried to synthesise to the maximum the empirical data, avoiding the incorporation in the text of more graphs, tables and illustrations than those strictly necessary for its understanding in a quick reading, so as not to lose the line of argument. For a greater penetration we have collected a selection of empirical and bibliographical facts in the appendixes, stating their references in the text. Probably, certain nuances will have been lost and the specialist reader or person who feels to be an actor in this same story would like to see certain information that we have not been able to include. Between exhaustivity and superficial simplicity we have wanted to choose a middle way that facilitates access to our conclusions for the uninitiated public, but that at the same time the reader interested in going in deeper can make his own elaborations from the empirical information that, of course, permit more readings than our own.

The doctoral thesis, written in Catalan and defended in the Faculty of Psychology of the University of Barcelona in March 1986, was directed by the professor of the History of Psychology in the same university, Dr. Antonio Caparrós. For its development, it counted on the inestimable advice of the team specialised in socio-historical studies through bibliometric processes of the University of Valencia, especially Francisco Tortosa, Luis Montoro and Josep Maria Peiró. I would like to express my gratitude to all of them, and to Professor Frederic Munné for his reading and critical observations of the manuscript, and to the whole tribunal who so generously evaluated the work. Equally, I would like to express my gratitude to all the university professors who are quoted in the text and who, subjects of this same report, agreed to converse at length with me and to provide me with originals and studies, some of them still unpublished, without which this study could not have been carried out.

ENRIC POL
Barcelona, July 1988.

Part 1
GENERAL APPROACHES

1 The science as a social organisation: Objectives and processes

1.1. Introduction

Within the framework of what has come to be called the sociology of science, we have sought to make our approximation to the evolution of Environmental Psychology (EP) in Europe with three basic sources of information: interviews with the central personalities of our story and the visits to the outstanding institutional centres that house them, the analysis of documents and the bibliometric analysis of the publications that have emerged from the international congresses in the tradition of the present day IAPS.

We have been interested in analysing how the power relationships that come about in the scientific community in which it is developed influence the evolution of a young scientific discipline (it is still being discussed whether the term domain can be applied to EP, but we will use it as being useful, although with the reservations necessary for its application), and how this is shown through the Invisible Colleges that are constituted, that in turn will be established from indicators already consolidated in this kind of study. Moreover, we analyse the relationship between the two disciplines that have showed themselves to be the dominant forces in this field: psychology and architecture.

Also, we have wished to see how the cultural substrata mark styles of development and differentiated theoretical perspectives, especially in the diverse linguistic ambits of Europe, and how one or another become dominant over the rest.

This is all centred on and adjusted to the European geographical ambit, conscious as we are of the difficulties and bias that are implied by isolating it from a wider framework in which the American weight is fundamental. However, this has allowed us to objectivise the theoretical influences and dependencies of the European whole. Also, working within a limited geographical framework has allowed us to approach the social and institutional reality with direct contact, as well as the selfsame protagonists of our story. This has furnished us with information never published previously in any form, of great value for the interpretation of the quantitative facts that emerge from the bibliometric analysis.

In an initial formulation, the objective of our study is to make an approximation - we do not intend to exhaust the subject, far from it - to the evolution of Environmental Psychology in Europe. We have sought to sketch out what is the actual state of this field of study which since the end of the fifties has been developing in our western cultural context and has as its first public appearance in Europe the conference celebrated in

Dalandhui, Scotland, in 1969, with the conscience of developing a new domain.

In this introductory chapter we are going to make a rough draft of the principal theoretical and instrumental references, as well as the sources used in our historiographical process, to finish by pinpointing operationally the objectives and their development processes.

1.2. Of science as a social organisation

As it has been sufficiently reported from the Sociology of Science, also called Science of Science (Bernal, 1964), it is not possible to interpret progress in a scientific field solely by considering a cumulation of advances and results of the investigations of isolated individuals. It is necessary to take the social interaction of the scientists and their communities into consideration as well. A profound relationship exists between the social and theoretical aspects (Crane, 1972; Carpintero and Peiró, 1980; Montoro, 1982). The direction that the development of science has taken requires a high degree of collaboration and exchange, a circumstance that has lead many authors to define modern science through one of its essential characteristics, being what Ziman (1968) defined as "a public knowledge". Today it is difficult to imagine the investigator as a personage isolated in his laboratory, without interacting and interchanging ideas with his colleagues in some way or another (Garfield, 1979; Montoro, 1982).

1.2.1. *Big science, Little science*

In this sub-chapter we will revise a series of notions and concepts that will serve us as a basis to tackle the study we propose. From the consideration of science as a social organisation, the use of bibliometry as a procedure, the spectacular growth of the science in our century, which has lead to the coining of the term "Big Science" as opposed to the old situation of "Little Science"; the sense, utility and limitations of the analysis of productivity and references as a way of x-raying the situation in a given period, and congresses as a communication channel, the study of which demonstrates not only a series of thematic and productive data, but also the relationships between the authors and the social structure that science takes on as an organisation.

In fact, it has been repeatedly affirmed: when science exists, it exists socially, publicly and in an organised way. This existence takes on multiple and diverse forms, from strongly formalised institutions to spurious associations or simply informal relationships between authors, without being given any kind of organisation. At the same time, they have different levels of structure. The professional associations, and especially the inter-professional associations on concrete subjects -this being the case of the IAPS (International Association for the Study of People and their Physical Surroundings)-, constitute an important kind of social organisation of the science at the same time as a scarcely formalised but fundamental channel in scientific intervention. Organisation and leadership maintain their unity within the scientific community (Merton, 1977) although, "At times this power structure has functioned to resist scientific revolutions" (Carpintero, 1980). So we can consider that the development of science is produced in concomitance with three factors (Peiró, 1980):

a) The theoretical approaches, which are developed according to their own internal logic.

b) The social and institutional determining factors (Peiró and Carpintero, 1980;

Crane, 1972; Merton, 1977).
c) The individual determinants (Cole and Cole, 1973).

These three factors will be the hubs that will inspire the development of our study. On the other hand, the enormous dimension taken on by science has brought Weinberg (1961) to define it as a "Great Science" and Price (1951) to formulate his theory of the exponential growth of science. According to this, science grows more rapidly than social phenomena, having a much smaller period of duplication than that of population. This can bring it to an almost infinite growth that borders on the absurd, if what is announced as a phenomenon of saturation that has not yet been reached does not come about. This fact has a series of consequences that cause repercussions in its social structure. We should highlight here that no author can have access to and dominate the whole of the information, resources, sources... Science, as a body of public knowledge (Ziman, 1968) cannot be realised except in communication and collaboration (Price, 1971; Garfield *et al.*, 1978; Garvey, 1979).

One of the facts that characterise present day science is, without doubt, the collaboration, the interaction between the selfsame scientists and groups, a circumstance that affects and contributes to its development (Peiró, 1981).

This collaboration has its maximum public expression in the joint signing of the studies, but it extends to an interchange that goes beyond the formal meetings. The interchange takes place in the gestation of investigation and in its development. More than ever, today we have at our disposal channels that permit communication between authors relatively easily, one of these being congresses. The study of the joint signatures in studies and of other signs of the existence of communication between particular authors will give us important information for the comprehension of the dynamic structure and functions of the scientific community in a given moment (Merton, 1977).

This structure of relationships, together with the consideration of other aspects related to exchange, has been denominated as "Invisible Colleges" by Price (1961). With this term, Price names the groups of scientists who, working in different places on similar themes, exchange information through diverse means other than printed literature, in particular *pre-prints* (Peiró, 1980). Peiró gathers the criticisms that Mullins (1968) made on this concept, in the sense that the relationships between scientists are very lax, diffuse and flexible, and therefore it is difficult to operationalise them for their study. On the other hand, Crane (1969) affirmed the convenience of substituting the term with "Social Circle", which in his opinion better reflects the kind of relationships that are established (Peiró, 1981: 54).

In any case, what is true is that a large quantity of studies have been developed on this theme, analysing the influence between invisible colleges, productivity, incidence in the distribution of investigation resources and their function in revolutionary periods of investigation. Indicators have been defined to study their different aspects like collaboration in jointly written articles, the master-disciple relationships, the influences (by acceptance or rejection) recognised through bibliographical quotations (Peiró, 1980).

The study of the references permits an approach to the relationships of influence, interaction, collaboration or convergence in the same sources of a group of authors. The custom of systematically and explicitly mentioning the previous works on which the arguments of a study are based originates around 1850, according to Peiró (1963), although in an explicit or implicit way we can say that it is possibly as old as knowledge itself.

For "references", seen from the perspective of who emits them, or "quotations", from the point of view of whom is the object, we understand them to be explicit mentions that are made of another author in a study, in a clear and systemic way, in footnotes or at the end of a study (Carpintero and Peiró, 1979). Evidently, references can have very different meanings: the recognition of an influence, the demonstration of knowledge of a field and in the manner of a "visiting card", contributing basic documentation, or on the contrary rejecting a study, discussing notions, concepts, ideas, or simply to "enounce" or divulge a study of another or oneself (Weinstock, 1971).

1.2.2. Conferences as a communication channel

We have already mentioned, in an indirect and collateral way, some of the functions of the congresses as communication channels. In this section we will try to recuperate them and make them clear through the ideas of the eminents. In his study of the International Congresses on Psychology, Montoro (1982) carries out an exhaustive compilation, which we have considered interesting to summarise here.

Conferences should be understood as an important channel of communication for modern science, both on formal and informal levels. Communication facilities have favoured -and are continuing to favour increasingly- informal contacts between investigators. This process culminates in personal contact, of which congresses are, but above all have been, important exponents. Ribot (1889) had already highlighted that while psychology had been a metaphysical construction, an individual study, psychologists did not need to meet and so they did not. The scientific orientation that it sought to take on from the end of the last century would, on the other hand, favour the growth of the need to meet, form groups and collaborate.

The international congresses, moreover, help to overcome the cultural and political barriers between the States (Angell, 1929; Claparéde, 1929). They also favour exchange and contact between the different tendencies and schools of thought (Richet, 1889) without seeking to provoke unity in doctrines (Mallet, 1978), but facilitating the resolution of controversies and eliminating misunderstandings (Michotte, 1959), facilitating models, not only of strictly scientific proposals, but also of social organisation.

Moreover, the psychological congresses have been useful as an element for making a formal balance in the situation of psychology in each moment, given that the latest results are reflected in them and that they reveal the principal interests of contemporary scientists, permitting the calibration of the maturity of its "scientific contexture" (Germain, 1948).

Bloch (1978) has pointed out their utility as a complementary channel for other media. The slowness of the editorial process, the saturation of magazines, the fact that there are really very few open and truly international magazines, that obviously do not in themselves contemplate the possibility of an open, agile and personal debate, mean that congresses play a very important role. Moreover, on their informal side, they allow old friends to meet again and to make new ones, redounding in possible posterior collaborations.

In spite of this important role in the development of the science, recognised as we have just seen by prestigious psychologists, there are few studies that have paid attention to the theme until quite recently. Apart from the aforementioned Montoro, it is necessary to highlight the study published by the APA (1968) en which the validity of international conferences was clearly demonstrated, both in aspects referring to formal and informal communication (Montoro, 1982).

1.2.3. The Environmental Psychology Congresses

The international meetings on psychology are neither a modern nor recent phenomena. It is necessary to go back to the last century, in 1889, to find the first documentary evidence

4

of the first congress in Paris, ten years after the foundation of the first Experimental Psychology laboratory by Wundt. This is not a casual event; various factors converged here, such as the expansion that psychology had undergone in the last part of the century, the need to find new channels of communication and to establish friendly links and maintain an open exchange between authors. Psychology began to be thoroughly accepted in universities, and academic programmes multiplied. Moreover, an important contextual event exists, which is that in other ambits of science this dynamic had been initiated successfully. Little by little, these meetings would settle in and new specialised branches would emerge and organise their own meetings.

The short life of Environmental Psychology has been marked out by a constant interchange between those concerned, in a long list of meetings that originated in the sixties. The logical low level of formalisation in its origins, the lack of specialised organs of expression, the social and technological determining factors and the intellectual and theoretical factors in a process of profound change that in psychology is the crisis of behaviourism, as well the social demand, give a vitality to this domain that finds its best and most immediate form of expression in the congresses.

The first meeting that we have a record of took place in the United States, in Salt Lake City, in a congress on "Architectural Psychology and Psychiatry" in 1959, but which some authors have placed in 1961 (the difference is probably due to the time that passed between the date of its celebration and publication). At the end of the sixties a strong dynamic of annual meetings was established, that would soon be constituted in the EDRA (Environmental Design Research Association). In Europe a similar number of smaller meetings were organised for the different cultural areas (see part II), that in 1970 would be crystallised into the so called First International Conference on the Psychology of Architecture in Kingston. Since then these meetings would be celebrated quite regularly by various of their cultural areas. In total, between 1969 and 1984 there would be eight international conferences (that will have been twelve by the time this English version is published), plus an uncountable number of small symposiums, meetings and conferences, constituting at the beginning of the eighties the IAPS (International Association for the Study of People and their Physical Surroundings).

In This study, we will concentrate on the eight international conferences, plus the British one celebrated in 1969, that while being of a limited convocation have an important transcendence and impact in their cultural ambit. So, we have sought to apply the techniques mentioned up to this point and explained later to the conferences that we have analysed (Dalandhui, 1969; Kingston, 1970; Lund, 1973; Surrey, 1974; Strasbourg, 1976; Louvain-la-Neuve, 1979; Surrey, 1979; Barcelona, 1982 and Berlin, 1984), with the intention of getting closer to the intellectual, social and organisational origin and evolution of EP in this geographical ambit.

It has been repeatedly affirmed that congresses, when they exceed certain dimensions, are not valid for theoretical debate, a circumstance which has lead some "eminents" not to participate except when occupying an important role in them (Moles, 1984; Pagès, 1984). Somebody has also said that the large congresses, so common these days, are something like a market or fair where merchandise, studies and authors are on offer, which are attended by many curious people, and where some close a commercial operation, being ideal places for climbing in the social ladder of the scientific community, whether it be in the Great International Community or in the same institution of origin. This is probably true. This is probably one of the reasons why an elevated percentage of the participants are little known authors, and the confirmed authors fail to attend if not under special conditions. But probably the contrary occurs too.

We do not wish to be too suspicious, but knowing the mechanisms of the social organisation of the science on the part of some groups, cases have arisen of the utilisation of the calling of congresses and meetings on specific themes (that have to exist, with genuine interest) as a mechanism for gaining "cohorts" or recognition by giving opportunities to new professionals, who on the other hand are in need of these opportunities. The exponential growth of the science, postulated by Price (1963), that has not yet reached the saturation point that he announced, and the same social dynamic that leads us to a more or less imperfect leisure society (Racionero, 1980) which requires the generation of activities that are not directly productive, strengthen and stimulate this phenomenon.

The conferences on Environmental Psychology have without a doubt covered both functions. However, due to their size, small in their beginnings and middling in the majority of the convocations, they have conserved to an important degree the function of exchange and theoretical discussion, as the controversy generated in 1979 demonstrates (see 3.7). Their analysis allows us to bring in new facts to evaluate what is expressed in the previous paragraph, the organisational configuration adopted by EP and the most relational aspects between the participants. Their scientific dimension and their condition as being one of the still few means of institutionalised communication in this field, permits us to analyse their documents, at least as a faithful reflection of what happens, both on the level of interests, productivity, professional profiles, collaboration and as a background, geographical distribution and collective identity.

1.3. **On the methodological instruments and of the sources**

1.3.1. *Bibliometry as an instrument*

Once we have defined this point of departure, next it is necessary to concentrate our efforts on how to tackle the analysis of the science as an organisation. Among the diverse historiographical, social and psychosocial methods, we have centred out attention on the bibliometric technique that allows us to deal objectively with a series of data susceptible to being taken as indicators of the aspects of the scientific domain previously mentioned for their posterior interpretation (Carpintero, 1980: 2), complemented with the analysis of the contents of the documents, the *in situ* study of institutions and direct dialogue with the "eminents" and/or key authors in the structure of EP.

Bibliometry has been defined as "the quantification of bibliographic information susceptible to being analysed" (Garfield, 1978: 180) with the application of mathematic and statistical methods (Huber, 1977). Its objectives can be summarized in two broad epigraphs (López Piñero, 1972):

1) The analysis of the size, growth and distribution of scientific bibliography.
2) The study of the social structure of the groups that produce it and use it.

The data base which we have worked on is made up of the publications that emerged from the nine international EP conferences that have been celebrated in Europe.*

The publications or acts of the congresses generally collect the whole of the studies presented, except in the three last callings analysed. In Surrey, 1979, the acts were not published, so we have worked with the repertory of abstracts. We have proceeded in the same way for Berlin, 1984, as at the moment of doing the computerisation of the

The congresses, conferences and symposia constitute a public manifestation of the structure of a scientific field, especially when it is recent and finds a communication channel in this kind of activity that agglutinates as many formal aspects as informal aspects.

The congresses, moreover, are a reflection of the concerns of the time and the community that calls them (Carpintero, 1979) and as Michotte (1938) said, their study constitutes a reflection which is a summery of the evolution of science that permits the elaboration of a rapid balance of the progresses made (Montero, 1982). And to a even greater measure if we consider that the number of studies published in the acts of EP is superior to that published in books and specialised magazines in Europe.

So the bibliometric procedure provides useful information, but it does not exhaust in itself the whole of the possibilities of socio-historic analysis in a scientific field. It provides quantified data that constitutes the base of an interpretation from other coordinates, which in this study we complement with another source of contrast: that of the contextual analysis in which they are produced and the vision of the protagonists themselves.

Carpintero (1981) gathers and synthesises the criticisms and accusations of limitations that have been brought against bibliometry, especially on the part of Endler and his collaborators (1975). He raises two problems to be taken into consideration, and which in our study are repeatedly revealed: the diffuse and polysemic meaning of some commonly used terms that hinder the task of ordering, classifying and recuperating the information; the fact that knowledge that is widely shared and taken for granted is rarely quoted. Otherwise, the ideas end up losing their personality and paternity.

In any case, bibliometry is a route -a good route- for obtaining data that is useful to us for the interpretation of existing relationships, for identifying relevant authors and works, the immediacy or obsolescence of the information used, at the same time as the organisation itself of the science (Garfield, 1979), as is shown in table 1.1.

1.3.2. *The contextual data of Environmental Psychology*

The bibliometric analysis of the EP conferences in Europe furnish valuable and rich information for the understanding of the development of this field of study, in its thematic aspects, of productivity and of analysis of its social structure. However, the study would be incomplete without an approximation to social, institutional and individual circumstances that, in accordance with the previous aspects, help us to understand the origin and evolution of the process followed. Moreover, the limitations that the selfsame bibliometric process implies, as we have seen, make it advisable to complement and contrast the resulting data with that proceeding from other sources (Carpintero, 1981).

Given the lack of similar studies in our field, except for some succinct revisions of the state of the question in a few countries that constitute the geographical area that we are

---

information the acts still had not been published. Moreover, in this case there are some notable differences between the studies described in the programme and in the abstracts. In Barcelona 1982, two monographic publications emerged from the congress, that only gather a small selection of the studies presented. On a formal level, and for coherence with the other congresses, we have worked basically with the publication. However, both in Berlin and in Barcelona we record in some of the tables the information of the programme as it is wider and more explanatory, being duly assertive in these cases.

TABLE 1.1. *Organisational elements of the science and their indicators*

Organisational characteristics	Scientific correlation	Representative Element	Indicator used
Aims	Discoveries	Areas of investigation	Dominant matters in the production
Specialisation	-Matters -Groups of investigation	Themes of investigation Invisible colleges and schools	Classification matters Co-quoting Collaboration Signatures
Authority and leadership	-dominant theories -Elite of authors	Paradigms Eminent Authors	Key concepts Indicators of eminence (visibility)
Members -participants -clients	Investigators, clients	Subjects and roles Institutes that support and gain benefit from science	Identific. variables Aid, financing Scientific policy of the government, industries
Training of members	Training Apprenticeship	Training plans Biographies...	Curriculum Masters-Disciples
Motivation of members	Rewards and *ethos* of the scientist	Prizes, salaries	Priorities in discoveries Systems of honours Economic reports
Communication and execution systems	Means of communication Investigation and evaluation institutions	Magazines, *pre-prints*, texts, meetings Academic departments	Editorial exigences, committees, requirements Indicators of institutions
Communicated Information	Publications (methods, theories, information)	Quoted Authors Concepts used	References Analysis of content
Group mentality	Scientific mentality	Attitudes, values The scientists habits	The psychology of the scientist

Carpintero, 1981

interested in, it has been necessary for us to investigate direct and indirect sources to elaborate this part of the work.

In this study we have tried to give an approximation to EP in the different European cultural zones, paying special attention to the contextual data, through the vision of the protagonists themselves. Historical and epistemological works are still very scarce, generally brief and of a very general nature. The European contribution is scarcely reflected, being very diluted in the whole dominated by the high American productivity. Few works of substance can be named, in their own right, in this justification. We have to keep on referring to basically the same group that we used as a base in previous works (Pol, 1979), fortunately with the inclusion of some colleagues from our country. It is worth saying that the majority of the introductions or works with divulgative pretensions make a brief historical reference, but do not go any further than mentioning platitudes. Those that do have the intention of contributing data, dates and reflections for the explanation of the history of our domain are scarce. It is necessary to continue to refer to the works of Craik (1970, 1973, 1977), Canter (1972, 1975), Craik and Canter (1981), Holahand (1982) and Stokols (1977,1978). We also have to consider the less widely divulged works, but of important profundity, of Graumann (1974, 1978), Kruse (1974), Graumann, Kruse and Lantermann (1985), and Kaminski (1976, 1978, 1983, 1984, 1985 and 1986). The works of Barker (1968), Pagès, Fourcade and Cafson (1974), Bagnara (1976) and Lee (1976) also contain some interesting reflections. More recently Lévy-Leboyer (1980) and Mikellides (1980) have been concerned with a certain historical reflection and Teymur (1982, 1984) with a more epistemological stance.

In Spain, comparatively, a notable interest has been awakened given the number of works that are concerned with the theme. From the pioneering daring of Llorens (1973), the brief but mordant reflection of Gómez de Benito (1976), the search for sources and the contextual analysis of Remesar (1981), Pol (1979, 1982, 1984) and some of the more academicist texts included in Jiménez-Burillo (1982), especially Pinillos and Carpintero, to the more recent works of Blanco (1983), Iñíguez (1983), Aragonés (1985), Hernández, Riba, Remesar (1985), the excellent epistemological reflection of Hernández (1985) and more recently those included in the compendium published by Jiménez-Burillo and Aragonés (1986).

To approach the reality of each area, from the double perspective that Caparrós (1982) considered as internal and external historical developments, and seeking to reunite the descriptive information that allows us to achieve a sufficient explicatory level without falling into the mere accumulation of dates, classifications, genealogies, scales... (Caparrós, 1980), we concentrate on the analysis of the origins, the roots, the social context, the institutional development and the thematic interests of the discipline, dealing with these aspects according to the level of information obtained about each area. This is information obtained basically from first hand documentation to which we have been able to have access, as well as our direct conversations with recognised "eminents" from the countries that we have considered as being fundamental, from an initial revision of the bibliometric information that we deal with in this study. This data, moreover, has the value of facilitating a contrast with the results of the bibliometric study, as Carpintero (1981) recommends.

We have dealt with the way in which each one reaches this domain, his experience of formation, the sources he considers essential, his evaluation of the social structure of this field and its institutionalisation process, as well as his intellectual links, the reflection that he makes on the incidence of the social context within which EP has developed and his evaluation of the future.

Except for some brief reports on this field in Italy (Perussia, 1983), the USSR (Niit, Kruusval and Heidmets, 1981) and Sweden (Gärling, 1982), and also in a way the volume published by Kaminski in 1976 for Germany and his article of 1978, at the moment of compiling and dealing with the basic information for this study (1985-1986), we have not find any other published work. We are aware, through the contacts we have been able to make, that a *Handbook of Environmental Psychology* edited by Stokols and Altman is in the process of elaboration (it certainly will already have been published by the time this study is published), that will collect different aspects of the theoretical and applied development of this field in some European countries. By courtesy of its authors we have been given access to the first essays of some of them, or we have been able to converse at length. This has been the case of R. Küller, of Sweden; L. Kruse and C. Graumann, of Germany; D. Jodelet, in France, and D. Canter and I. Donald in Great Britain.

However, in the cases where it has been possible and we have thought it necessary for the complexity of the situation and the level of development, we have complemented this information with a series of interviews with the most important authors who have been, and are the protagonists of this story. Apart from the authors mentioned in the previous paragraph, we have been able to visit and interview Terence Lee, Ian Cooper and Alan Edge in Great Britain. We had an appointment arranged with Alan Lipman but on arriving in Cardiff we were confronted with the disagreeable surprise of his serious illness that had become apparent the day before. We have had to be satisfied with contact by letter. In any case, it was possible for us to converse at length with his assiduous collaborator Howard Harris.

In the francophone area we have been able to converse on repeated occasions with Jules G. Simon, of Louvain-la-Neuve, Belgium; Perla Korosec-Serfaty, of Strasbourg, and also with Abraham Moles, Robert Pagès and Gisella Fourcade, Claude Lévy-Leboyer, Anna Gottesdiener, Paul-Henry and Marie José Chombart de Lauwe.

In Sweden, apart from Rikart Küller, we have spoken to Sven Hesselgren; and in Italy to Mirilia Bonnes, of the University of Rome. As far as the Spain is concerned, the data that we present constitutes a synthetic re-elaboration of a report elaborated by Florencio Jiménez-Burillo from Madrid, José-Cecilio Sánchez-Robles from Valencia and by myselves from Barcelona.

Finally, as far as a point of view exterior to Europe is concerned, we have been able to converse with Daniel Stokols, in Irvine, and Kenneth H. Craik, in Berkeley, California. We have recordings of all the conversations. However, for a question of discretion and respect towards the interviewees, given the frank nature of the dialogue and the fact that practically all the authors that we are talking about are active in EP, we will not publish the transcriptions.

Altogether, these sources constitute the basis for the data used for the elaboration of this study.

1.4. **On our objectives and development process**

Starting from the premise that the science advances from its internal logic, but also from social institutional and individual conditioners, we have thought it necessary to specify our objectives in the following points:

1) To describe the context of the emergence of Environmental Psychology in each of the cultural areas defined, attending to their social, institutional and personal conditioners.
2) To analyse the impact of EP in society.
3) Given the interdisciplinary origin of the M-E field of study, to analyse the evolution of the relationships between architecture and psychology, as well as the professional profile of the authors who work in this field and the institutional framework that supports them.

4) The most outstanding authors.

5) The thematic field of interest.

6) Intellectual and theoretical roots, sources and dependences.

7) The social, organisational and scientific structure of the community of investigators, invisible colleges that are formed and how power relationships are established.

8) The differentiated characteristics that it adopts in each context of the European cultures and if, finally, it is possible to speak of a European Environmental Psychology (EP), as well as the evolution that can be foreseen for it.

These eight points are articulated between themselves from the partial analyses that we will see next.

In Figure 1.1 the kinds of analysis that have been carried out are outlined, as well as their articulation and sources. Although the apparent structure of the figure gives an impression of centrality to the notion of "Invisible colleges" (which in fact it has, as it arises from the analysis of the whole), our interest is not exhausted in this, but rather we put emphasis on the partial analyses which we consider can furnish information of interest for the scientific community that works in this area.

In the first place, at the base of the figure, following the order of realisation, we have the process of emergence of EP in the various European cultural areas, a study realised from data collected in the direct interviews with the recognised eminents and visits to the most important institutions. From this source, apart from the socio-historical context of the emergence of the discipline (in a broad sense, we must insist), fields of interest, the most productive authors, etc., we have been able to tackle a first approximation to the dominant nuclei recognised by the eminents, which has allowed us to establish, together with other indicators derived from scientific collaboration, the principal nuclei of power. At the same time, we have been able to detect some first groups of authors who possess the conditions of "Historical Classics" and "Functional classics", starting from the explicit recognition of the eminents.

The four remaining levels of analysis, productivity, thematic analysis, geographical analysis, that of references, emerge together from the bibliometric analysis of the nine conferences previously mentioned.

To tackle these analyses in the first place a detailed study was carried out of each conference, considering the social and institutional circumstances of each convocation. Then we concentrate on the analysis of productivity which in fact constitutes the source of the data on the *items* exhibited here.

The most productive authors emerge from this, and they allowed us to go into the analysis of the general scientific collaboration, concentrating on the most productive nuclei. Equally, the impact of the works presented in the conferences mentioned was analysed, both within their own framework, that is to say, to simplify their impact within IAPS itself, and in the international community, measured through its presence in the Social Science Citation index that Garfield started. This,together with a part of the analysis of data transmission, allows us to detect the nuclei of collaboration between the "eminents"

In the analysis of the geographical structure, we consider collaboration according to source, thematic interest according to the country, the origin of the authors and the quotes. All of this also concentrated starting from bibliometric analysis.

Through the analysis of the quotes we can detect who are the most visible authors, that is to say those who receive the highest number of quotes and have the greatest interest in the Environmental Psychology (EP) community that is expressed through the conferences under consideration. This has allowed us to detect which authors constitute the

Figure 1.1. *Synopsis of work procedure*

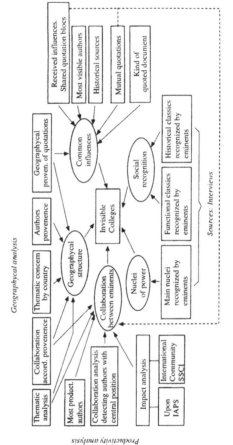

historical source of the discipline, that is, a second means of approximation to those who comply as "Historical classics" and "Functional Classics".

Finally, we complete the figure with the analysis of the common quotations and mutual references between the most important authors, being able to analyse the common received influences.

Starting from a second level of analysis, that is, collaboration, nuclei of power, social recognition and mutual influences in function of the geographic variables, we have been able to determine the structure of the principal "Invisible colleges" in this field in Europe.

2 The definition of a space for environmental psychology

It has been repeatedly affirmed that EP is a new field of study. If we have to be loyal to reality, this affirmation is rather inexact. It is true that the impulse that it received in the sixties and seventies had not been felt previously with the same intensity. However, it is necessary to consider that surroundings or environment had always been present in the background of philosophic and psychological thought, even if they could be mistreated by the latter, except in isolated cases such as the German psychological thought of the beginning of the century.

In this chapter we will make a first succinct approximation to the theoretical framework that constitutes the substrate over which EP is developed, and we will see how it is being progressively defined to the point of reaching certain levels of formalisation and institutionalisation, which gives it degree of acceptance as a disciplinary ambit, starting from an indeterminate space hinged between diverse disciplines.

2.1. A psychological framework for a history

When trying to place a date on the origin of EP, it is customary to refer to a process of convergence of interests of different disciplinary fields, situated at the end of the fifties and a part of the sixties. At the most, "Classics" or "roots" are sought *a posteriori* in some isolated or marginal approaches of classical psychology (Lévy-Leboyer, 1980).

Certainly, according to what we have been able to confirm, present day EP began in this period but was not constituted in an absolute vacuum. There have always been studies that have considered the relevance of the surroundings, although in a more or less marginal way. In fact, all of the psychological schools also contribute to the recent development of EP, directly or indirectly.

Man from his origins has questioned his relationship with the physical surroundings. Whether it were with his struggle to dominate a hostile natural environment, adapting himself and adapting it, or whether it were to dominate an urban environment, created by himself as an optimum setting, or at least more favourable, that had also become hostile to him. In fact, the biblical tale of Genesis constitutes a fable, with all of its typical elements, in which the first efforts of man to break a stable balance is expressed, to accede to knowledge and dominate the rest of the elements of his ecological niche, which constituted a revulsive that exercised its influence on all branches of knowledge, and that in psychology would take on special virulence in the controversies between hereditists and environmentalists. This line of thought of the so called natural sciences would form a constant spur throughout the history of psychology, to our days.

On a more strictly psychological level we already find the environment, the surroundings in the approaches of the first laboratory of experimental psychology. For Wundt, the object of psychology is the immediate experience. The person who experiences something is a live organism that responds to external stimuli which produce sensations in him. The appropriate method for a psychological study has to combine all kinds of physiological experimentation, the auto-observation of the subject, and an analysis of the cultural products of the human mind. That is to say, introspection, experimentation and cultural analysis.

With Watson's behaviourist conception the objectives would change and be focused on the prediction and the control of conduct. On a methodological level, Watson advocates the observation of the environment as a substitute for introspective analysis. In this scheme, and with the great methodological emphasis that characterises it, neo-behaviourism would conceptualise the surroundings as a complex series of stimuli, that is to say, as events external to the person that modify his conduct. The emphasis on the control of the experimental situation would make the reductionism into which it falls dilute the environment as a system and all that remains is a group of denaturalised stimuli, that would move it away from its first approaches and make it unviable as a framework for development of EP. Critical authors from the centre of the same behaviourist paradigm would not take long to raise the point that it is not possible to understand the S-R relationships without referring to the intervention of cognitive processes. In agreement with these mediational models, the significance of the environment and its impact should be measured by mental processes, through which the properties of the place are reinforced. Tolman was to play a decisive part for posterior EP. The crisis of what has been qualified as the Behaviourist Paradigm and the emergence of the Cognitivist Paradigm would mean an important opening for the authors that were placed within this tradition, towards the consideration of the environment as an active element to be considered in psychological processes.

Two of the authors from the Germanic area that have had greater transcendence in its posterior development in the United
States have been Kurt Lewin and Egon Brunswik. Lewin's concept of Personal Space, for example, which emphasises the constant interaction between the forces of the environment and the individual, and that leads to the vision of conduct as a function of personal factors perceived from the environment, together with some aspects of the so called "Chicago School", would be gathered and developed by Roger Barker in his Ecological Psychology, readopting of terms of Lewin's proposal.

Brunswik (1956), on his side, would concede more specific attention to the perceptive process, through which the individual comes to know his surroundings. The valuation of what he denominates "distal" and "proximal" stimuli would be important as determining factors for behaviour. Later, Gibson (1966) would highlight the importance of the perception of the properties of stimuli and the circumstances in which the perception is produced. Gestalt, however, had already brought about an awareness of the relationship between stimuli and context.

For the British Bartlett (1932), early cognitivist with a strong impact in his country and the later cognitive developments, the perception of simple or complex objects is only possible if one has a conjunction of cognitive structures that permit the processing, and of course distortion, of the information that is received from the surroundings (Canter and Stringer, 1975). Canter points out that it is starting from these currents when the possibility

15

arises that perceptual psychologists begin to take an interest in and to examine the environment, now outside the laboratory.

Elsewhere, the developmental studies, whether from a genetic or analytic perspective, gather in some ways the impact and function of the environment in the building process of intelligence, through the action and transformation of the surroundings, as is shown in the works of Piaget, Wallon and his disciples, and/or its affective investiture in Spitz and Klein. Psychoanalysis has been very little concerned, in an explicit way, with these aspects. There are, however, some efforts that should be described. Charles Madge (1951) made an attempt at approximation to the study of the public and private space; Marc Olivier (1972) carried out an evaluation of the home from this perspective; The most outstanding was the German Mitscherlich and the wide number of studies he made on the urban environment.

With this revision we do not intend to be exhaustive, far from it. We are only trying to point out some key connections with the development of psychology as a global discipline, as well as to emphasise that the emergence of Environmental Psychology is not a contemporary and sporadic phenomenon, but rather that the sixties and seventies eclosion is the consequence of a reality that had been settling in over a long time and that finally found the necessary conditions to crystallise, as will clearly be seen from each European cultural area throughout the second part of this book.

In any case, and as a form of orientation for the reading of the later chapters, we can schematise the emergence of Environmental Psychology in two stages or two births with different nuances and historic moments.

2.2. The first birth

In this study we show, and emphasize, how the present day situation has to be understood from the seed sown from a first Environmental Psychology that developed at the beginning of the century and that would be divided and disseminated by the effects of the two world wars. Hellpach, as Kruse and Graumann describe (1984, *pre-print*), collecting the influence of the notions of "Umwelt" from its definition within ecology by Haeckel (1866) and Uexkül (1909 and foll.), and for the influence of meteorobiology of great popular response at the time, published *Geopsyche* (geopsychological phenomena) in 1911, that openly tackles -and as far as our information goes, for the first time from psychology- a wide spectrum of physical-environmental phenomena in behaviour.

In 1924, after the first world war, a *Manual of Biological Methods* was published, whose third volume, compiled by Hellpach, was entitled "Psychologie der Umwelt", that's to say, Environmental Psychology.

To this author, the names of the Muchow should be added, especially Marta Muchow, who in 1935 developed the notion of *vital space*, which has been divulged and is generally attributed to Kurt Lewin, when, with different modulations, it had a widespread circulation in the epoch (without a doubt, this would be a subject for a thesis in itself). Together with Lewin, the Hungarian Brunswik, trained under the influence of the "Circle of Vienna" and with them the Californian Tolman, their American host, they would exercise an influence on posterior American and European psychology.

The vicissitudes of the war, the social determining factors, have attributed to these authors - especially to Lewin - the honour of being one of the most recognised keystones of the modern development of EP, leaving the contribution of the whole of that movement that overcame the biological and the strictly psychological in the dark. Without a doubt Lewin could not be understood without the climate experienced in the times of Hellpach and Muchow, but also of Marie Jahoda, of the sociologist Simmel, of the Gestalt itself, or the impregnation of the psychological and sociological contents of the Bauhaus (Mies Van de Rohe's famous school of art and architecture), but this would also reach the rest of Europe.

French anthropology and ethnology, the approaches of Le Corbusier; those of Chapman in England, the thinking of Spanish Ortega y Gasset and of Eugeni d'Ors, the Catalan *noucentisme*, and later the groups of the architectural avant-garde like GATPAC (Artist and technicians for the progress of contemporary architecture) with Sert, in spite of their opposing aesthetics, would receive the influence of this German period, although in these contexts it did no take root openly in psychology.

The second world war, for the origin of some, for the ideological links of others, would practically lead to the disappearance of this field of interest, which as we have seen in Europe, apart from Germany, had impregnated the other social sciences rather than psychology.

Moreover, the predomination of behaviourism and of neo-behaviourism would not allow a field to progress which outside the manifested conduct, based -as early as this- its studies on the socio-cognitive processes and the experience that the individual has of the surroundings.

2.3. The second birth

It was not be until the period of change that we referred to previously (the end of the fifties and during the sixties), in which the intellectual and social factors necessary and sufficient converged that there would be a new emergence in this field of interest.

The optimism of progress, the economic growth, the profound changes in the structure of production, and the urban concentration in cities in full reconstruction, would be factors that would create great problems for architects, technicians and urban planners, and with the awareness of social responsibility that characterised the period would lead them to look for answers in the social sciences.

In turn, psychology would live through a period of paradigmatic crisis, in which the dominant behaviourism and neo-behaviourism crumbled and gave way to more open theories, which would permit what Caparrós (1977) called the "resurgence of perception" (a very dear theme for the architects of the moment), the recuperation of the internal processes of knowledge starting from the elaboration of models (generally cybernetic), the revaluation of genetic epistemology and the slow but growing recuperation of phenomenology within psychology.

Within this context the first Man-Environment (M-E) studies were to begin in a parallel manner but independently in the different geographical areas.

In 1954 we can find Terence Lee's studies from social psychology in Great Britain, without being conscious of working on EP, still within the ambit of neo-positivism (as Lee himself recognised [Lee, 1984, personal interview]), but strongly influenced by Bartlett; the first works on the perception of architecture by the Swedish architect Hesselgren, that sought foundations in the phenomenological psychology of Hering; the socio-anthropological studies of P.H. Chombart de Lauwe and the urban sociology with a Marxist slant of Lefebvre in France; or in Spain, some isolated studies of the prohibited and repressed sociology (at that time), or at least of social psychology like that of *Del campo al suburbio* (from the fields to the suburbs) by M. Siguán.

As early as the sixties, the studies on tangible surroundings would begin to be frequent (the "case studies"), basically directed towards improving the urban habitat and the working environments on a functional level, for a better and greater productivity with a clearly technocratic background.

In this context the approaches of the first meetings or symposia emerge in different areas, such as the British ones (1963-1967) or the Swedish one (1968). The labour of that first EP was seen to bear its fruit in them, in spite of not being in a conscious or recognised way (even today). The references to Lewin and Brunswik are frequent, and even the participation of some of the survivors of the diaspora occurred, like the Austrian Marie Jahoda (Lee, 1984).

The convergence of interests between the different disciplines was to form a common space, hinged between them, that for some authors (Joiner, 1982; Hernández, 1985) had not yet acquired the conditions to merit the category of "domain". It would take on a specific nature in itself, acquiring its own identity and a progressive level of social, institutional and academic recognition, but of difficult disciplinary classification. Psychology has wanted to heel the dimain towards itself. In fact it is clear in the denomination with which it is known today and identified in the international scientific community (see 10.4).

2.4. From inspecificity to configuration: the emergence of the IAPS

Against the background of the comments of the previous section, in the following chapter we will describe the collaboration of architects and urban planners with psychologists, at the end of the fifties and during the whole of the sixties, to tackle together a series of specific problems. On the other hand, other social sciences increased their interest in the relationship between man and his surroundings. Circumstances that place us before a meeting of interests between architects and urban planners, as designers of the space where man develops; sociologist and geographers studying man and his activity in the most visible aspects, while he constitutes one more element of the landscape, without mentioning more than accidentally the aspects of interaction between human beings; anthropology, interested in the production of the surroundings, in as much as they form a part of material culture; ecology in its aspects of biological interaction and of profound transformation of the environment, as well as quite a long list of etceteras of fields that border on what are called the environmental sciences, and in those in which the psychologist enters, if not rather late, then once the interest was awakened from other fields.

These interests created a common space, an interconnection zone between the different fields that would configure itself as being specific, acquiring its own identity and a certain social, institutional and academic recognition, but of difficult disciplinary classification although psychology has intended to appropriate it, throughout a process in which it has acquired a certain preponderance. In fact, this is quite clear in the selfsame denomination with which it is known or identified today in the international scientific community.

This interdisciplinary contact would not take long to overcome the merely individual and sporadic collaborations, and soon would enter the progressive construction of a social structure that facilitated and fortified exchange between investigators. The first sectorial meetings would go on to amplify their domains, entering a continued dynamic of meetings that would find the height of their expression in the International Congresses. As we have seen in point 1.2.3., first it would be the United States and later Great Britain that would initiate a series of conferences as a better formal and informal communication channel, and that would set the standards of the development of the new domain.

As we shall see in the next chapter, in spite of the fact that at times there is a gap between the dates, the origins are very similar, being parallel but independent in the majority of European countries. Even so, in our study we take the 1969 Dalandhui conference as a point of departure, strictly British and with great transcendence in the future of EP; an international convergence would not take place until the following year in Kingston. Kingston, 1970, was organised by a different team, but which had been present the year before. In this conference the celebration of the next one was agreed upon, in Lund (Sweden) in 1973, which like the previous ones was celebrated under the name of the Conference of Psychology and Architecture, this time including the qualification of "international". Without intending to do so, a working style and a minimum social structure had been created in this new born domain. Plans were spoken of to form an association, but the proposal was rejected given the great quantity existent in all the participating disciplines, and it was decided to continue with the same procedure of naming a linking committee that would organise the subsequent conferences, maintaining the common denomination of IAPC (International Architectural Psychology Conference). Thus the Strasbourg conference was called three years later, and in 1979 the Louvain-la-Neuve conference.

However, this process was seen to be altered in England. In 1972 Canter and Lee created the first academic syllabus that would take the name of Environmental Psychology, in Guildford, Surrey, following the current that was imposing itself in the United States. Canter, who had been the spirit of the Dalandhui conference and Lee called the International Conference on the Built Environment in 1974 as a promotion for their syllabus, which generated a certain ill-feeling within the community that had begun to be formed around the IAPCs, which already represented an important part of those interested in the theme. Later, in spite of the fact that members of the two groups participated in the linking committee that emerged from the 1976 conference for the preparation of the following one in 1979, it was inevitable that with only one week's difference two congresses were celebrated, one in Louvain-la Neuve within the tradition of the IAPC's and another in Guildford, Surrey. The climate between the two groups began to turn sour.

The situation presented brought about the imperious necessity to build an association of European ambit to start with that would coordinate these activities. Without a doubt, the calling of the conferences had been used to take up advantageous starting positions in the power structures of this association, predictable from the beginning, that for the strong individual links of their members were destined to mark profoundly the orientation of the development of this domain.

In fact, in the background of the question there were deep disciplinary and theoretical divergencies. Disciplinary, inasmuch as the Guildford group represented the preponderance of psychology over this field. The IAPC group was composed of a much wider spectrum, architects and psychologists formed the majority groups, but there were others with a certain representation, especially those from the social sciences like sociology, anthropology and geography. Theoretical, inasmuch as the Guildford group was explicitly identified with the British empiricist tradition (Lee, 1984; Canter, 1984); while the nucleus of the IAPC's, through their very constitution, showed a greater openness to other postures. In this sense, a great controversy blew up in 1979 in the Surrey conference between positivists and phenomenologists that would last until the end of 1984, and that would openly and publicly continue in the pages of the magazine *Architectural Psychology Newsletter* (see 3.7). Apart from the typical discussions between both points of view, the appropriation that was intended from positivist psychology and the occupation of key places in the submerged power structure were denounced -especially in that referring to publications- of the field of study. It also coincided with a moment of economic crisis and of disenchantment with the possible contributions of psychology to the practice of design, above all from the architect's point of view. In fact, however, it was no more than the "environmental" version of the much trumpeted and disputed crisis of social psychology, which appeared here with a certain delay.

In the context of this tense atmosphere, now in Louvain-la-Neuve, a commission was named of twelve representatives of different countries and professions to study the convocation of the fifth IAPC for three years later (speaking of Italy as a possible seat), as well as to initiate the contacts necessary to build a more permanent organisation that would coordinate these activities.

Driven by Küller, Fatourus and Canter, the commission met in London on the 26th of September of that same year, 1979. Of the twelve members elected in Louvain-la-Neuve six attended: Küller, a Swedish psychologist and architect; Fatourus, a Greek architect; Canter, a British psychologist; Simon, a Belgian psychologist; Kemp and Sue-Ann Lee, British psychologists.

Also in attendance were the Japanese architect Masao Inui, the Israeli architect Peled, the Dutchman Prak, the German psychologist Espe and the British psychologist Griffiths. This last group, although they had not been elected, were interested in the project.

The discussion took place during the whole day in a hotel en London, and revolved around the analysis of the evolution of the domain, the tensions that had been generated between individuals, groups and disciplines, and on the name to adopt. After a forceful debate the creation of the association that would take the name of IAPS was agreed upon. With this denomination, it was sought on the one hand not to lose the dynamism nor the image forged over ten years, maintaining an acronym and logo similar to those used until then (IAPC). On the other hand, that the name of the association should not be identified with a concrete discipline with an end to maintaining to the maximum the formal openness, but at the same time that the previous acronyms should be remembered (IAPC and ICEP for the Surrey group). In this way the sibylline name of the "International Association for the Study of People and their Physical Surroundings" was agreed upon, deciding to maintain the acronym IAPS for all languages. The objectives of the association were succinctly defined in the following points:

1) To facilitate exchange and contact between investigators concerned with the relationships between man and his physical surroundings.
2) To stimulate investigation for the improvement of the built environment, for an improved welfare.

All this through:

a) The stimulation of exchange at international level.
b) The organisation of regular meetings and seminars.
c) The publication of a bulletin of information and the acts of the meetings.
d) The maintenance of contacts with other similar associations.
(IAPS, 1985).

Thus, it was agreed that the Architectural Psychology Newsletter, fruit of the first meeting in Dalandhui in 1969, would become the association's official bulletin, foreseeing a change of title. This change was carried out later, progressively, given the image and implantation achieved in the previous period, starting with the addition of a subtitle, which since the has become the present day "IAPS Newsletter".

In the meeting a provisional executive committee was chosen that was to take charge of the functional management with a four year mandate, formed by Rikart Küller as the president, David Canter as secretary-treasurer, Sue-Ann Lee as being in charge of the newsletter and publications secretary, and Dimitris Fatourus and Jules G. Simon as spokesmen. A desire for deliberation and equity was clearly visible in the committee between members proceeding from the IAPC and ICEP groups.

We should say that the constitution of the IAPS would not mean the end of the tensions, but rather, in any case, their redirection towards an internal debate. Neither would it end with the disciplinary discussions. An indication of this is that in spite of the formal constitution in 1979, the 1982 Barcelona Conference would be wholly organised by the local committee, with a minimum implication of the executive body of the association, too busy in resolving its own internal problems. It would not be until the eighth conference in Berlin, 1984, that a total implication of the executive would come about, assuming full responsibility for the organisation, where on the other hand it proceeded to the preceptive renovation of its powers.

In this initial stage the role of the British groups was fundamental, especially that of Guildford. David Canter was the absolute leader, with an efficient management, without which the development of the organisation would probably have been different and certainly less effective. On a personal level, in very few years he would manage to reach a clear position in diverse ambits of the social organisation of EP: create the EP programme in Guildford, recuperate the leadership initiated in Dalandhui, with the constitution of the IAPS create and direct the *Journal of Environment Psychology* in co-edition with Kenneth H. Craik of Berkeley and a panel of consultant-writers composed of the leading world figures. His contribution, clearly based on his management capacity, but which has not ignored a copious production, has been important for the consolidation of EP.

With this concentration of the powers that be, matchless in the European Environmental Community, it would be strange that what had been constantly reproached did not come about: in spite of the formal denomination of the association, the real domain has been in psychology and especially in the tradition of empiricist and markedly positivist psychology. Even though on a formal level the duplicity of the conferences had been eliminated in Europe, on a disciplinary and theoretical level the situation has gone through critical moments. The psychological predominance meant that other disciplines felt themselves excluded, discussed and mistrusted, when they did not in fact abandon the association. In the heart of psychology itself, the tendencies argued violently. Some of their leaders deserted, other taking core positions that could lead to a foreseeable opening up of theoretical perspectives. These situations would place us before the renovation of the executive in 1984. We should add that the strong anglo-saxon domination created tensions with other cultural ambits, causing some dissident and/or secessionist movements, such as the aborted creation of GEEM [study group for mediterranean surroundings] on the part of the latin countries in the Barcelona Conference, after the drawing up of a critical document, or the attempt to form a Francophone section within the IAPS.

Before going into the new period we would like to mention a quotation that we consider as being very significant. Referring to the conference on "A model for social Action in Environmental Design: Nine questions to bridge the gap", Duncan Joiner, 1982, said:

> *David Canter seemed to be trying to place us all within a recognised and defined discipline. A laudable effort, but he forgot one of the most significant points demonstrated during twelve years of Environmental Psychology, that each discipline jealously retains its information and its constructs [Joiner, 1982: 27].*

In spite of this notably extended critical spirit, it is widely assumed that EP owes an important part of its actual reality to David Canter.

What we could qualify as a second stage was created in the Berlin Conference in 1984. At the moment of writing this study there had hardly been enough time to draw up a line of action. The elected committee was formed by four architects, the Swiss Gilles Barbey as president, the Britons Martin Symes as secretary and Peter Ellis as treasurer; the Dutchman Niels Prak as spokesman, with the psychologist linked to Kingston Polytechnic, Sue-Ann Lee, continuing as editor of the *Newsletter*. It could be considered as a renovation committee for the personal orientation of its members, still continuing with a British majority. It had gone from four psychologists and an architect to the reverse. Moreover, Gilles Barbey maintained strong links with the central European group, eminently phenomenological, formed by the triangle of Strasbourg (Korosec), Heidelberg (Graumann and Kruse), and Lausanne (Barbey).

On the other hand, the growing participation of other non-European countries is seen to be reflected in the constitution of the consulting committee or full committee, with a strong American specific weight. Moreover, the General Assembly of 1985 had taken place in New York, within the framework of the EDRA annual conference, and the 1986 Conference took place in Haifa, Israel. Altogether this would form an ambit of implantation that surpassed the geographical limits of the first European theories, to reach a strong position of world association.

To sum up, this is the organisational framework in which the period that we are studying was to develop, and which will have to serve to integrate and/or rebind the sectorial dates that we will be furnishing for the cultural areas or conferences in the following chapters.

2.5. Summary and conclusions

In this chapter we have revised, within the framework of psychological thought, the role that is given to the environment from some of its most significant perspectives. We have seen how, starting from an unspecific field of study, hinged between various disciplines, it would be defined until it reached certain levels of formalisation. The constitution of the IAPS was accomplished through a process marked by the different conferences that were celebrated in Europe, to which we have made an initial approximation.

Environmental Psychology is still fighting for its own definition, not from nothing, but rather it is recuperating a tradition of treatments, at times marginal, from psychology, and finds sufficient conditions for its emergence in a given moment, supported from other disciplinary fields, progressively endowing itself with a bureaucratic infrastructure, with magazines and periodical publications, with institutional programmes and associations which define a recognised field of study, even though it is not overly integrated. The evolution of the organisational process that presents itself seeks to be a framework of global reference that permits the integration, in a global whole, of the sectorial data that we shall be furnishing in the following chapters.

Part 2
THE ORIGIN AND DEVELOPMENT OF ENVIRONMENTAL PSYCHOLOGY IN THE PRINCIPAL EUROPEAN CULTURAL AREAS

In the preceding chapters we have described and analysed some common characteristics of the development of Environmental Psychology within the geographical framework that we are concerned with. Once we have situated some theoretical antecedents and important events that would set milestones in its development, now we shall concern ourselves with how the emergence of Environmental Psychology came about in the nain cultural areas, taking into account its context, its distinctive characteristics starting from their own cultural traditions, the process of institutionalisation, and as far as we have been able, the present situation and participation in the dynamic of the conferences described.

The main source of information for this revision has been documental analysis, interviews and the direct experience of the principal centres of interest. As a complement, data referring to each area emerging from the analysis of the conferences will be used.

The areas analysed are Great Britain, Sweden and the Scandinavian countries, Belgium and Francophone Switzerland (from Luxembourg we only have evidence of a single study), the Germanic area, basically the Federal Republic of Germany although with some precise references to Austria, the USSR, Italy and Spain. In the epilogue to the second part, we will make a reference to some areas that we did not consider in the 1988 version, like Holland, Portugal, Greccc or Turkcy.

With these partial analyses in mind, we will go on to evaluate Europe as a whole in the third part of this book.

3 Environmental psychology in Great Britain

3.1. The origins

In the United Kingdom all of the authors consulted coincide in indicating one name as a pioneer of the discipline: Terence Lee. In spite of the fact that in reality the dates of the initiation of his work do not differ particularly from the first French or Swedish studies, the key difference is in the fact that T. Lee has continued his work, identifying it and strengthening Environmental Psychology after what we have called the second birth or formal birth of the discipline took place.

Considering that Environmental Psychology is an emphasis on physical surroundings, built or natural, evaluating attitudes, experiences, feelings ..., for T. Lee (1984) it would not be true to say that this had not been studied before. It is necessary to point out that psychology, however, had paid very little attention to environmental aspects. The only interesting material that was found when he began in the fifties were some sociological studies, the work of Lewin on vital spaces and Tolman on maps. Later he discovered other pioneering studies in psychology: Howard and Templement on spatial orientation; Horwitz, Duff and Stratton also on orientation and on the body as an amortization zone. We should point out the pioneering work of Charles Madge, a psychoanalyst who worked assessing sociologists and planners, and who in 1951 had already published an article on public space and private space, in which the home was compared with the body of the mother. The field in which Lee began to work was quite unusual for the psychologists of the time; neighbourhood problems and the planning of new towns in the reconstruction of post-war England; - we see, on the other hand, how it was a liberating social factor quite common in the other pioneering countries in Europe -. The first question that the planners posed was the suitability or not of building self sufficient neighbourhoods (a notion introduced in America since 1919). This was the subject of his thesis, presented in Cambridge in 1954 and directed by F.C. Bartlett. "[...] but he did not denominate this as Environmental Psychology, because he could not have imagined it" (Lee, 1984). Later he went on to study educational problems caused by the closing of schools in small villages during the process of educational centralisation that was carried out in England during the fifties. He studied how the change of surroundings, the integration into an urban environment to a large centre and scholastic transport could affect the children. "I began to understand that this was psychology of space, but basically it was Social Psychology." Later he would develop this in the little book *Psychology and*

Environment in 1976. For Lee, the date in which this field became thoroughly accepted was 1963, when in the annual conference of the British Psychological Society (BPS) he organised a symposium on the theme (Lee, 1976). There had been, however, some other previous meetings in the core of the Scottish branch of the BPS in 1960. (Lee, 1984).

3.2. The social context

The strong planning and constructive activity that was registered in the country during the postwar period brought the planners to propose the investigation and trial of new solutions for urban problems in general and habitational problems in particular. The quantity of methods that were used for this task and the lack of success of some actions meant that architecture opened out to the social sciences, especially to sociology and progressively to psychology, which in some cases would be integrated as key disciplines in academic curricula, and their professionals were required as consultants to resolve various questions.

> *There was great expectation for the answers that other disciplines could supply. Now, however, there is a profound disenchantment among architects with respect to psychology and sociology in Great Britain [...] my profession says that these disciplines do not offer us anything. In fact they make life more difficult [Cooper, 1984, personal interview].*

Ian Cooper is an architect, a professor in Cambridge, who has worked actively in the field of Environmental Psychology and Sociology. If this is a genuine reflection of the opinion that exists in a goodly part of the profession, which is translated on an institutional level to the reduction to virtually nil in the support that RIBA (Royal Institute of British Architects) had offered since the end of the fifties for psychosocial studies in relation to architecture, the justification from psychology would obviously be quite different.

It is not true, as all the psychologists interviewed agree, that psychology has not made important contributions to architecture. Today there is a substantial body of knowledge, although it is true that much of it is irrelevant. The problem is that it did not exist when the first demands were formulated (Lee, 1984; Canter, 1984; Mikellides [ed.], 1980; Edge and Donald, 1984). Another question is the posture of the architect who hopes for concrete prescriptions for immediate application that obviously psychology will not be able to give him (Lee, 1984; Edge, 1984).

Independently of this anecdotic discussion here, which we will take up again in a later chapter, the problem from the perspective of context is the type of social and demographic development followed by Great Britain since the middle of the sixties. Like the whole of the western world this country has seen how the growth of population has come to a halt, and therefore urban development as well. Moreover, the economic crisis of the early seventies was a severe blow to the expansive dynamic of construction in the country, especially the massive construction of social housing (Canter, 1984; Lee, 1984; Donald and Edge, 1984; Cooper, 1984).

3.3. Institutional development

Environmental Psychology in Great Britain grows particularly on three axes: the public and/or professional institutions, academic programmes, but above all on the meetings, symposia and congresses that play a fundamental role and are a driving force in the expansion and consolidation of the discipline.

We have already spoken of a far off meeting in Scotland and of the symposium that T. Lee organised in 1963 in the annual conference of the BPS. It would not be until 1969, however, that we can find the first monographic conference (in whose organisation T. Lee was not involved) in Dalandhui, forum of the dialogue between the sixty psychologists and architects gathered on what was then called Psychology of Architecture, as is reflected in the *Newsletter* that they decided to create. The most significant debates took place with respect to the necessity to form a theoretic framework, but above all, for the application of concrete methods. Just one year later, in 1970, what is considered as being the First International Conference on Psychology of Architecture (IAPC) took place in Kingston. The attendance and active participation was triplicated and the range of topics diversified. George Kelly's Theory of Personal Constructs was considered in a special way as a possible theoretical framework.

In 1974, Canter and Lee, from the still recently created syllabus on Environmental Psychology in the University of Surrey, called a conference on "Psychology and the Built Environment". Mikellides (1980) evaluates it as a conference in which for the first time, as opposed to the previous ones, there was a greater emphasis on psychological aspects than on architectural aspects, paying special attention to the methodological questions.

In 1975 in Sheffield a conference was called with the generic title of "Education for the Urban Environment", where the training of designers and the role of the different professionals was emphasised. The sociological perspective came into play and the ideological problems invaded the scene, demonstrating a certain crisis in the relationship between psychologists and architects (Mikellides, 1980).

In 1979 we find the last British conference of a general and international nature, also in Guildford, Surrey. The relationship between psychology and architecture was maintained, but it has become problematical. There has been an identification, or an auto-attribution in this process, of the title "Environmental Psychology" by the group with empiricist and positivist tendencies, institutionally more active than the phenomenologists, ethnomethodologists, symbolic interactionists..., that apart from in the controversies maintained in the pages of the *Architectural Psychology Newsletter*, it became evident in the scientific sessions of the congress.

Without a doubt, the process followed by this field becomes clear through the conferences we have referred to. We will itemise some of the most significant milestones for the understanding of this evolution within the British context.

3.4. The second origin

Previously we spoke of a double contemporary birth of Environmental Psychology in Great Britain. In fact, Terence Lee is one of the origins, but in spite of some pioneering studies from the fifties until Dalandhui, he considers himself as being a Social Psychologist or an Educational Psychologist, and works in Dundee. The other, independent, is the 1969 Conference in Dalandhui, Scotland, where there was an incipient dynamic of interdisciplinary study.

The Dalandhui conference was personally promoted and organised by David Canter. Canter studied in the University of Liverpool, where he received his doctorate in the same year of 1969 with a thesis on the evaluation of offices, directed by Bailey. Bailey was in contact and had participated with the organisers of the first conferences in the United States, to be precise in the Utah conference, where Bryan Wells had also taken part. Canter -as he himself told us in 1984- had always been interested in Art in general and has written some plays. He wanted to unite art, aesthetics and psychology. On looking for fields to investigate, the only possibility he found was that of joining a group from the School of Architecture of the University of Liverpool, the Filkington Research Unit, with which he began a work of investigation on offices, with special emphasis on the aesthetic aspect. "It was an unusual group, there was an architect who had congregated a geographer, a physicist and a psychologist." The psychologist was Bryan Wells, who was finishing his doctorate on the evaluation of offices. There, Canter organised an important national exhibition on Art and Science.

Bryan Wells and David Canter joined the Department of Psychology of the University of Strathclyde. Once there, Tom Markus, another important name for the understanding of this moment of the story, a recent arrival in the Department of architecture, proceeding from the School of Cardiff where he had already been a professor, asked for psychologists who wished to join his team on the evaluation of buildings, the BPRU (Building Performance Research Unit), which had just been founded. All this came about before 1966. That same year Wells abandoned the field that was already called the Psychology of Architecture.

In 1969, D. Canter was in the School of Architecture of Strathclyde University, Glasgow, and from there he was able to organise that weekend conference in the university's country house that is called Dalandhui. Twenty people were expected and sixty attended, among them T. Lee. This conference, in a sense, is the point of reunion between the two origins mentioned previously. A series of key events in the development of Environmental Psychology would emerge from here, either directly or indirectly:

- It was decided to create a *Newsletter*.
- The psychology courses were created in Kingston Polytechnic.
- The Kingston Conference was called (1970).
- The embryo of the future programme in Surrey was formed.

The Kingston Conference, in 1970, was closely linked to the sensitivity of Basil Honikman, who after Dalandhui called on Sue Ann Lee to take charge of the creation of the new psychology courses in the Polytechnic, and offered its infrastructure and economic support for the creation of the *Architectural Psychology Newsletter* which would be published from then on. Sue Ann Lee (1984) had been studying Geography and Psychology in Dundee with Terence Lee (with whom she is not related despite her surname). The *Architectural Psychology Newsletter*, now the official bulletin of the IAPS since its foundation in 1980, has played a fundamental role in the development of the discipline, both for the possibilities of connection and exchange between authors, and for the information and dialogue in its pages. Proof of this is the intense debate that we will refer to later.

Sue Ann Lee's work is marked by its functional aspect, at the service of the environmental community. There is complete agreement between all the groups on the recognition of the correct management and editorial criteria followed, which have played a fundamental role in the development of Environmental Psychology in Great Britain and in Europe in general.

The fourth milestone that emerged from the Dalandhui Conference was the creation of the Environmental Psychology Programme in Guildford, Surrey. When D. Canter returned from Japan (1970-1971), he rejoined the Glasgow group. Rapidly T. Lee, who was head of the Department of Psychology in Guildford, called on him to form part of it. Canter laid down the condition of being able to establish an M.A. course in Environmental Psychology, which Lee accepted. In 1972 he joined his new department and in 1973 began the first master's course, plus various investigation projects. Immediately, they prepared the First Surrey Conference, partly as propaganda to attract pupils. (Lee recognises that with this conference some of the first tensions with continental Europe began to appear [1984].)

This academic programme has been, and is, very important as it was the first (and at the moment, according to our information, is still unique) complete programme on Environmental Psychology in Great Britain. There are isolated subjects or important parts of syllabi in various places (Cardiff, Glasgow, Birmingham, Oxford, Portsmouth, Sheffield, London, to name a few examples), but no complete programmes. Another reason why it is important, transcendent and controversial is the terminological question. In Guildford, the term Environmental Psychology was adopted in Britain for the first time. Until then the denomination Architectural Psychology had been used, both in Great Britain and the Nordic countries of continental Europe and even in some American nuclei (Utah, for example). The term Environmental Psychology, which had been launched by Proshansky, Ittelson and Rivlin of the New York team and some of those from California (Craik, Appleyard, etc.), was much wider for Canter, it was not restricted to the built space, and had a worldwide image. Moreover, according to the evaluations of some authors, it permitted the demarcation of such an ambitious project from the rest of the British and continental European dynamic, where discrepancies and tensions were beginning to emerge.

In 1975 Peter Stringer joined the Guildford team, as Griffiths had done a little earlier. Stringer, the most "sociologist" and political of the three (Lee, 1984) and the least positivist, he would leave for Holland in 1979, going on later to Belfast where he continues to work on questions of the people's participation and social representation in the urban setting.

In 1979, the *Journal of Environment Psychology* was created, through the particular impetus of Canter in Guildford and K.H. Craik in Berkeley, California, basically under Canter's direction. The *Journal*, without institutional links with the IAPS (International Association for the Study of People and their Physical Surroundings; the contacts for the formation of which began that same year, and of which Canter would be the general secretary and principal promoter), would be an unofficial scientific mouthpiece, at least during the period between 1980-1984, the year in which the changeover in the post of general secretary came about. Guildford, Surrey had become the nucleus of the most important institutional power in Great Britain and in Europe in general.

In ten years, Guildford has had more than a hundred graduates. Among its students, around 50% are foreign (Spanish, Belgian, German, Greek, Turkish, Israeli, etc.). In the beginning the architects had a certain specific weight, but this has gradually been reduced, while that of other social sciences has increased. Its graduates make up what Canter calls the third generation.

3.5. The three generations

According to David Canter (1984), we can distinguish the clear existence of three generations within British - and in general western - Environmental Psychology, and probably a fourth. Every five years a new generation would emerge. Korosec in France distinguishes two: the one which was formed by itself through the direct reading of books and investigation, and the one formed through the classrooms. She had already made this qualitative distinction between the first and the second generations in 1969.

The first generation would be composed of the pioneers who, without being conscious of working in this field, laid the groundwork for the future: T. Lee, Bailey, Brian Wells, John Langdon, Allan Lipman and Basil Honikman would be the British names, together with Sommer, Hall, Proshansky, Ittelson and Rivlin in the United States. It is a generation that has been concerned above all with methodology more than with the theoretical angles and which received a formation closely linked to the relevant names of psychology in general.

The second British generation was composed of the young participants in the Dalandhui Conference. Still being pioneers in the work, they received influences from the previous group and began to be aware of the fact that they were defining a new field, that would lead them to the institutionalisation of programmes and academic subjects in the general curricula of psychology and architecture, and would give them a certain level of organisation, such as the first foundations of the IAPS discussion between sociologist and architects within the British Sociological Society. They would be provided with the first organs of expression like the *Architectural Psychology Newsletter*, and later on the *Journal of Environment Psychology*. This is the generation of David Canter, Byron Mikellides, Sue Ann Lee, Steven Tagg, Ian Griffiths, Alan Edge, Martin Symes, and due to his British training, the Valencian Tomás Llorens. On a European level it would be the generation that accepted certain methods as being relevant, it was largely concerned with theory, with the role of the environmental psychologist and to a lesser extent with application.

The third generation is fully concerned with application, without underestimating the theoretical and methodological aspects. It has been formed inside the classrooms, either through the Guildford programme, as in the cases of Ian Donald, David Young or the Valencian architect José Cecilio Sánchez Robles, or through having been sensitive to isolated subjects of Environmental Psychology or Sociology within a general training programme, as in the case of Ian Cooper and Howard Harris of Cardiff, to mention just two names that now have some resonance.

A possible fourth generation, for Canter, will be formed by the present day students.

3.6. The roots

British authors recognise common roots with those of other countries. Lewin and Tolman, the influence of Gestalt, are the bases of the references given, but without granting them any special relevance. What Lévy-Leboyer (1980) labelled as "roots *a posteriori*" is fulfilled. In fact, the two tendencies that have been taking shape, and that we will analyse later on, show differentiated theoretical references.

T. Lee recognises his debt to Bartlett, who directed the study of his thesis, and through him, to Galton. Bartlett (1932) was to conceptualise space explaining the forms into which we divide it, not only into physical units but also social units. The social system is isomorphic with the physical system. This can permit the comprehension and the prediction of behaviour, a subject of great interest to Lee. "Behaviourism has had very little influence, whereas cognitivism, whether it be of Neisser, Piaget or Bartlett, contributes a more interesting framework for the understanding of the spacial behaviour of the individual". (Lee, 1984)

One detail that Lee mentioned in passing in our conversation, without giving it more importance, was Marie Jahoda's participation in the Reading Conference. This Austrian authoress, had begun to develop studies on unemployment in Marienthal in the thirties, and continued her work later on in the U.S.A. This Germanic connection could give the key to the unrecognised influence of German Environmental Psychology of the first third of this century (see 6.1) on British development in the sixties. Marie Jahoda published her American studies in the *Journal of Social Issues*, a review that has collected the first pioneering studies and other highly influential works in the development of Environmental Psychology, such as that of L. Festinger (1950).

Peter Stringer brought Social Psychology and the Political Sciences to Environmental Psychology. In 1969 he introduced George Kelly's "Personal Counselling". It is precisely in the line of Kelly's personal constructs that Canter had based some of his latest proposals. In any case, Canter (Canter and Craik, 1981) situated the most remote antecedents in the psychophysics of Fechner, in the eighteen-sixties. He also recognised the great influence of the American geographers of the forties and fifties. Within the field of more recent psychology he points out the importance of the works of Festinger and his collaborators (1950), Barker and Wright (1955), Mintz (1956), and T. Lee (1954). Conceptually, he recognises the importance of the molar perspective contributed by Tolman. Personally, in his thesis (1969), he accumulated the influence of the first works of Pressey (1921), Broadbent (1955, 1958, 1964, 1967), Markus (1967*a*, 1967*b*) Maslow and Mintz (1956), Wells (1964, 1965*a*, 1965*b*), Langdon (1963, 1965, 1966), and methodologically of Osgood and his collaborators (1957) and Thurstone (1947). Lipman and his collaborators recognised that they were in special debt to Marxist Sociology and the phenomenology of Foucault and Bachelard.

If we concentrate on what results from the bibliometric analysis of the IAPS conferences (which we will synthesize globally in the third part of this book), then we find a great dispersion of authors who are at the base of the development of Environmental Psychology, inasmuch as their remote studies generate and inspire posterior investigation, fulfilling the condition of "Historical Classics" and/or "Functional classics" (See 10.5). To be rigorous, however, we can speak more of "historic" and "functional" more than "classics" as far as they fulfil objective criteria of classification starting from the publication dates of the studies mentioned, but more rarely a wide common acceptance occurs (authors in which an elevated number of quotes converge). In spite of this we will mention them here, inasmuch as they are indicative of theoretical origins and orientations which the present day authors base themselves on. *

Among the quoted authors that fulfil the condition of historical classics in the studies presented by Britons in the conferences analysed, we should highlight the search for fundamentals in the developmental and cognitive approaches of Bartlett and Piaget on the representation and image of surroundings, in Sargent, Duncker, Cohen and Nagel for the thought processes, resolution of problems and scientific methods. A second block, which takes a greater interest in personality, makes references to Freud, Plant, Usher and Hunnybun; the impact of the social and cultural dynamic is based on Sorokin, and significance and symbolism on Ogden and Richards and Firey and Bender. Finally the question of well-being and comfort, strongly weighted towards the architectonic components is based on Ruskin, Houghten and Yaglov, Gershun, and Bedford, among others.

Taking the "Functional Classics" into consideration would widen this list to include authors who are still active, and some of whom act as "founding fathers", apart from authors of who take theoretical frameworks of reference, incardinated with psychology in general, social psychology in particular,other social sciences, town planning, architectural and industrial design. The concern for urban issues is what assembles the greatest number

---

The complete references to the studies of the authors mentioned in this section as "historical classics" and "functional classics" are gathered in the bibliographic Appendix.

of references in this category of functional classics. However, a great dispersion of sources persists. The author with the greatest impact is Chapman and his studies of the home and social status. Also Kuper from urban sociology, dealing with life in the city, and Rossi analysing why families change their residence. The analysis of the social system finds its sources in Parsons, Whyte, Loring, R.K. Merton, Strauss and even the novelist George Orwell.

In the field of proxemics there are references to the first editions of the pioneering works of Hall and Sommer, with the impact of distance in the process of social interaction according to Gullahorn also being analysed. Making it more specific in the group, Sprotts's work receives some quotes, but the work that has most impact is the classic by Festinger, Schachter and Back.

Paying attention to the importance of personality, the references punctually turn towards Adorno, Cattell, Duffy, Laing, Maslow, Eysenck, and with more emphasis to Goffman and Spinley, but especially to Kelly. This perspective brings us to another important field as far as references are concerned, such as thought processes and cognition in general where we are referred to Bartlett, Lynch, Mintz, Morris, Osgood, Whort and Vinacke.

Bennet, Campbell, A. Edwards, Mackworth, Siegel, Townsend, Walkers and J.Y Young contribute the experience of their works to the methodological notions. The epistemological preoccupation seeks its sources in Popper's philosophy of science, the sociology of Merton and also in Poincaré and Wittgenstein. Concentrating on the ecological perspective the support of Barker is sought, and from architecture they go to Gropius and Neutra.

The references to studies on merely physiological aspects (from the period we are analysing) are very scarce, being limited to Euler and Schnore. In some cases including this aspect but being especially centred on thermal comfort and the effects of artificial ventilation, we find Bedford, Chrenko, Pepler, Teichner and Webb. Centred on illumination the references are directed towards Moon and Waldram.

Globally, among the functional classics a much more specific weight of the studies of theoretical interest is appreciated in their conception than a recuperation of concrete empirical studies.

3.7 The controversy

Once the basis of the development of Environmental Psychology in its constructive aspect has been established, in this section and the following ones we shall see some of controversial aspects and critical dissensions, normal in any scientific field, which will help us to understand the actual state of things.

In 1979 a strong controversy was unleashed from the critical evaluations expressed in various articles by A. Lipman and repeatedly defended in conferences and papers (Lipman and colleagues, 1979, 1982, 1983) and that were especially developed in the pages of the *Architectural Psychology Newsletter*. The critical perspective of A. Lipman, professor of Psychology and Sociology in the Welsh School of Architecture in Cardiff, and his disciple and collaborator H. Harris and in a less belligerent way I. Cooper, already raised the following points.

Basically the legitimacy and validity of the dominating positivist perspective was questioned, strengthened from the power structures of Environmental Psychology (institutional programmes, associations, publications) in Great Britain, and adopted in the majority of the applied studies.

The criticism, as in the crisis - paradigmatic or not - of other fields of psychology, is centred on the asocial character that the discipline has taken on. Lipman and Harris (1980) argue that from the methodological assumptions, the central dogma of positivism treats people as if they were objects, social relationships as if they were determined by laws, and moral problems as being resolvable with technical solutions (Lipman and Harris, 1980, 1980*a*, 1983*a*). Cooper (1984*a*) remarks that this is not the first criticism of this kind, as J. Daley (1968, 1969, 1971) had expressed similar doubts in this direction a decade earlier. With the consent of the "ingenuous expectations of the designers", they take the role of "technical servant".

For T. Lee, on the other hand, "Science has to be apolitical, anti-ideological", and he reveals to us his disagreement with the approaches of the commitment of the scientist, that he calls generically "Marxist". For Lipman and Harris (1980) the evaluation of space has to be carried out as a social recourse within the social reality. The very construction of the instruments used reflects the presumptions of the theoretical perspectives of the investigator. It is not possible (Harris, 1977) to investigate directly causal relationships between thought and elements of the physical environment. " These relationships are transformed symbolically".

The reactions did not take long to appear. D. Young (1980) expressed in the *Newsletter* that "[...] in his arguments there is no indication of how non-positivist Environmental Psychology can be developed" (Young, 1980, 24).

I. Cooper (1984*a*) indicates that the alternative was already outlined in the same work of Harris and Lipman (1980*b*, 1983*b*). MacDonald (1980: 32) expresses it like this:

> *[...] hearing people talking about their problems with respect to Architecture and the Environment [...] as a significant point to begin with. Really, they often give indications of what to investigate and how to investigate it.*

The use of more open techniques that do not impose a structure and a direction of response, free descriptions, are defended as being considered more adequate. They were already successfully used (Cooper, 1984*a*) in the investigation on the environment carried out by Liverly and Bromley (1973), by Harris (1977) and by Cooper himself (1979).

The debate was wide. It originated in 1979, and in 1984 it was still possible to find argumentations in one direction or another. The list of controversial names is long. For five years, David Young, Alan Lipman and Howard Harris had been supported or refuted in practically every issue of the *Newsletter*. It is significant, however, that neither T.Lee nor D. Carter had intervened directly in this controversy.

One interesting aspect of this controversy is its public airing in the pages of the *Architectural Psychology Newsletter*, apart from the moments of friction in the bitter debates that arose in the congresses, especially those in 1979 in Surrey and also in Barcelona in 1982.

The identification of the power structures with the positivist tendency, especially in Great Britain during the period 1976-1982, implied difficulties for the "dissidents" to win official subsidies for investigation, and even lead to their exclusion from some of the scientific media for "not fulfilling the scientific requisites for publication."*

The consequences of this controversy -that on the other hand is nothing new in the

---

* *We possess documents that prove these facts, and not making them public answers a compromise of discretion with respect to the sources. It should be considered that all the authors are active today.*

field of psychology and sociology- have been the auto-exclusion of some authors from the field of Environmental Psychology. The criticism, now supposedly external, is that Environmental Psychology has become narrow minded in its objectives and conception. It has sought to appropriate a field of study that in principal was shared, interdisciplinary, due to social and personal conventions (Harris, 1984). R. and J. Darke reflect on the consequences for us:

> *Environmental Psychology is abandoning its first disciplinary base, wide and eclectic [...] for some more exclusively psychologicalist methodological proposals, and all this is provoking the disillusionment not only of the most critical psychologists in this field (see Menzies, for example, 1979), but also of those that have relationships with other traditions of sociological, political, economic and architectonic etc. investigation. The space abandoned by the displacement, lead by Canter and his colleagues, for a more rigorously psychological knowledge, is crying out to be occupied [Roy and Jane Darke, 1981: 8].*

There have been some alternative efforts to occupy this space, as early as immediately after the conferences of July 1979. In the following month of November a meeting was organised in Oxford Polytechnic under the name of "Environmental Psychology in Action" where around fifty people assembled, all of them British, who had felt excluded by the dominant approaches in Surrey. Among them were historical names in the development of Environmental Psychology, such as Brian Goody, Roy and Jane Darke or Michael Menzies. One of the results was the rapprochement towards sociology and architecture, that would result in the formation of the sociology-architecture debate groups in the British Sociological Association, and which in a first instance were also expressed through the *Newsletter*. All this lead to a second theoretic-critical debate between this new group and Howard Harris, in which they would place themselves in a double perspective of marxist materialism and symbolic interactionism.

Another result of these confrontations was in part the colloquium that was celebrated in Paris in June 1981 (see 5.6) with the participation of Peter Stringer, Peter Ellis and two more delegates from Great Britain, where there was wide discussion of this controversy.

The paradox is that at the present (1986) we find authors who, while rejecting the label of Environmental Psychology, are still working in the same line as they were before, with approaches closer to those dominant in other cultural areas, and whose field of study has been homologated by the international scientific community as Environmental Psychology.

3.8 The crisis

Although until now we have seen and commented upon the internal crisis, we will now go on to inspect the identity crisis with respect to its social function.

We had commented previously that one of the critical points was the role that Environmental Psychology had taken in the interdisciplinary whole that was concerned with the M-E relationships. This, as we have pointed out, has created certain ill feeling and has even made some authors shy away from it. T. Lee (1984) rates this fact most seriously. The interdisciplinary work was in danger.

On the other hand, the very origin of Environmental Psychology as Psychology of Architecture in Great Britain emerged as a demand from architecture. With the economic crisis and the crisis of expectations in the psychological contribution, this collaboration was seen to be affected gravely. Lee (1984) understands that:

> *We are not providing what the architects are looking for [...] for two reasons: one because the architects are looking for a manual of concrete formulae instead of understanding the interaction of people with their environment. In a second place because the architects and planners are in decadence for economic reasons [...] and I think that with the present crisis we are in danger of disappearing altogether, because people do not see it as a differentiated thing [...] [1984: 8. Recorded personal interview. The transcription and original translation is ours].*

The critical sector, as we have seen, was much harder in its evaluation and considered that the problem was that there had not been - and there could not be - relevant contributions from the "official" perspective adopted (Lipman and Harris, 1980; I. Cooper, 1984a).

Canter (1984) denied emphatically that there was a crisis, affirming on the contrary that it was an expanding field. However, in 1984 he adopted for his professorship the name Applied Psychology and not that of Environmental Psychology. In this way, he says, he sought to cover a wider field, not being restricted to designers and psychologists. Alan Edge and Ian Donald (1984) made a similar evaluation of the situation. They considered that "the economic cutbacks and the social pressure to carry out applied investigation mean that all the departments are moving towards Applied Psychology". Lee considered this tendency worrying.

Alan Edge (1984), from his position as professor in Birmingham, but above all as a professional in Applied Psychology, considers that the changes in the social issue (there are no longer problems of overpopulation, nor programmes of massive construction of public housing, and there are on the other hand problems of unemployment and crime) have blurred the function of the architectural psychologist. The problem consists in redefining exactly what is the clear offer of environmental psychology. EP has a definite function that the architect cannot cover, which is the possibility of obtaining more information from people so that it can be used by the architect in the reflection for the conception of his project. He does not cite, however, the widening of perspectives that the environmentalist contribution could signify for the analysis of the new social situation of unemployment and delinquency, as has already occurred in other geographical ambits.

All of these problems have a possible solution, according to Lee (1984), if the environmental psychologists carry out good investigation and are capable of explaining it. In the interdisciplinary field, the psychologist has to be more methodical than other professionals. The crisis with relation to architecture is a strong internal crisis within this discipline: "[...] a part of its crisis is due to not having learnt from the contributions of Environmental Psychology." The relationship that must continue to be maintained is both on a level of formation and of consultants (Lee, 1984; A. Edge and Ian Donald, 1984; Mikellides, 1984).

It is clear from the controversy and the crisis that a situation of strained relations has been reached among the professions that were found in the origin of EP as a pluridisciplinary science, when its very beginnings were, in part, the answer to a demand external to psychology.

In any case this process explains the interest in knowing just how the professional profile of the authors who work in this field has evolved, inasmuch as it can provide orientations with respect to the tendencies of its possible later development.

3.9 The professional profile

The pluri-disciplinary origin of the nature of our field of study and in particular of the active participation in the conferences that we are studying allows us to analyse the percentages of participation of the principal professions that converge in them. Considering that we have not been able to determine the profession of 10% of the British participation, the dominant professions are psychology with 43% and architecture with 38% of participation. This majority of psychologists has been maintained over the years with fluctuations. The proportion was inverted on two occasions and was equal on two more, coinciding with critical events that could explain the fluctuation, such as the tensions between British and continental groups, and between British groups in themselves that we have already mentioned in this chapter. We should highlight the decrease of psychologists in the conferences that were celebrated outside the British Isles, and a relative decrease in British architects since 1979 (table 4, appendix).

This information indicates to us that the tensions mentioned previously do occur and mainly affect the psychologists, while the architects are principally affected by the crisis of their profession instigated by the economy, fashion and the crisis of relevance of EP itself. It is important to point out, however, that Great Britain contributes 15 of the 25 architects with academic training in social sciences. In comparison with other countries, this indicates a sedimentation of the interest in this field among a notable nucleus of architects. The launching of the Guildford programme has played an important role in this, along with the incorporation of social science subjects in various polytechnic and architecture schools.

As far as other professions are concerned, their presence is punctual. It is necessary to emphasize the low participation of sociologists, especially if we take into account the creation in 1979 of the discussion group previously mentioned between architects and sociologists in the seat of the British Sociology Society, which clearly places this absence on a level of power relationships rather than an object of interest. This happens in the same way with other professions that by rights should be present, for the sake of the emblem of pluri-disciplinarity. The participation of geographers (there has been no participation on the part of British anthropologists) even though it had always been limited, practically disappeared in the last five conferences studied. British biologists and ecologists have never participated in these conferences either.

One important reason that we feel partly justifies these absences is in the initial definition that the field took as Psychology of Architecture, which was to mark the psychological slant that it would take, in spite of the fact that in the formalisation of the IAPS it was sought to avoid the term "psychology" to make the association wider and more comprehensive.

We will leave this point here to pick up on it later in other sections, especially in chapter 11.

3.10. Other important facts for the understanding of the present situation

Other information that could allow a better understanding of the actual situation is related to the productivity and impact of the British authors.

The specific weight that this state has within the European context comes about due to its notable position as the first country most productive in the conferences analysed. Its contributions come to a total of a quarter of the studies presented (see chapter 11 and table 13 in the appendix), although in relative numbers this productivity has lost force in the last few years. In any case, of the ten most productive authors in the heart of the IAPS, five are British, with Canter occupying the first place, Lipman the third, Stringer seventh, Lee ninth and Mikellides tenth. We should consider, however, that of the nine conferences analysed, four were British and to these a goodly number on "minor" symposia have to be added, not for their qualitative evaluation but for the breadth of the convocation.

Other indicators reinforce this position of British leadership. The European authors with most impact (those of the greatest impact are the Americans, as we will see in 11.4) are also British. The list of authors that we have just mentioned changes if we pay attention to the number of times they are quoted in the IAPS conferences, with Lee going up to second place. If we analyse their visibility in the *Social Science Citation Index* (SSCI) Lee goes on to occupy the first place, which reinforces his image of "founding father", Canter takes second place, followed by Lipman, with Honikman appearing fifth and Mikellides in eighth, both far from the preceding authors as far as quotes received are concerned.

To all this we should add that the selfsame IAPS is legally based in Great Britain. On the other hand, the cultural and linguistic links with the U.S.A. reinforces the influence of its publishing centres with a double head office, which facilitates the domination of the market and therefore, also in part, the control of scientific diffusion and hence of the theoretical orientations that it can take in a given moment. However, as we have seen, the perspectives that coexist (more or less controversially) are multiple, in spite of the existence of a dominant group with a powerful institutional infrastructure that together with dynamism, proficiency and productivity has converted the group of Surrey University, Guildford, into the nucleus of one of the European "Invisible Colleges" (see 12.4).

Finally, before concluding this chapter we would like to mention briefly the dominant themes (see tables 16 and 17, appendix). As in the whole of Europe the greatest centres of interest are related to the evaluation and the experience of the environment, mainly treated from a cognitive perspective, referring to the general environment, followed at some distance by the studies applied to the habitat or housing and spaces for children, especially the school environment. The essays on architecture, design and planning are also outstanding, along with user participation in design and the relationships between architecture and the social sciences.

3.11. Summary and conclusions till 1984

Great Britain is one of the pioneering countries in EP and one that registers the most activity in this subject, as the conferences that have been realised with international transcendence demonstrate. With influence from contemporary North American EP, its development is connected with the first empiricist tradition, especially with Bartlett, through the first studies of Lee, and also Bailey, Wells and Langdon. The phenomenological point of view is present through Lipman, Daley and Harris, among others. A strong public controversy emerged between both of these tendencies when, after a period of peaceful coexistence in the first profound interdisciplinary moments, with the process of the progressive institutionalisation of psychology, the positivist approach took on a preponderant role. The constitution of the EP programme in Guildford, Surrey, directed by D. Canter, who in turn would be the general secretary and principal promoter of the IAPS during the first four years, has been fundamental for the development of this field.

In Great Britain we find the highest level of institutional formalisation of the whole of Europe, but also the greatest disillusionment and the deepest crisis: scepticism (I. Cooper); the search for a space (A. Edge); the crisis in the relationship with architecture and the establishment of new parameters of intercommunication (B. Mikellides); a negation of the crisis (Canter); refuge in academicism (T. Lee); desertions (A. Lipman). British Environmental Psychology has been too strongly marked by its applied origin in the Psychology of Architecture and by a debt to the selfsame empiricist tradition, that in spite of the more formal than real attempts at opening (Psychology of Architecture went on to become Environmental Psychology in Guildford), and for the same institutional structuration with a clear intentionality to dominate and control the "Environmental Psychology framework" means that approaches and areas of intervention in other geographic zones (U.S.A., France, Germany, etc.) are left within the field defined as such (the environmental implications of delinquency and vandalism, the environmental implications of unemployment, the penetration of other perspectives and/or schools of psychology for the environmental perspective, etc.), were auto-excluded in Great Britain. It also explains why, apart from personal and institutional protagonisms, the clashes between the differential labels and titles occurred.

Once the professional profiles of the authors active within the framework of the IAPS, the "historical" and "functional" theoretical references understood as roots, the dominant themes and in part the social structure of the scientific community have been analysed, we have shown how British EP presents its own personal profile differentiated from other European cultural areas that are under its influence to a lesser degree, while it plays a leading role both for its dynamism, social and institutional consolidation and for its key position in the international editorial context. All this allows British EP to exercise a powerful influence on the rest of Europe at the same time as maintaining a high level of Anglophone autarchy.

3.12. Epilogue: 1984-1992

During this period, Canter* has detected a drop in the interest of architects in the contributions of psychology, frequently due to disinformation. Moreover, due to budget reductions environmental psychology is no longer taught in some schools of architecture. Sue Ann Lee and Jonathan Sime coincide in this same opinion.

The irruption of the Prince of Wales's opinions on architecture in 1987 were to encourage -and vulgarise- a great public debate on the subject, but the environmental psychologists, according to Lee, practically took no part in it. Only Canter says that he was frequently required by the press, radio and television to give his opinion on preferences and effects of colour, people's emotions, some designs, skyscrapers or the development of new towns, as well as being called to participate in some studies on environment and delinquency and collaborations with the police on the theme.

Some conflicts of interest and competencies have come about with urban, human and social geography, to elaborate evaluations of resources and facilities. On the other hand the interest in environmental psychology is being revealed from what Canter calls 'more commercial disciplines', such as marketing projects, the evaluation of the types and distribution of services and products in particular places like supermarkets, for example. A decrease in interest in design and architecture has been detected on the part of the Surrey students along with an increase in questions of a larger scale, like landscape, tourism or environmental protection.

For the writing of this chapter the interviews with David Canter, Sue Ann Lee and Jonathan Sime in the summer of 1992 have been most useful to me.

Within the framework of the teaching of architecture in the University of Kingston (previously Kingston Polytechnic), Sue Ann Lee highlights the favourite themes of interest of her students as being themes related to perception, community architecture and social control with relation to the physical structure. Similar programmes within the framework of architecture and planning are maintained in Oxford, with Byron Mikellides, in the Bartlett School in London, where Martin Symes was teaching until he obtained the professorship in Manchester, where Necdet Teymur is also working. They all play an important institutional role in the IAPS.

Canter and Sime coincide in highlighting the interest of the British public administration in the participation of environmental psychologists in the consideration and resolution of problems of overcrowded settings where emergency situations can arise, like football stadiums or the underground (Canter, 1980, Canter, Comber and Uzzell, 1989). Also the interest of businesses in organisational themes, security, perception of risk and management. There are some European research programmes under way.

One characteristic that Canter has pointed out is that the concern for environmental problems, both for natural resources and design quality, has spread more than might be apparent from the demand of specialists. In many different areas there are people making evaluations of behaviour in concrete settings, of the creation of new landscapes and their impact or of hospital buildings, for example. People who at times have done some courses on environmental psychology in their basic training, but have not specialised in it in the university, but frequently also people who do not know that what they are doing is environmental psychology. What was done in the seventies and eighties, sporadically as specialisations, is now beginning to be absorbed in the ordinary professional work of the clinical, school, organisational or social psychologist. This has meant a consolidation of the ambit of work of environmental psychology, at the same time as a reorientation of its professional development. The problem is that frequently, its formation is insufficient to be successful with its proposition.

4 Environmental psychology in Sweden and the Scandanavian countries

Environmental Psychology in the Scandinavian countries has developed particularly in Sweden, and we only find punctual studies in Norway, Denmark, Finland and Iceland (Norbert Schults, 1967; I. and H. Gehl, 1970). Language (T. Gärling, 1982) has been an important barrier to a greater expansion in these countries than there has been in Sweden.

4.1. The origin and social context

In Sweden, as in the rest of the European countries, we find the pioneering studies in the mid-fifties (Hesselgren, 1954, for example), but it would not be until the mid-sixties that a certain movement would start that was to culminate in the progressive institutionalisation of the discipline. Küller (1984c) places the date of origin in 1965, on the basis of the work of two authors: Carl-Axel Acking and Sven Hesselgren, both architects. Also, just as it happened in Great Britain around these dates but in a totally independent way, as the work of that country nor any other had been heard of, Environmental Psychology started off as Psychology of Architecture from the requests for help from architects especially to work on themes related to perception.

The oldest pioneering work, and with a notable impact, was the work of Sven Hesselgren. He began to be interested in this field more than fifty years ago, Hesselgren tells us (1984), when he was an architecture student and he was interested in colours and the sensations they produced. This author is not a psychologist, but he has a broad autodidactic formation. In his first studies he sought to develop the phenomenological postulations of Hering on the sensation of colour. Later he found the works of Katz, with whom he maintained contact. Collecting these aspects he presented his thesis on "The Language of Architecture" in 1954, published in Swedish on this date and in English in 1969. Professor of architecture in Ethiopia between 1961 and 1965, on returning to Sweden he devoted himself to investigation with the economic support of the Swedish Council of Building Research, and he integrated two psychologists into his group, Gärling as assistant and Küller as collaborator. "I found that it was very difficult to converse with psychologists [...]" (Hesselgren, 1984), says kindly. During this period he re-elaborated his work on the perception and sensation of colour. He wanted to show that architecture could be discovered through the description of the perception of the environment. It was published in 1975 with the title *Man's Perception of Man Made Environment*.

After his investigation period, towards 1975, he began to give courses on the Psychology of Architecture for architects and civil servants in different towns and cities. In 1981 he began to write his latest book, *On Architecture. An architectural theory based on psychological research* (1987), published in Swedish and English.

Carl-Axel Acking, from his professorship of Architecture in Lund, organised a first conference on the Psychology of Architecture in 1967, that since then would be celebrated every two years as a meeting of Swedish ambit and in some of the editions as a meeting covering the Scandinavian ambit (Gärling, 1982). Rikart Küller, with a degree in Psychology and later a doctorate in Architecture, who had joined Acking's team in 1966 as a psychologist in the School of Architecture, tells us (Küller, 1984c) that after the first conference, in 1968 they saw the need to create a Swedish association for Architectural Psychology. A little later, at the end of 1969 they heard of a conference that was to be celebrated in 1970 in Kingston, and decided to go and present the results of the four years of their investigation on colour, perception and the semantic differential. In England the idea was approached of organising the next meeting in Sweden, in 1973. This is how the dynamic of European international meetings originated, without there being any organised association to promote them. In 1973 the single Swedish presence in the American EDRA conferences came about, with the attendance of Küller and Acking.

Later on Sweden would play an important part in the development of Environmental Psychology in the European ambit, on assuming the presidency of the IAPS in the person of Küller in its first period of 1979-1984. He was the person with the highest number of votes in the assembly that decided the creation of an association of international character in 1979, within the framework of the Fourth Conference on the Psychology of Architecture, celebrated in July 1972 in Louvain-la-Neuve, Belgium.

Sweden was not as intensely castigated by the second world war as France, Germany or Great Britain were, and therefore the problems arising from urban reconstruction did not take on the same dimension as they did in these countries. This explains in part why Psychology of Architecture was developed more as an investigation on problems of architectonic design, rather than an answer to the concrete social question of reconstruction. Moreover, the demographic density is lower than in Central Europe or the British Isles. Later on, with the oil crisis and the emergence of ecological problems, studies in this terrain would begin to be developed, and Architectonic Psychology would widen its field to the environment in general.

4.2. The institutional development

According to Tomy Gärling's report (1982), the work in Environmental Psychology (or architectural psychology), apart from the papers published in the proceedings of the Conferences or the international magazines, has included two regional magazines that have aroused interest for their quality: *Environmental Systems*, that gathers studies proceeding from other Scandinavian countries, and the *Scandinavian Journal of Psychology*, apart from the support and publications on the part of the Swedish Committee for the Research on Construction (Gärling, 1982).

Since 1967, as we have said, every two years in Sweden there has been a National or Scandinavian Conference on Environmental Psychology or Psychology of Architecture, of which the proceedings are only in Scandinavian languages.

There is also broad support for investigation. The Swedish Committee for Research in Humanities and Social Sciences has established a professorship, occupied by Brigitta Berglund, which has been awarded to the Department of Psychology of the University of

Stockholm. In the Lund Institute of Technology there is a programme for graduates in the Environmental Psychology Investigation Unit, with Rikart Küller as head of the group, in the School of Architecture. There are also programmes for graduates and licentiates in various universities. Gärling cites the following: Department of Environmental Hygiene, in the National Institute of Environmental Medicine, with Ulf Berglund; the Department of the Theory of the Form, in the Royal Institute of Technology in Stockholm, with Harriet Rydh, the Department of Psychology of the University of Göteborg, with Lars Sivik; the Department of Landscape Planning, in the Swedish University of Agrarian Sciences, with Gunnar Jarle Sorte; and the Human Laboratory, in the Swedish National Institute for the Research on Construction, with David Wyon. The development of Environmental Psychology in Sweden is very wide in comparison with other fields of Psychology, and is clearly recognised. Moreover, there are no signs that this interest is declining (Gärling, 1982).

4.3. The generations

In the same way as in Great Britain, in Sweden we find a precursive generation, not very numerous, in the persons of Hesselgren and Acking, and that we can qualify in this case as of the first generation, that would merge with the posterior generation, formed by the people who have characterised the institutionalisation process: Ulf Berglund, Brigitta Berglund, Harriet Rydh, Lars Sivik, Gunnar Jarle Sorte, A. Härd, J. Janssen, Tomy Gärling, G. Edberg, David Wyon, and Rikart Küller as the most visible leader, among others. A third generation would be the one formed in the classrooms by those mentioned above.

4.4. The roots

The first author that Hesselgren (1984) recognises as being indebted to is Hering (1878) with the analysis that he made on the sensation of colour. Later it would be Katz, who opened the door of perception of landscape to him. He has also taken an interest to a large extent in Eysenck's treatment of personality (1941). In the same way as with Langer (1942), Ogden and Richards (1949) with their studies on significance. He took Härd's Natural System of Colours into consideration and, finnaly, he refutes Eco's *The Absent Structure*.

Carl-Axel Acking and Rikart Küller have been very interested in Gibson's proposals (1950) on perception, Arnheim (1954), Bird, Köhler and Hesselgren. However, Küller especially recognises his debt to Berlyne, with whom, moreover, he shared various studies before his death, and to Eysenck and Osgood. The transactionist perspective of Kilpatrick and the other members of his group, Aims and Ittelson, influenced him at some time, as well as Valentine, Lewin and his Field Theory. In any case, he sees himself as being more within the Germanic-Anglo-American tradition than the French tradition (Küller, 1984c). Tomy Gärling shares the influences of the aforementioned even when his references name studies now more fully identified in the world of Environmental Psychology.

Perception sought its foundations in Arnheim, Bruner, Gibson, Ekman, Heyl, Hungerland and Johansson. With reference to the study of colour Hesselgren becomes one of the most frequently quoted authors. The analysis of significance finds its principal source in Osgood. Ogden is referred to from the context of language, and Langer and Cassirer for more philosophical approaches. Kelly is quoted In other perspectives, along with Lévy-Strauss's anthropology, Lacey's psychophysics, and in the methodological aspect the quotes are founded on Attenave, Torgerson and Guildford, as well as Long and Moreno.

This information is confirmed without adding any new authors to the bibliometrical analysis of the "Historical Classics" (see 10.5). As for the "Functional Classics" the list gets longer. In general the references are scrupulously related to the themes that they develop, and their field of origin is wide and universal.

4.5. The fields of interest

Both through our bibliometric study (tables 16 and 17, appendix) and what Gärling describes in his report, the fields of interest in which Environmental Psychology has developed in Sweden show a notable concentration in perception and its cognitive and physiological aspects, in contrast to Great Britain and France, There is, moreover, a notable presence of experimental laboratory studies. In the description that Gärling gives us (1982) it is possible to emphasize the construction of an environmental simulator in the Lund Technological Institute for Acking and Küller.

Gärling distinguishes three large areas of study: on the sensorial bases of the processes related to the environment; the significance of the environment; and the cognitive processes and the surroundings.

In the first area we can refer to the work on perception, colour and texture, with Küller (1971), Acking and his various collaborators (1972, 1976), Gärling (1969, 1970a, 1970b, 1972a, 1972b,) Sorte (1981), Härd (1975). Referring to exterior spaces we find the works of V. Berglund (1981) and B. Berglund and her team (1971, 1973, 1975, 1976, 1977). On interior environments, a field particularly well developed due to the influence of the climate, Wyon (1970, 1973, 1974, 1975); we should also emphasize the work of Berglund and his collaborators (1982a, 1982b) and Küller (1981).

In the second area it is necessary to highlight Hesselgren in particular (1954, 1967, 1969, 1975), Küller (1972, 1973), Sivik (1974a, 1974b, 1974c, 1974d, 1974e, 1975), Sorte (1971, 1973, 1981), Acking et al. (1973).

Pertaining to the third group in the first place we find general studies related to the cognitive process with reference to the environment, by Gärling, Küller, Sivik and Sorte (1974, 1976); emotional reactions to the environment, with Küller, (1976a, 1976b, 1978, 1979, 1980); orientation, memory and spacial-temporal sequencing of daily activities, with Gärling and his team (1974, 1979, 1981, 1982), Böök (1981), Acking (1976), Gustavsson and Manson (1980), and the categorisation of environmental information, Gärling (1976a, 1976b, 1980a) and Janssens (1976).

For Gärling (1982), the fact that many of these studies laid the groundwork for the establishment of a bridge between Environmental Psychology and personality studies constitutes a promise for the future due to its infrequency in this domain. The interactionist and cognitivist approaches have gradually been occupying an important place as a theoretical framework of Swedish EP, in the way of understanding how the person perceives and interprets his or her environment (understood in a broad sense that includes both the social and the physical). So Sweden has made some important theoretical contributions in this field.

Although many of the authors have not developed a personal theoretical scheme, a great number of the studies are projects carried out on a grand scale, with a well established empirical foundation, that furnish well founded data for future theoretical interpretations.

4.6. The crisis

In spite of the fact that the economic crisis had not been felt with the same intensity as it had in other places and had not greatly affected the budgets for investigation, EP in Sweden suffered a moment of crisis concerning its applied aspects and its relationship with architecture. This crisis had a different, minor intensity, however. We recall that one of the characteristics of our field in this country has been the genuine interdisciplinarity, the collaboration in mixed teams. On the other hand, the contributions of Psychology have basically been directed towards the understanding of interaction, or better, the perception of the environment, and to a lesser degree to the solution of concrete design problems.

In spite of all this, Küller (1984c), observes that twenty years earlier the architects had been very interested in psychological questions, ten years ago their interest was centred more on sociological questions, and now they are basically interested in aesthetic issues and the theory of architecture. They are neither interested in scientific, psychological nor sociological investigation. Anyway, five or ten years on from now, he says, there will once again be a balance. This lack of interest is due, apart from fashion, to the fact that investigation cannot bring concrete solutions to specific design problems. "We can respond to some questions, we can give explanations, but the responsibility of design is still with the architect" (Küller, 1984c).

4.7. The professional profile

The Swedish participation in the European Environmental Psychology conferences presents a perfect overall balance between architects and psychologists. It began with a majority of psychologists in Kingston, 1970, and was inverted in the following conference (table 8, appendix). This is logical as the Lund Conference (1973) was hosted by the Departure of Architecture of the Polytechnic in this city. This global balance is a faithful reflection of the situation in which this field of study has developed in this country, as cooperation between the two professions. This is reflected equally in the institutional links of the authors.

4.8. Other relevant data on participation and impact

In the conferences analysed Sweden is placed within the bloc of important producers, occupying fifth place with 6.3 % of the studies presented. Its active participation is regular and stable, with a logical upwards jump in 1973 in the Lund Conference (table 13, appendix). The only author who is placed within the group of maximum producers is Küller, who figures in second place on the list, having been present in all the international conferences except for the first in Surrey. On the other hand, he is also the best situated author (sixth place) as far as the number of quotes received in the core of the conferences is concerned, followed by Hesselgren in twelfth place.

Analysing the impact of the most productive authors starting from their visibility in the SSCI, Küller is the only Swede who appears, although with a very low index of visibility. As in other British cases that we have seen in the previous chapter, being the compiler of the proceedings of the Lund Conference published by an editorial with a strong American implantation has reinforced his international projection, apart from his institutional role played as president of the IAPS, and apart from the undoubtable interest of his personal work. This should not obscure the role of the authors with a productivity relative to the their particular area, like Gärling and Acking, and above all Hesselgren as far as theoretical contributions with a notable impact are concerned.

4.9. Summary and conclusions till 1984

In the Nordic countries EP has only developed with any strength in Sweden, where it has achieved a defined status, recognised and integrated in the global framework of both psychology and design. Its growth process has come about more as Architectural Psychology, initially centred on perceptive issues, and later widening its range to cognitive and interactionist perspectives, rather than urban problems, stemming from pressure from social groups. Its orientation has turned to reinforcing the links with classical psychology in general.

Sven Hesselgren is the Swedish pioneer ("Functional Classic") who has had the greatest impact in the European community. Küller, Acking and Gärling have also played an important part, achieving certain impact.

Within the international context, in spite of the fact that Sweden lead the creation of the IAPS, it has not played a belligerent nor controversial role. Neither have tensions appeared within the nuclei of power within the country itself.

Due to its situation as a country within the Anglophone cultural area of influence and the character of English as its second unofficial language, they have had easy access to all the anglo-american literature, and also a certain impact on publishing the most important studies in English from within Sweden itself, although the majority of the production is in Swedish.

4.10 Epilogue: 1984-1992

In Sweden and Denmark, there are some new programmes in environmental psychology. But for Küller (1992, personal interview) there is basically a continuation of what had gone on before.

There continues to be an interest in cognitive psychology, in which Tomy Gärling's Umeä group is maintained as the central nucleus (Gärling *et al* 1985, 1991, 1993) centred on the perception and cognition of the surroundings, the impact of memory, the familiarity and configuration of space in orientation systems (Gärling *et al* 1986); the appreciation of distances in large scale cognitive maps (Säisä *et al* 1986) and in small scale ones, such as pedestrian areas (Gärling and Gärling 1988). It has also been concerned with the cognitive aspects of the evaluation and preferences of residential spaces (Gärling , Garvill *et al.* 1991, Lindberg *et al.* 1992), and the perception of risk with relation to children (Gärling 1985, Gärling *et al* 1989, 1990). In September 1992 Gärling left the Department of Psychology of Umeä and moved to the University of Göteborg.

There is still a great interest in the architecture of colour. There is still interest in research on sound, air quality and related themes in Stockholm. In Lund the interest continues in the semantic description of the urban environment, aspects related to driving and traffic, the impact of light in aesthetics, but above all in welfare and health and especially in buildings for old people.

In the buildings for elderly people the Lund group, lead by Küller, takes the health of old people and their requirements into special consideration for the design of healthier environments, connecting in this way with biology and medicine. The group is closely linked with neuropsychology, psychophysiology, the physiology of human beings, seeking to form a solid foundation for the theories that are being developed in psychology of the environment.

A new field of research for this group is that related to traffic as a cause of atmospheric pollution, noise and changes in people's use of time. Küller and Laike (1992) consider that as well as restrictive measures, the problem presents a psychological dimension of attitudes and changes in behaviour with respect to forms of transport that should be considered adequately if the problem is to be tackled seriously.

In Denmark the work of the Danish Building Research institute is outstanding. Ambrose and his collaborators (1989, 1990, 1991) are developing an interesting programme of management and implementation of ecological measures in human townships and evaluating the experience of the users of Danish experimental housing. We should highlight the celebration of a colloquium on the theme with the participation of more than twenty specialists, in September 1991 (Ambrose & Christiansen, 1991).

Much of the research is coming out of the laboratory and becoming more 'like life'. Many field studies are analysing real people in their real surroundings (Küller, 1992). There is interest in more complex designs, including more diversified aspects of the social environment and of health, and not only of the physical environment. They are beginning to include medical perspectives in the studies. A noticeable incorporation of cultural aspects in the interdisciplinary studies is coming about.

Küller considers that environmental psychology in Sweden is becoming more complex, more holistic in essence. But it still trying to maintain a strict methodology and to work in a systematic way. 'Furthermore, we try to develop theories, the best theories possible to cover broad aspects of the interaction among people and their surroundings,' (Küller, 1992).

As for the ecological aspects, some studies have been initiated recently, but I have the impression -as Küller says- that the interest in this problem of the world crisis has fallen. There is interest in the conservation of energy, but it is not as strong as it was ten or fifteen years ago. Biological aspects are being considered in the research through medicine and the affectation of health.

5 Environmental psychology in France and the Francophone area

In 1973, K.H. Craik used the term "Environmental Psychology" in one of the first revisions of the environmental field as a "useful term, inclusively and theoretically neutral" to refer to the human-environment relationships, emphasizing its perfect interdisciplinary character. If that could have been true for the America of the moment, it could hardly be so for some European countries and especially for the Francophone area, both in that referring to the labels and in the manner of understanding interdisciplinarity. This is clearly revealed in the interviews with relevant authors from the area (see 1.3.2), which together with Jodelet's report (1984 *pre-print*) and a manuscript (the script of one of his own conferences) lent by Korosec, have served as a basis for the elaboration of this chapter.

5.1. The origins

The M-E paradigm had a rapid development in the seventies in France, which Jodelet (1984 *pre-print*) does not doubt in qualifying as multidisciplinary. It will be necessary to clarify the meaning of this declaration.

If it is true that in some moments an interdisciplinary collaboration and/or contact among architects, town planners, psychologists, sociologists, geographers and anthropologists etc. came about, the truth is that in France, the development, more than being a collaboration, was parallel among the various disciplines, though of course frequently feeding off common sources. Korosec (1984) remarks that while working in monodisciplinary teams within social psychology, she had fed from philosophy and sociology on a formative level. For Pagès (1984) only after '68, in a short period, was authentic multidisciplinary study realized, fruit of the spirit of fusion that emerged from those historical dates, but is now "directed" inwardly from each discipline. He tells us that it was a very enriching period, and that we have to develop and experience everything that we learned. Elsewhere P.H. Chombart de Lauwe (1984) began his interdisciplinary studies in 1949 (a study on Paris). From his anthropological and socio-ethnological position, he called on sociologists and economists to collaborate with him. The collaboration with psychologists would come later, especially with M.J. Chombart de Lauwe. So Environmental Psychology in France is a pluri-disciplinary field, but it is not completely accurate to qualify it as interisciplinary.

Moreover, it is necessary to point out that psychology did not consciously enter some of the M-E approaches more than tardily, well into the seventies, except in some exceptional case. Although we find, as early as in the mid-sixties, some sociological approaches defined with the desire to deal with space rather than the environment (Moles and Rohmer, 1964; Fisher, 1967), and even a decade earlier the first ethnological and psycho-social studies of P.H. Chombart de Lauwe as we have seen; the purely psychological investigation -or better, psychology conscious of being environmental- did not appear until more recently. This came about as an Anglo-American influence (Lévy-Leboyer, 1976, 1980; Korosec, 1984) or as homologation with the European dynamic of exchanges in symposia and conferences, in the beginning as the Psychology of Architecture and later on as Environmental Psychology, and in part as a recognised effect of the celebration in Strasbourg of the 3rd. IAPC in 1976.[1]

In spite of this, there has been a strong dynamic of studies within the other social and human sciences for thirty years. Jodelet qualifies them as comprehensive in her approach, and remarks that they have been especially favoured by the institutions. The environment was not defined in clearly material and spatial terms, but they took the political, social and cultural dimensions into consideration. Even from the psychological perspective the subject was not only studied in the bio-psychological dimension, but also in relation to its membership of the group, its localisation and social practice, values and belief, that are expressed looking at the context of daily life (Jodelet, 1984 *pre-print*). This meant that the frontiers between diverse disciplines are rather blurred.

The autodefinition of Environmental Psychology as a pretended nucleus, or at least as a comprehensive and neutral term, with which we introduced this chapter, created a certain distrust in other social sciences. They do not deny, on the other hand, the right of psychology to make its contributions from its own identity as such (P.H. and M.J. Chombart de Lauwe, 1984). However, the treatment that the social and human disciplines follow in the consideration of the M-E relation is focused on the subject, and at least in psychological aspects, with the frequent use of merely psychological methodology. It is for this reason that in spite of the fact that Lévy-Leboyer says that it was artificial, with Jodelet we can affirm that Environmental Psychology began very early in environmental studies.

5.2. The social context

Just as in the rest of Europe, the development of studies on the environment were marked by the postwar period and urban development. Until the mid-sixties the environmental studies concentrated their interest on understanding the changes in social life with relation to urban space. Later the centre of interest would move towards the system of production of space and the power, space and social control relationship that would culminate in May 1968 and its consequences. Now interest is concentrated on public participation, social, pacifist and ecological movements, especially due to social pressure in the face of the progressive degradation and depletion of environmental resources and the effects of the crisis.

---

[1] *In all the interviews that I have had with French authors, this has been considered a key date for the autodefinition of Environmental Psychology.*

5.3. Nuclei of development and their theoretical references

If we have seen that it was difficult or forced to speak of an Environmental Psychology with a long tradition, in spite of the psychological perspectives adopted by other social sciences which have been concerned with M-E, to analyse the panorama from within the selfsame field of discipline can be even more confusing. In spite of converging sources, we find notable differences not only in the approaches (very often not so disparate) but also concerning to nomenclature, the label that is adopted and the "founding fathers".

Jodelet points out that there is a common desire to maintain the links with the French tradition itself, and in a wider sense the European tradition. Durkheim, Weber, Marx, in the sociological tradition; Heidegger, Bachelard, Baudrillard, Lyotard, Foucault, in the philosophical tradition; Lévy-Strauss and Duran in anthropology; Barthers, Greimas in semiology; Lewin and Freud in psychology.

P.H. and M.J. Chombart de Lauwe, from the Centre of Ethnology and Psycho-sociology (CNRS) in Paris, reject the environmentalist label and define themselves as being within the framework of psycho-sociology and ethnology in their work on representation, attitudes, necessities and aspirations with relation to the city (P.H. Chombart de Lauwe, 1956, 1976, 1982, 1982) and from the perspective of the development of personality and socialisation of the child in relation to the environment (M.J. Chombart de Lauwe, 1976, 1979). They recognise the influence of Halbwacks from sociology, Mauss from ethnology, Sivadodn, Wallon and Bastide from psychology and Le Corbusier from architecture. With a proximate perspective we find the studies on the experience of life in the home, the emotional investment of space, combined with a clinical and sociological treatment, in a Marxist and psychoanalytical theoretical framework by Lugassy (1970) and Palmade and his collaborators (1970-1977).

Pagès and his group (Pagès, Fourcade, Gafson, 1974; Fourcade, Lesieur, Pagès, 1976) from the Laboratory of Experimental Social Psychology in the Sorbonne, in their studies on the impact of the environment on behaviour, through social codes, developed in his Theory of *Emprise*, critically integrate some of the approaches of anglo-american Environmental Psychology, but reject the label. The term *environnement*, he says, is not purely French and does not express the content of our field of study well (Pagès, 1984). He also rejects the term Psychology of Space, quite prevalent in this country, as "space" has a connotation of emptiness that it is necessary to fill, and for the social psychologist this vacuum never exists, this "void". He defines it as Experimental Ecological Social Psychology for two basic reasons: because Ecological Psychology is necessarily always social, and because Social Psychology always has to be ecological; that is to say, it has to consider the framework wherein the action develops. Moreover, the identification with the term "ecological" is made fully consistently, both when it includes the relationship not only with the abiotic environment but also with the biotic. He considers that it does not enter the rubric of Environmental Psychology, as it is connected with a much longer tradition and with different sources. It takes, he says, the sources of the movements and the social thought of the 19th. century. But it would be necessary to go back to the epicurians, to Hippocrates, to Greece. Human geography has also been important, especially Blache, with relation to the German variety. He considers that in fact, the first investigations in Experimental Social Psychology with architectonic and urbanistic implications were realised around 1865-1870, on the planning and construction of the Palais Social in the city of Guise, which was to house the first production cooperative.

In spite of the terminological coincidence he does not consider the approaches of Barker as being too important, but he does on the other hand consider those of his predecessors Lewin and BrunswiK as being particularly so.

The nucleus of Strasbourg is one of the pioneers. Moles (1984) considers himself as being the originator of the term "Psychology of Space" as early as 1964, a widely accepted term which we can confirm by revising the francophone bibliography. His disciple of the time, who now works in Metz, Fischer would not take very long to publish his *Psychology of Space* (1967) and to form the "Group d'Études de la Psicologie de l'Espace" promoted by Perla Korosec, a real driving force in this branch, organiser in the 3rd. IAPC and the visible head in the heart of the European scientific community, particularly in the IAPS. The basic consideration is of space as territory -as a labyrinth according to Moles in some of his work- that includes people as a whole as an anchor point and field of liberties, to a greater or lesser degree structured in concentric circles, with a central nucleus in the individual and around which things, people and the world are structured (Moles and Rohmer, 1984). The basic background of Lewin, the theories of Gestalt and the German phenomenology of perception stay in his approaches. Phenomenology, he tells us, is an immediate approximation, of spontaneity, of identification with the environment, and gives results that even though they can be not different, they do go in another direction from the experimental ones (Moles, 1984). The term Psychology of Space, however much it appears similar to that of Environmental Psychology, has different origins.

> *[...] When we began we did not know about the work of Hall, Sommer, Proshansky and the others. We had the impression of starting a totally autonomous field of studies [...] When we learnt about them, around 1970, we had been working for more than four years and already had a regular formation [...] (Moles, 1984). [Personal recorded interview. The transcription and translation is our own.]*

Perla Korosec, for her part, recognises the importance of the role that urban sociology and human geography has played in the formation of this domain. People like Ledrut, Tournier, Margeraux or Castells play a very import role, but especially important is Henri Lefebvre and his approach to the relationship between space and politics, the city, the habitat, the generation of space, and above all the notion of appropriation of space, that later on she would develop from the phenomenological perspective and that would constitute the central theme of the 3rd. IAPC Conference in Strasbourg in 1976. The continuous contact in seminars and studies with colleagues from Heidelberg (Graumann, Kruse) and from Lausanne (Barbey) has been of key importance for its definition. After a visit to the USA where she was able to confirm the basically clinical origins of the American environmental psychologies, in contrast with the psychosocial origins of the French ones, "I decided to restore a European tradition, which is the tradition of phenomenological psychology" (Korosec, 1984).

Claude Lévy-Leboyer, proceeding from the social psychology of labour and organisations entered Environmental Psychology belatedly, as a result of assignments to study the effects of noise in the city. Of the group of authors that we have revised he is the only one to fully assume the name of "Psychologie de l'Environnement" of anglo-american origin, and points out the importance *a posteriori* of Lewin, Brunswik and Tolman (Lévy-Leboyer, 1980). For this author the discipline is not as well established in France as it is in Great Britain and the United States. Moreover he does not concede importance to the sociological and anthropological antecedents of the French tradition (Lévy-Leboyer, 1984b). Perhaps for this reason the revision *Psychologie et Environnement* (1980) has been translated into English with notable acceptance, greater than that of the other nuclei analysed until now. This fact could be considered as an indication of the power relationships that exist in the International Environmental Community.

Claude Lévy-Leboyer, The last nucleus that we will revise is the one that works from the psychosocial perspective in Paris, linked to the Maison des Sciences de l'Homme, formed by Sergei Moscovici, Denise Jodelet and their team. They tackle the question of the environment from the point of view of social representation, paying special attention to the cognitive and symbolic process that underlies spatial behaviours. From here they take into consideration the political sphere implicit in the social representation of community participation and attitudes faced with the environment (Jodelet *et al.*, 1983, 1984; Jodelet, 1984 *pre-print*). The interaction of the group or culture with their environment is mediated by the common values that are social, in the sense that they are generated or built through the social process, and can be studied in themselves, independently of the individuals that practise them.

Outside the France it is necessary to name two nuclei with a certain relevance: Lausanne and Louvain-la-Neuve.

In Lausanne, Switzerland, in the Institut Politéchnique Federal, we found the group lead by Gilles Barbey. He has maintained a close collaboration with two more nuclei, Strasbourg and Heidelberg, with which he shares the same philosophical substratum and theoretical perspectives. Since 1984 Barbey has been the second president of the IAPS and Korosec is a member of the consulting committee. All this means that this triangle (completed by Kruse and Graumann) is outlining itself as an important nucleus of central European power for an immediate future.

Louvain-la-Neuve, the Francophone university of this Belgian city, presents two aspects: the Psychological aspect, with the visible figure of Jules Gerard Simon, from the School of Architecture, who also with a phenomenological perspective, was the principal person responsible for the organisation of the 4th. IAPC, and an untiring fighter for the breaking of the Anglo-Saxon monopolitism of the IAPS, from within the first Executive Committee; in the School of Sociology, with a marxist urban sociology perspective, we find Jean Remy and Lylian Voyé.

Finally, although he does not belong to the European area, we would like to mention Jean Morval. From Montreal, in Quebec, he maintains a strong and logical cultural link with the European Francophone area, apart from a personal link with Belgium. Morval has published one of the few systematic introductions to Environmental Psychology in French (1981), also from the perspective of social psychology, but in this case with certain behaviourist touches.

In spite of this confused panorama of identifications and pseudo-tendencies we can find the first bibliographical and conceptual revisions quite soon, as early as the beginning of the seventies (Gaillard and coll., 1971; Gruska, Mazelrat, 1975; Lecuyer, 1975, 1976*a*, 1976*b*). The defunct *Institut de l'Environnement* played an important part in this. It is curious, however, that in spite of the strong recognised influence of American Environmental Psychology and the its very active tradition, there is no translation of the texts that have become classics, as in the case of other contexts, such as Spain for example.

In spite of the acknowledgment of receipt expressed by the French authors, their autodefinition, the theoretical aspect, the epistemological foundation and the methodological development explain that there is not a full identification with the parameters of Anglo-Saxon Environmental Psychology, which is strongly criticised from the theoretical and ideological point of view (Pagès, 1984: Jodelet, 1984 *pre-print*).

5.4. **Other tendencies**

As far as the non psychological social sciences following Jodelet (*pre-print*) are concerned, next we will attempt to synthesise an orientative panorama.

In the first place it is necessary to speak of the studies of sociological tendency, the most stimulated by the French administration, according to Godelier (1982). With some initial studies in the fifties, with their root in Durkham, Weber and Morin, and spurred on by American urban sociology, it borrowed a goodly part of the methodology of social psychology. It took into consideration social practice in relation to space, the meanings of the M-E relationship and analysed the answer to urban planning on the part of the citizen. Inside it we find a culturist orientation (Raymond and Haumont, 1966; Haumont, 1972) that considered that determinate social types and cultures correlated with determinate types of ways of life and ways of using space. Moreover, construction was a crystallisation of the cultural model, an expression of social customs.

Another tendency within sociology is the Marxist tendency, on which Lefebvre (1968, 1970, 1974) exercised a considerable influence. He related the urban phenomenon as qualitative to the quantitative growth of production. He pointed out the contradiction between the social production of the environment and social practice. He established the theoretical unity between mental space and social space, and introduced the notion of appropriation. Space is never neutral, and knowledge has to be used to change the urban way of life. Together with him we should mention Ledrut (1968), Kopp (1975), Ragon (1975) and especially the structuralist Marxist Castells (1973, 1975) who criticised the theoretical and empirical axes of environmental investigation (Pol, 1981; Hernández, 1985). It is necessary to mention, also within sociology, the systematic perspective (Touraine, 1969; Gaudin,1979) that analyses the environmental questions as a result of a complex network of social and institutional, economic and political interacting systems.

To find another study group we have to look to the field of anthropology. The analysis of spatial symbolism by Lévi-Strauss had a strong impact. Moreover, this tendency was seen to be reinforced by the expansion of ethnology and ethnohistory in France and Europe. In principle we find an interest for the rural landscape, for the symbolic system of space as a study tool of attitudes and values (Rambaud, 1969; Courgean, 1972; Vincienne, 1972) or the symbolic marking of the home, to facilitate the reproduction or creation of intra or inter group relationships, through the concrete representation in space (Bourdieu, 1970).

We should emphasize the urban anthropological tendency which takes methods from ethnography and the Chicago School for the study of microsocial environments (buildings, stairways...) in small groups (family, groups of youths) with a participant method of observation (Pelletier, 1975; Althabe, 1977).

Finally semiology, which analyses the city as a non-verbal system of significants related to the cultural system. The city has suffered a transition from a closed system, of slow change of significants (archaic or medieval societies) with key systems of signification (political, religious, social), to an open system and of rapid change, hyposignificant, that leads to the pollution of stimuli (Choay, 1965, 1972; Sansot, 1972).

5.5. The roots

The authors interviewed recognise their debt to a wide and varied spectrum of predecessors who have already been specified in the treatment and exposition of the lines of work that we have just finished. So now we will go on to mention briefly the authors that emerge from the bibliometric analysis (see 10.5).

As "Historical Classics", apart from the studies on anthropology and urban sociology, the genetic perspective is predominant with reference to psychology, with Piaget and Wallon, and some references to James's functionalist psychology, Hull's behaviourism and Freud's psychoanalysis. Referring to the "Functional Classics" the authors quoted can be grouped together into three large blocs: the first, profoundly marked by the social sciences in general; the second, more strictly psychological; and the third, composed of specific studies.

Within the first bloc we include the subgroup with a philosophical outlook characterised by Bachelard, especially with his poetry of space, and by Malraux. A second subgroup with a more socio-anthropological outlook is formed by P.H. Chombart de Lauwe, Bloch, Gurvitch, Dumezil, Mitrani and Hall. The second block, more strictly psychological, is formed by references to psychoanalysts, with Freud, Lagache, Lacan, Klein and Spitz. Those interested in childhood also quote Piaget and Wallon, and finally make references to Festinger and Kelly.

In the specific works mentioned the differential studies between the sexes by Jones are quoted, Erikson's behavioural studies and Radcliffe's anthropological work. Mancipoz's analyses of social movements are also quoted, along with those of the evaluation of environments of Ladol and Smitch.

In synthesis, the strong philosophical, social, psychoanalytical and genetic components are quite definitive differential characteristics in this area, which reflect what we have seen in other sections of this chapter in their choice of "Historical Classics" and "Functional Classics" .

5.6. The institutional development

As far as the institutional aspect of the discipline is concerned (university, education and investigation), Jodelet's report gathers and analyses the existence of 214 programmes centred on the M-E relationship. Of these, 12% have a psychological or psychosocial perspective, 18% are socio-historical or socio-ethnological, 31% legal, economic or political, 26% of human geography and 13% of physical and ecological geography. In 1984 there were sixty-two universities and higher education institutes that include curricula on environment. Twenty-five do this in their departments or laboratories of psychology or social sciences. Jodelet concludes that the social sciences and psychology are more deeply concerned with investigation and consultancy on environmental questions than it could possibly seem, only placing value on the universitarian curricula, thanks to the governmental initiatives to promote interdisciplinary programme. Jodelet counted 820 publications on environmental themes between the end of the sixties and 1981.

In spite of the non-publication of texts on philosophical themes on the environment in the last period, the authors of philosophical character are still appearing in the references of investigators, as the prevalence of the general studies demonstrates the tendency towards the development of reflections of a global and humanistic kind, which are quite characteristic traits of French thought.

As for the relationship between the studies of the social sciences and psychology, after a moment on the downturn during the seventies, the psychological perspective goes up again. Jodelet attributes this inflection to two reasons: one that is intrinsic, related to theoretical questions of the same discipline, and another extrinsic, through a social demand related to the ideological moment. She relates the intrinsic reasons to the influences and the posterior crisis of behaviourism and the experimental perspective. Later, at the end of the seventies, a revival took place because of the new impetus of applied psychology and the influence of Environmental Psychology from other countries. She continues in this way:

After 1968, the inclusion of psychoanalysis in the higher educational curricula, the great emphasis made in the symbolic processes to explain behaviour, and the influence of the Marxist perspective, help to understand the field of interdisciplinary investigation - in which psychology has to play an important part - and modify some of the perspectives of Environmental Psychology [Jodelet, 1984 pre-print, p.24].

Also in social sciences structuralism and Marxism gradually widened the field of environmental investigation, facilitating a greater convergence with psychology. With all this, and in spite of Jodelet's optimistic analysis, the work of interdisciplinary collaboration is quite minimal, in the light of the data that we have at our disposal.

On the other hand, the Francophone area is prolific in meetings, symposia and congresses. Here we will only quote three that have a special significance and explicit identification with Environmental Psychology.

The first conference to be described is that of 1976, the 3rd IAPC, in Strasbourg. Organised by Moles and Korosec, with the collaboration of Graumann and Kruse from Heidelberg and Barbey from Lausanne, among others, it signified the irruption of professions like sociology and anthropology, completely absent until this point in the "Psychology of Architecture" conferences. This same tendency was consolidated in the 4th IAPC, in 1979 in Louvain-la-Neuve, organised by Simon, Remy and Voyé, where the first indications of the British controversy appeared (see 3.7) and where the first steps for the creation of the future IAPS would be taken (see 2.4).

The third meeting that we refer to was the restricted colloquium that took place in Paris in June 1981, organised by Peter Stringer, Serge Moscovici and Denise Jodelet under the title of "Towards a Social Psychology of the Environment". Twenty social psychologists gathered in this meeting, proceeding from France, Great Britain, Holland, Germany and North America. The existing tendencies in this field would be analysed in the light of the tensions manifested in the last conferences. P. Ellis (1981) resumed them in this way:

1) The empiricist Anglo-Saxon Social Psychology, with its roots in Watsonian behaviourism, and that sees society simply as the sum of its individual parts.
2) The rationalist European Social Psychology, which sees Durkheim as its progenitor and that is principally linked to the "social representation of the world and its influences on the norms of social conduct" (P. Ellis, 1981: 44).[Our translation.]

The conclusions of this meeting were without doubt clarifying for the moment in which they occurred, when the controversy between the two British groups, a sign of what was happening in EP in general, was still fiercely burning, not to say at its most intense point. The definition of referential paradigms, linking Environmental Psychology to social psychology was not new in itself. We only have to remember the origin of EP in a goodly number of the cultural areas that we are analysing. What was new, the other hand, was its formal definition in a symposium and to explain from here the conflictive situation and the sentiment of crisis that was being lived in the field. Moreover some of the bases of the central European tradition were pointed out, which as we have seen, were shared to some extent at least in the Francophone area. However, this does not redound in an increase in collaboration nor in the joint institutional activity that could be emphasised, sue to the "peculiar" structure of relationships (or non-relationships) between the groups that arose, at least in France.

5.7. The professional profile

The belated entry in the dynamic of meetings (1973) was realised mainly through psychologists. Psychologists and architects share percentages of 41% for the first and 16% for the second. This was maintained in all the conferences, with the exception of Surrey, 1979 (table 7, appendix).

As a unique case between the most productive countries (in the group of average producers Belgium is repeated) in France the sociologists have an important specific weight (16.7%) after their incorporation in Strasbourg, 1976. This is understandable from the dynamic that we have seen in this chapter. EP in the Francophone area was mainly developed from the social sciences. On the other hand the psychologists' links to architectural centres and the selfsame collaboration between social scientists and architects in the conferences were very reduced.

5.8. Other relevant information

The strictly French participation in the conferences places this country in the fourth place in the *ranking*, according to what can be seen in table 13 (appendix). However, if we consider the whole of the Francophone area, that is to say, if we add the active Belgian and Swiss participations and the single participation from Luxembourg, it surpasses the American presence and goes on to second place. We find the Francophone participation notably concentrated in the conferences that have been celebrated in its own cultural area and in Barcelona, the three in which the French language has been at least co-official and has had an active presence (in Berlin, 1984, it was merely formal). So the linguistic factor is shown to be fundamental in this area.

The only author who is placed in the group of maximum producers is Perla Korosec. As outstanding authors in relation to the same area we have to mention Barbey, Bernard, Pezanou, Remy and Voyé (see 11.1).

In the most visible group of authors, solely Piaget appears, in eleventh place, and the Institut de Sociologie Urbaine in sixteenth place (table 11.3). Korosec remains in fifth place among the most productive authors in the order of the impact that they create, quite far from those that head the list in quotes received. On the other hand, the SSCI appears with only three quotes to works not related with the conferences (table 21, appendix). The impact of the authors from this area is fairly limited.

The majority of the themes in the work in this area are centred on territorial behaviour and especially the appropriation of space, followed by the evaluation of the environment in both cases, as an application to space in general and to the home in particular. The cognitive and anthropological perspectives have a notable presence (tables 16 and 17, appendix). This is not surprising given that their highest participation came about in Strasbourg and Louvain-la-Neuve, where the central themes were the appropriation of space and the conflictive experience of space, respectively. In any case this is coherent with their tradition and with the data that we have seen throughout this chapter.

5.9. Summary and conclusions till 1984

In synthesis, we see how the Francophone area, and very especially France, displays a tremendously complex panorama that is difficult to schematise. There is a long tradition of work that concerns itself with the M-E relationship from all the perspectives and disciplines possible of the social sciences, architecture and urbanism but which does not accept any disciplinary label that would include them all. In psychology development occurs within social psychology, closely tied to and identified with this discipline. There is a certain common reference to Environmental Psychology, but almost always as an exterior framework from which to differentiate itself. Another characteristic is the individualism and personalism with which each tendency has developed, with strong emphasis on particular origins and the ignorance, disregard or even contempt for the work of other groups.

The Strasbourg nucleus is the one which offers an image, and maintains a "homologated" presence, more assiduous and assumed, participating in the formal structure of the European environmentalist scientific community, fully and openly identifying with phenomenological perspectives, of which it has made a symbol of differential identity.

Without a doubt, the situation of unhearing rivalry between the groups described means that the contribution of French "Environmental Psychology" (for want of a better name) to the International Community has a specific weight quite inferior to what should correspond to its conceptual wealth.

5.10. Epilogue: 1984-1992

In the last eight years, the interests in the Francophone area have not changed particularly. They are still centred on the urban issue, facilities, housing and working organisations, as a general tonic, with the two basic tendencies already explained: the most experimentalist centred on the perception and evaluation of the environment and the most phenomenological, more concentrated on the appropriation and experience of space (Jodelet, 1987 explains it completely, systematically and measuredly). The institutional changes in this period are the most relevant.

According to the reports, documentation and interviews, apparently some of the bitterness has been removed between different groups, which we referred to throughout the chapter. However, analysing the revisions published, we can see how exclusions from tendencies or groups are silenced, minimised or maintained, according to who is preparing the report. However, there does seem to be more collaboration in research, a certain agglutination around institutional groups and entry in international interdisciplinary programmes, like those of the M.A.B.

After 1989, the investigation units of "Architecture, Urbanism and Society" of the CNRS linked to the University René Descartes-Paris V, were merged into a single "Laboratoire de Psychologie de l'Environnement", directed by Claude Lévy-Leboyer. It has notable members in Yvonne Bernard and Gabriel Mosser, together with around twenty more teachers and investigators (Report, 1991). It is organised into three thematic fields. The first, on insecurity, noise and affectation of children by environmental stress; the second, on social representation and conduct, centred on space, on identity and the working world; the third on theoretical and methodological problems. This is clearly reflected in the report of the laboratory mentioned and in Bernard and Lévy-Leboyer's revision (1987) on the monograph that the magazine *Psychologie Française* dedicated to environmental psychology.

One of the most relevant developments has been the disappearance of the Strasbourg nucleus. Even before the death of Abraham Moles, from the beginning of the eighties the group had been breaking up. The difficulty and the conflicts in the succession to the leadership signified the disintegration of the principal group of the 'Psychology of space'.

By 1985-86 there was nobody left to take charge. Korosec had left for the United States and the people who had more or less worked with Moles approached space in a very precise manner. They gradually joined the university or professional practice in a very conventional way, according to Korosec (1992, personal interview). With the dismembering of the psychology of space group, the most formally structured syllabus of formation in environmental psychology disappeared. In section 5.6 we gather from Jodelet's pre-print (published in 1987), that there are around twenty-five environmental programmes with a psychological or psychosocial perspective. Despite this apparent profusion, Bernard (1991) remarks that there are no formal courses. In any case, they are given under the heading of social psychology.

In comparison to this disappearance, there has been the creation of a second purely experimental environmental psychology laboratory on the part of the CNRS in Strasbourg. It was an old bioclimatology centre which is now an environmental psychology unit. It works on stress, confinement, noise, temperature and similar themes.

Another innovation pointed out by Korosec is that the ministry that takes care of facilities, construction and housing has incorporated environmental psychologists and sociologists in order to launch the research plans financed by the French state on the environment. Yvonne Bernard has initiated a line of work on security in the city within this framework (Bernard, 1992) and Marion Segaud has launched a programme on the question of urbanity (Segaud, 1992). Moreover, the Ministry finances research in a precise and disperse way which is commissioned to professional collectives.

Korosec is still working on public spaces and housing (Korosec-Serfaty, 1991*a,b*). She recognises many changes in working methods, above all for the valuation that she gives to the contribution of history to research. With her move to Quebec, to French speaking Canada, she shares her time between the University and the activity as a consultant in architecture and urban design. This fact goes beyond the anecdotal, as it permitted her to accede to the post of Superintendent of Public Space and Green resources of the city of Montreal. It is interesting in that it is a post traditionally occupied by an environmental psychologist. Being a managerial post, it can generate commissions and projects that seriously approach social aspects.

Korosec (1992, personal interview) defends the consideration of Quebec as being within the valuation of the Francophone area for its strong orientation to finding links with Switzerland, Belgium or France. The link sustained in shared works between Gilles Barbey, Perla Korosec herself, Yvonne Bernard and Marion Segaud, Vittoria Giuliani and Mirilia Bonnes are notable. In this way, Switzerland, France and Italy are becoming an important axis of understanding. This group does not maintain sustained bonds with Belgium. In fact, the Belgian participation in the IAPS conferences has diminished greatly, apart from precise participations like that of the sociologist Jean Remy, of Louvain-la-Neuve, an invited lecturer in Thessalonica. Korosec also highlights the ease of communication and the bonds with authors from other countries like Holland and Spain.

The universities of the cities of Quebec and Montreal where the tradition of multi-disciplinarity in urban work is maintained are remarkable. In Quebec, with Denise Piché, essentially action-research is being carried out, on space and women and on elderly people. In Montreal, as well as work on local urban spaces, there are research programmes on underdeveloped countries, on housing in Trinidad, in Mexico or some studies in Asia for example. In 1991 an important international congress was held there on housing. The previous conference had been in Paris.

In France, the participation of psychologists in the evaluation of environmental impact is quite minimal, except for urban contexts, where some studies on the noise of motorways by Lévy-Leboyer's team can be found. On the other hand it counts on a long tradition in Canada, for the kind of legislation that exists in the country.

To finalise this section, it should be mentioned that the term psychology of space has lost ground to the term environmental psychology. The disintegration of the Strasbourg group has probably had a lot to do with this, as well as the homological tendencies with the international community.

6 Environmental psychology in the Germanic area

Environmental Psychology in Germanophile Central Europe reveals two births differentiated in time. In fact, the first German Environmental Psychology formed one of the sources from which a goodly part of posterior developments have been nourished, not only in the German area but also in Europe and especially America.

6.1. The first origins and antecedents

We find the first formulation for a field of study with the specific name of Environmental Psychology as early as 1924 on the hand of Hellpach. The second would correspond to the seventies.

Both Kaminski (1976) and Kruse and Graumann (1984), fundamental sources of our approximation to this cultural area, point out a necessary terminological differentiation between the two fundational moments. If Environmental Psychology was spoken of explicitly in the twenties (Hellpach, 1924), in the seventies there seemed to be preference for the term "Ecological Psychology" or "Ecopsychology" as something wider, while "Environmental Psychology" would refer only to a more technological branch of the former (Kaminski, 1976, 1978). For Kruse and Graumann it would serve, moreover, to differentiate the two historical moments.

Environmental Psychology or Ecopsychology in the Germanophile area displays roots in ecological biology, meteorobiology, urban studies and phenomenology. It is in the framework of ecological biology where we find the notion of *Umwelt* formulated just as it would be used with variations later on. From the establishment of this notion within ecology in the framework of biology by Haeckel in 1866 and Uexkül in 1909, it jumped to the social sciences, being picked up on particularly by the Chicago School (Park, Burgess and McKenzie, 1925; Hawley, 1950) passing on later to Mukerjee's social ecology (1940).

The term *Umwelt* takes on a variation that differentiates it from the environment, it would be this in as much as it is experienced and action is taken on it (Kruse and Graumann, 1984). Also in the 19th century, there was interest in scientific analysis of what had popularly been taking root, such as the incidence of meteorological cycles, landscapes, etc., in human behaviour. This lead to the emergence, at the beginning of the century, of meteorology and geopsychology, that would also form a part of the *background* of the first Environmental Psychology, as Hellpach recognised (1911).

The urban studies and the criticism of civilisation or culture began with the consequences of industrialisation. They contributed the French tradition of Compte and Durkheim to German Environmental Psychology. The first urban sociology studies, with a strong psychosocial vision of the problems, were by Simmel in 1903, Thurwald in 1904 and Sombart in 1907, and they were concerned with the impact of the new kind of city on the individual.

Phenomenology is the most recent of the sources that constitutes the substrate of German Environmental Psychology. It has brought a metatheoretical framework to the study of the Human-environment relationship (Kruse and Graumann, 1984).

Obviously, there are other fields that have also made contributions to German Environmental Psychology, but have less historical weight and more contemporary impact, like anthropology, ethology, geography, systems theory or architecture.

In 1924, after the first world war, Abderhaden published a *Manual of biological methods*, whose third volume was entitled *Psychologie der Umwelt* ("Environmental Psychology") edited (compiled) by Willy Hellpach (1877-1955). It divided the environment into three circles, the natural one or the "Geopsychological factors" the community or the "Psychosocial factors" and the "Built world", that later on would be specified as an object of "technopsychology". In 1911 he had published *Geopsychological Phenomena*, which was translated into French, Dutch, Spanish and Italian, and that with the title *Geopsyche* can still be found in Germany. In this study he analysed the influence of a wide range of physical-environmental phenomena in behaviour. From climactic and astronomic phenomena to built spaces, artificial climates, etc., at the end of the thirties Hellpach would also take an interest in typically urban phenomena like density; crowding and over-stimulation, continuous change, pressure and state of alertness, concluding that the perceptive style of the urban citizen is quite distinct from that of the non-urban citizen. The urban environment is ambivalent, liberating on one side, it permits independence but it leads to isolation.

Another important author that Kruse and Graumann indicate, who is not particularly well known in the international community either, is Martha Muchow, disciple of Stern and Werner in Hamburg. In 1935, together with her brother, she published *Vital space in urban children*. She recuperated Stern's concept of "personal world", centred on the dimensions of "personal space" and "personal time". In her study, Muchow distinguished three dimensions of the vital space of the child: the space that the child lives *in*, the space that the child experiments *with*, and the space *where* the child lives. Her promising work would be left inconclusive with her voluntary disappearance because of the second world war.

In Austria we found another author who would have transcendence in European development, especially due to her participation in the first British meetings: Marie Jahoda. Taking into consideration the environmental aspects, in the thirties she began a study on unemployment in Marienthal. Later she moved on to the United States where she would join the project known as "Columbia-Labemberg", defined by the *Journal of Social Issues*.

We should also speak about the work of Kurt Lewin, before emigrating to the United States, on vital space (K. Lewin, 1931, 1935), that in some way would be a reactive factor for the later studies of E.C. Tolman and in part, for many of the first studies of American and European Environmental Psychology. We will also quote the Hungarian Egon Brunswik, trained in Vienna under the influence of the "Circle of Logical Positivism", who proposed the differentiation between "central" "proximal" or "distal" stimuli. On Tolman's invitation he would fix his definitive residence in Berkeley,

California, exercising a notable influence.

Another forerunning side of the fields of study related to modern Environmental Psychology was psychotechnics, a term introduced by Stern in 1903 and that now takes the name of ergonomics, defined as the systematic analysis and the optimisation of the person-machine relationship, and in a wider sense the person-place relationship at work.

6.2. The second birth: the social context and institutionalisation

If until now we have seen the most notable milestones of the first birth of Environmental Psychology in Germany, we will now revise the coordinates of its second birth. Kruse and Graumann emphasize as factors that favoured this process: 1) The effects of the post-war reconstruction of the German cities, 2) the growth of environmental questions, 3) the growth of anxiety for "Ecological validity" within the predominant experimental psychology, 4) the growth of Environmental Psychology in the United States of America.

In this period one of the pioneers in the analysis of urban issues was Alexander Mitscherlich, with titles like *Psychoanalysis and urbanism* (1963) and *The inhospitality of our cities* (1965), some of them translated to French and Spanish and that can still be found in the publishing market of these countries. Anyway, such key cases of concern for urban problems from psychology were not frequent in the fifties and sixties. Just as in the rest of the western world, there are more and more students of architecture who feel the need to know more about psychology and sociology and ask for the inclusion of these subjects in their training programmes, but who are confronted with a lack of adequate answers, through educational deficiencies in this field, especially from psychologists.

The decisive impulse for the development of the second Environmental or Ecological Psychology, as they prefer to call it, came about through the arrival of the influence of American Environmental Psychology. This second birth was a progressive process of sedimentation, that would culminate in 1974 with two manifestations that mark its public origin.

On one hand Kruse presented the first doctoral thesis within this "new" domain. In her thesis the authoress presented a revision of the pioneering American EP and revealed the lack of theoretical foundation, emphasizing the possibilities that the notions of "vital world" and "vital space" offered from phenomenology as a possible theoretical framework.

The second manifestation, that allows us to take 1974 as a fundamental date, was the celebration of the biennial conference of the German Psychological Association (Deutsche Gessellschaft für Psychologie) in which a symposium on Environmental Psychology was included, conducted by G. Kaminski, in commemoration of the fiftieth anniversary of the work of Hellpach. It was in this symposium that it was decided to reserve the term "Umwelt psychologie" (Environmental Psychology) for the study of the influence of factors like noise and pollution, etc., and use the term Ecological Psychology for the more fundamental study of the Human-environment interaction (Kaminski, 1976). In any case, the dominant tendency has been not to create a new disciplinary field, but to penetrate all the areas of psychology with the ecological perspective (Kruse and Graumann, 1984). Two more colloquia would take place the following year, in February and October, organised by Kaminski and Graumann respectively, on "The ecological approach in psychology" (Kaminski, 1976).

In July 1984 in West Berlin the 7th IAPS Conference took place, organised by a team directed by Martin Krampen and in which 27% of the 311 papers which appeared in the programme were German. Without a doubt this meant a point of convergence between the selfsame tradition of the country and the international community, that would have positive effects in the medium term on the dissemination of the German contribution which was particularly little known in this domain.

Environmental Psychology or Ecopsychology has undergone a progressive process of institutionalisation, forming part of the environmental investigation programmes that in principal were totally dominated by the natural sciences.

The present day activity of the discipline has been marked by a strong dynamic of investigations and symposia, which do not always reach the rest of Europe due to the question of language, but in part are seen to be reflected through their participation in the international conferences. For an exhaustive knowledge of specific studies and actual fields of interest we refer you to Kruse and Graumann's study in Altman and Stokols's *handbook* (1987), and the international and multidisciplinary Bibliography included in the period 1970-1981, collected by Kruse and Arlt (1984) and Kaminski's previous works (1976). Next, however, we will present an orientative synthesis.

6.3. The fields of interest

A first area of investigation that Kruse and Graumann indicate to us, is that concerning architecture and urban planning. As we have pointed out already, the issues raised by rapid urban development in the post-war reconstruction and the urban crisis of the sixties would generate a goodly number of investigations on the experience of users, needs and satisfactions, environmental perception and cognition. The contribution of the psychoanalyst Mitscherlich has special relevance, as we have already pointed out, along with Berndt's criticism of the ideology of urban planners in 1968. At this point, however, we should emphasize the collaboration between sociologists and urbanists, which began a decade before that of the psychologists. As psychologists it is necessary to highlight Frankel, in Nuremburg, who worked on the evaluation of the habitat from techniques of semantic differentials, and on participation.

Another field of interest under development referred to the school environment, as an important nucleus of promotion in the "Institut für Shulbau" in the University of Stuttgart. The street as a scenario has been studied from various perspectives, notably by the group from the University of Augsbourg, with a Barkerian perspective. Delinquency and the school as principal objectives of vandalistic behaviour constitute the interest of the University of Münster. Recreational spaces in the open air form the centre of interest of a wide group of authors (Bierhoff, Schmitz-Scherzer, Nohl and Kruse).

The problem of environmental degradation has an important resonance in the present day German ecopsychology. Noise pollution, both from aircraft and from that produced by urban traffic, are analysed by Schönpflug and Schulz and their group in Berlin, and in some cases together with smells (Katska), are treated as causes of environmental stress.

Density and *crowding* have begun to be analysed much more recently, with an initial revision by Kruse (1975) and some empirical works by Schultz and Gambard (1979-1983), while still being themes with a certain tradition in the field of social psychology.

The conservation of energy, the risks of technologies, environmental protection and environmental education have also been the object of a large number of studies.

From within classical psychology itself a theoretical reorientation has taken place following the behaviourist fashion of the post-war period. It has gone from theories on behaviour to what Kruse and Graumann (1984, *pre-print*) call theories of action. The term "action" is understood as the most appropriate concept for the complexity of daily life, while "behaviour" has been left restricted to activities of organisms in the laboratory. This discussion on a theory of action would be recuperated in various studies by Kaminski (1973, 1978, 1983) and others.

In a broader sense, although less formalised, we find the perspective of the M-E interaction which was contributed by phenomenological psychology (Graumann, 1974; Kruse, 1974; Fischer, 1979).

From a perspective of developmental psychology with an ecological orientation, the work of Bronfenbrenner (1976) found wide and positive resonance. The impact of different systems such as the family climate, economic conditions, attitudes and styles of the exercise of the function of parents (Engfer, 1979; Lukesch and Schneewind, 1978), the "school climate" (Dreesman, 1982), etc., generated a large quantity of considerations and studies. Some of the approaches that Muchow had already made in the thirties on vital spaces were recuperated here, which we have referred to previously.

Elsewhere, German participation in the conferences was centred on the evaluation of the environment in general. The studies from the perspective of territorial behaviour and appropriation of space appear in a second place, and as a third perspective in order of importance we find environmental cognition, although with little specific weight in the last two. The environment in general is its principal object of study, followed at a distance by the interest in housing and the urban environment. We should mention the fact that on two occasion they were concerned with natural space, something quite infrequent in the conferences, as we have seen (tables 16 and 17, appendix). As a whole, German participation in the conferences does not reflect the thematic wealth of its development according to the reports used for the elaboration of this study. We will concern ourselves with this in sections later on in the book.

6.4. The roots

We have already seen in the previous sections what in fact constitutes a rich tradition of its own in this area. However, apart from some studies with a special interest in historical investigation, it is not frequent to find quotes from them in present day studies. In any case, in some studies on ways of life and the prospectives of the city of the future we find references to Simmel, together with Park, Wirth and Burgess from the Chicago School, and one or two architects and urbanists of the time, forming part of the reduced block classifiable as "Historical Classics".

As for the "Functional Classics", in the same way there are few references, dispersed and punctual, except for Osgood who collects more mentions. However, Kelly, Piaget and Maslow have to be mentioned as authors shared with other cultural areas (see 10.5).

Shannon and Weaver are quoted for their mathematical theory of communication, and Siegel for methodological and statistical aspects. Lindblom and Ashworth are quoted in the analysis of the urban environment, as well as those already mentioned as "historical". Finally, references are made to Hatt and Klose in the concern for the natural environment.

6.5. Other relevant information on the participation in the conferences

Germany, in overall numbers, is the second most productive country in Europe, but only due to the Berlin Conference. Its presence in the previous conferences was quite diminished (table 13, appendix). In spite of its long history in the field of EP it did not enter the conference circuit until Lund, 1973, through the sole presence of an engineer. It would continue in Surrey, 1974, through the organiser of the Berlin Conference, Krampen. The first institutional link would come about in the participation of the nucleus of psychologists from Heidelberg in the organising committee of the Strasbourg meeting in 1976, a nucleus with which it already maintained intellectual links. Despite the fact that the authors who played a more notable role in the social structure of the conferences were psychologists, the better known majority is of architects (table 6, appendix). We should point out, however, the high percentage of those who do not specify a professional or institutional link. This known professional distribution can probably explain the lack of concordance between the themes treated in the conferences and those exhibited in the reports, which we have already mentioned in section 6.3.

Martin Krampen enters the group of the ten most productive authors, but we should make it clear that with seven studies he has only participated in four conferences, three of which he himself organised. So it is not a particularly significant presence. He does not enter, on the other hand, the group of the most visible authors in the conferences, gathering five quotes received from his collaborators in other studies. On the other hand, in the SSCI he attains a certain visibility (tables 11.1 and 11.2).

The German participation in the analysed conferences, although massive, is infrequent and therefore still not particularly representative of the line of study that comes about in this country.

6.6. **Summary and conclusions till 1984**

In synthesis, we find before a quiet developed domain within the Germanic area, more for the research aspects than for training programmes, while not having a particularly strong international impact.

In this cultural area we find the first formal development of an Environmental Psychology on the hand of Hellpach and Muchow, which was lost because of the war, but whose seed is in the origin of posterior approaches in the majority of the countries which goes beyond the cliched and almost exclusive reference to Lewin. The second birth would take place formally at the beginning of the seventies, with names like Kaminski, Graumann, Kruse or Krampen, among others. In the first quarter of the century the name of Environmental Psychology was adopted. In present day approaches, however, the term Ecological Psychology is preferred.

The study and recuperation of this first Environmental Psychology could be a most important Germanic contribution, which modifies the coordinates of actual Environmental Psychology.

6.7. **Epilogue: 1984-1992**

I have spoken of this period in Germany with Lenelis Kruse from Heidelberg. Together with Graumann and various other collaborators this centre is still shown to be one of the most active nuclei. Another notable nucleus is the already well known Tubingen group, with Gerard Kaminski.

The vision of this period of environmental psychology in Germany presented to us by Kruse differs slightly from the observations that other interviewees have made. Both for the kind of analysis, and for the kind of expectations and themes that he foresees will develop.

A first element of evaluation is that there have not been significant modifications in the number of groups working or nuclei of research. The expectation of the seventies was that environmental psychology would to grow a lot, and this has not been the case. In those moments the interest of the architects was revealed, and therefore the research was centred on urban and constructive themes. Today it is still concentrated on he same themes, but it no longer interests the architects. They are too occupied with their 'art' and with realising their own private 'utopias'.

In any case, the work groups seem to be maintaining themselves. Some grow and some shrink. For example, Tubingen decreased due to Kaminski's retirement. In reality -Kruse considers- the question has to be what is happening to the posts. In fact there is only research for public institutions, and environmental psychologists basically work in planning.

In any case there is a certain diversification of fields of study. For example, some of Kaminski's students are acting as consultants in traffic problems and in transport. Kruse considers that one of the fields in which environmental psychology should be more active and which has a scarce presence at the moment is the naturalistic or ecological field.

One of the problems in Germany is that there are no interdisciplinary syllabi in the universities. There are syllabi of psychologists, architects' syllabi, biologists' programmes, but not integrated syllabi. Some precise psychosocial programmes have been created by naturalists, especially by those who were to go to the Rio conference in 1992, but little else. Of course, it could just be a beginning. In this way he considers that now the ideal companions for environmental psychology would be biologists, ecologists, geographers...

As far as environmental education is concerned, it is being carried out by pedagogues or educators and ecologists or biologists. He does not agree with the way they carry out environmental education because they are closed off, not connected with other groups. They only connect with some technicians. There are aspects linked to attitudes and behaviour, for example, which require another kind of treatment, more psychological.

In this way, the reunification of Germany presents some interesting aspects for environmental psychologists. The principal problem for them is environmental cleansing. There are green reserves that are now being used indiscriminately and destroyed. The naturalists do not know what to do nor how to save them. It is necessary to define criteria for the conservation of the environment and to carry them out. This requires the joint work of economists, social scientists and of naturalists. In this case the psychologists are well situated to make valuable contributions.

There is a combined commission between the Ministry of the Environment and the Ministry of Science, Technology and investigation, which is not devoted to education but to the consideration of the resolution of these problems, changes of behaviour and attitudes. It has money for research into natural problems (changes of climate, etc.) and to help to draw up environmental policy. It is made up of twelve members, six natural scientists, four economists, one environmental technician (engineer) and one environmental psychologist (in representation of the other social sciences).

This means, above all, that there are people who are able to consider the role of the social sciences in this task. This was unthinkable five years ago, but now it is in the minds of people. There are problems like those of rubbish and residues in which the solution is not in the incinerator, but rather it is a case of reducing the tons of waste that are produced and on the other hand to recycle. This has a great deal to do with behaviour and the creation of social habits.

The manual edited by Kruse, Graumann and Lanbermann (1990) is coherent with this approach. Due to the reasons of differentiation between the first and the second environmental psychology (see section 6.2) it is called *Ökologische Psychologie*. A key book with entries by topics (like an encyclopaedia), it provides a spectrum of what is relevant in Environmental Psychology, including references to green Psychology, as well as recycling, health, noise, pollution, etc.

To finalise, I would like to point out that Lenelis Kruse has been the only author (authoress) interviewed who has clearly and spontaneously considered the need for Environmental Psychology to be more directly implicated in naturalistic themes and that, moreover, the role of the social sciences has been systematically discussed more than environmental psychology.

7 Environmental psychology in the ex-USSR

To revise the development of Environmental Psychology in the Soviet Union was difficult and daring for us, due to the difficulty posed in finding available and verifiable sources. Our work would have to be based on, by necessity, the report prepared by Tomas Niit, of the Department of Sociology, Jüori Kruusvall and Mati Heidmets of the Environmental Studies Unit of the Pedagogic Institute, all three from Tallinn, in the Republic of Estonia, published in 1981 in the *Journal of Environment Psychology*. To add to the difficulties, these authors point out, the literature on the subject is written in ten different languages, although 90% of it is published in Russian. Moreover, the term "Environmental Psychology" does not have a great acceptance, except in Estonia, in spite of the existence of works of interest within this field. The term Psychology of Architecture, on the other hand, has had greater circulation among architects since the sixties.

7.1. The social context, antecedents and influences

As in the rest of Europe (and also in North America) architects were the first to take an interest in the built H-E question in the Soviet Union, well before the psychologists. Names like Malevich, Melnikov, Ginsburg, and others, from architecture, approached problems of perception of the form, colour, space, composition and psychological aspects of residential buildings as early as the twenty and thirties. In fact, their peculiar political organisation (without wanting to go into value judgements) and especially the social organisation, has lead to situations, above all with relation to housing, in contrast with the western world and its own traditional habitat, which converts it into a tremendously suggestive involuntary field of experimentation and analysis. Moreover, the housing policy followed by the model of soviet development required an elevated level of public planning, something that would soon cause a problem to be resolved on which to reflect, similar to that which has constituted one of the first explosive elements of Environmental Psychology in the rest of Europe.

On the other hand, despite the relative isolation of the scientific community of the countries in the soviet orbit in some aspects, recognised influences have filtered down that range from J.J. Gibson's ecological theory of perception to E.T. Hall, R. Sommer or the more recent D. Stokols, among others.

7.2. The institutional development

While in the rest of Europe the conferences, congresses and symposia have played a driving role in the study of the M-E relationships, facilitating the contact between investigators and the creation of a certain community conscience, in the soviet states this has not been the case.

In general, each author works independently and isolated from the rest, of course, within the well established scientific branches, with which he identifies and that include the environmental perspective in any case. Only belatedly has there been an attempt to break this tendency, with the organisation of some conferences.

In 1979 we find an initial conference on "The Development of Ergonomics in the Design System" celebrated in Borzhomi (Georgia). In 1980, a symposium was celebrated on "The typology of the world of things", organised by Zelenov in Gorki, and in 1981 T. Niit, M. Heidmets and J. Kruusvall organised a congress in Lohusalu, Estonia, on "Man and environment: Psychological Aspects". This dynamic has probably continued, but we have no references to this. T. Niit has participated in two European conferences, Surrey 1979 and Berlin 1984, in the latter as an invited lecturer. In 1985 there was a proposal on the part of the IAPS to all the European conference organisers to transmit an example of the corresponding acts to this last group in order to facilitate experience and future interchanges.

The process of institutionalisation of our field of study is on the other hand, anterior to this. In 1973 the Investigation Unit in Environmental Psychology was established in the Pedagogic Institute of Tallinn, Estonia, with Jüri Kruusvall and Mati Heidmets, where themes on the relationship of man with the built environment have been tackled in collaboration with various institutes of architectural investigation, both referring to the urban question and housing.

The M-E field of study was established under the perspective of different disciplines in other institutions not specifically psychological, without neglecting the psychological aspect. The Central Institute of Investigation in the Theory and History of Architecture, in Moscow, with Glazychev, Minervin and Rappaport, among others, its homonym in Kiev, with Evrienov, Marder and Tkachikov, an the Moscow Central Institute of Investigation and Experimental Planning of Residential Construction, with Kartashova, Orlov, Ovsyannikov, Koloskov and others, took an interest in social aspects of architecture. In the Moscow Institute of Architecture, with Nechaiev, they were particularly interested in psychological problems related to the creative work of architects. In this institute the courses on social factors went beyond the merely ergonomic contents - the most habitual-, incorporating social aspects of residential design. Psychological and social aspects were also treated in the Institute of Investigation of Urban Planning in Leningrad, with Borisevich and Kaganov, and its homonym in Moscow, with Privalov, Shvedou and others.

7.3. The fields of interest

Soviet psychology, in contrast with western psychology, has counted on an ideal theoretical framework for a development that would surpass the mere gathering of empirical data, the psychology of action. Human activity is the principal factor that moulds mental development. Objects and the environment, social or physical, play an important part which is recognised inside this process. One notable author following this line is Leontiev. This approach contributes a valid theoretical framework for the development of a theory of the environment. There is, however, a strong emphasis on the part of other authors (Glazychev, 1981; Savchenko, 1981) on the differentiation of scientific knowledge of the environment as "positive" and objective, and of the subjective as intuitive, and on the necessity to develop empirical studies. In this way they criticise the studies on the personalisation of space "so widely developed in western Environmental Psychology" of subjective knowledge of the environment. T. Niit, J. Kruusvall and M. Heidmets (1981), in the study that we have used as a foundation for our revision, on the other hand, lament the lack of exploitation of the existing theoretical models - both in eastern EP and the western - and expound on the nonexistent reason for thinking that this model would emerge from the collection of empirical data in studies on concrete environments. The only possibility was to create the theory, verify it and finally adapt it, and extend it to specific environments. "In our opinion the nucleus of this theory is the study of human activity and its relationships" (Niit, Kruusvall, Heidmets. 1981: 163).

The fields of interest of the empirical studies that we can find in the USSR do not differ greatly from those of the western world, apart from registering their influence. A pioneering field was that of the perception of the environment, especially of colour which, as we have seen, began to develop in the twenties and thirties on the hand of the architects. It is still being worked on particularly in relation to light and architecture (Feodorov and Koroyiev, 1954, 1961), in its influence on the perception of scale (Kirillova, 1961; Belyayeva, 1977), also aesthetics and semiotics, from the perspective of the theory of information (Azgaldov, 1978; Seredyuk, 1979, among others), or from the phenomenological point of view (Rappaport, 1981). However, there are very few studies on the image of the environment and cognitive mapping (Shemyakin, 1962; Sokolov, 1971; Raudsepp, 1981) applied to the study of central areas of cities or schools.

Another group of studies is made up of those related to spatial conduct, influenced by Hall, Sommer and Stokols, among others. This was developed at the end of the seventies (Valsiner and Heidmets, 1976; Niit, 1978, 1981c; Ojala, 1980; Solzhenkin, 1980), some of them centred on *crowding* (Niit and Lehtssar, 1981; Niit, 1981a, 1981b) or on the appropriation of space (Heidmets, 1979, 1980).

One of the most interesting fields, from the western perspective for the peculiarity that it represents in the social development system, is that concerning the home and residential environments. The Russian Revolution signified an important change in these aspects. The traditional Russian house was made up of a single large room, with a stove in the centre, that conditioned the functional distribution of the space. There were no small rooms, something only typical of castles, convents or aristocratic residences. The use of rooms as private spaces is a relatively recent phenomenon, say Niit and his collaborators. In the first years of the soviet regime the construction of buildings with small independent family flats was initiated. The urgent general housing situation and the cost of the small apartments lead to the construction of the so called "community flats", in which each family only disposed of one room. It would not be until the end of the fifties that single-family flats were built again. According to Niit and his collaborators:

> *[...] the growth of urbanisation made the psychological necessity of privacy in the home more urgent, with an end to compensate the growing quantity of social stimulation outside the home [Niit and coll., 1981: 167].*

It seems that the single-family flats built from the fifties onwards have not resolved the problem of individual space either, due to their small size and low quality. This set off, as early as the sixties, studies whose objective was to improve living conditions. Now they are building houses with more rooms (4-5) and that are larger. The architects have put some design solutions into practise, but it is calculated that only 40% of the homes use the space in the way foreseen by the architects (Orlov, 1981).

This question has generated a large number of studies from very different perspectives, such as the historical, anthropological, sociological, psychological, architectonic, etc., considering the way of life, the socio-demographic composition of the family, regional peculiarities, the degree of urbanisation of the environment, and the impact of television (Kartashova, 1980; Kruusvall, 1980). The problem of available space in the home has lead to different psychosocial-environmental interpretations. On one side we find the posture that defends the necessity of a private space for individuals and for families, as "the subjective *crowding* in small houses is compensated by psychological isolation" (Niit, 1981*a*) and this would lead to passivity in the home. On the other side, Kruusvall (1980) seeks to demonstrate how the personalised division of rooms does not only come about in relation to the need for privacy, but also that above all it is an expression of the type of social interaction that occurs between the members of the family itself, with the dominance of individual aspects over functional ones. More than endeavouring to care for the design of individual rooms - he concludes - what is essential is to have special consideration for the spaces of common use.

As a result of the interest in this problem, a standard questionnaire has begun to be applied in different towns, to learn about the ways of life of families and the evaluation of the building of new homes and residential complexes. As well as the studies on neighbourhoods and homes we can also find work on specific environments, cultural centres, recreational areas (cinemas, clubs, dance halls, etc.) (Savchenko, 1980; Bakshtein, 1980; Heidmets and Kruusvall, 1979, 1980; Niit, 1980*b*). Among the latter a differential typology was established according to activities: those of "relaxation and interaction" and those of "entertainment and formative development" (Kruusvall, 1979*b*). The aspects of interaction with natural spaces have also been treated, using questionnaires or differential semantics as basic techniques (Lepik, 1981; Rips and Lepik, 1981).

7.4. The participation in the conferences

The participation in the conferences of the Soviet Socialist Republics has hardly been more than testimonial. Only four studies have been presented on the part of two authors. Niit participated for the first time with a paper in the Surrey conference in 1979 and would be invited to give a lecture in Berlin, 1984. After his stay in Surrey he would prepare and publish the report that we have used as a basis for this chapter.

7.5. Summary and conclusions till 1984

In synthesis, we can see that Environmental Psychology has developed in the USSR as an influence of its western homonym, on a peculiar set of problems but without attending to the theoretical framework that the psychological tradition itself offers, even when its approaches admit and consider the contextual and environmental aspects of human activity, especially Leontiev's Psychology of Activity. There are a considerable number of centres where this field is institutionalised in the large capital cities and especially in Estonia.

The interest is centred particularly on perception, urban problems and housing, reinforced by the peculiar social structure of the country which requires a high level of planning.

7.6. Epilogue: 1984-1992

To speak of the ex-USSR has become a complex question at the very least. In any case, as we have seen throughout the chapter, the principal activity in environmental psychology is concentrated in Estonia, in Tallinn, the capital of the now independent republic. To contrast the present situation I was able to speak to Thomas Niit, one of its most notable members, as well as working with the proceedings of the last conference celebrated in 1991 in Tallinn, still with participation from the ex-USSR (Niit, 1991).

Environmental Psychology has obviously been affected by the conflictive political situation and the economic difficulties that come with it. In any case the centres of interest do not seem to have changed too much.

The principal topic of research since 1978 was and still is Housing. The problem of how, in the USSR, very disparate ways of life in very different and highly separated geographical areas, has had to adapt to very similar buildings (in the line of what we have already revised). Over the last few years there have been some small projects on perception of neighbourhoods and the limits (frontiers) of the neighbourhoods, the study of the use of complex residential zones, and the participation of people in the building process.

As a matter of fact, one of the last projects to be carried out between 1990 and 1991, financed by the extinct USSR, was on the participation in the construction of Youth Housing Projects, realised by Niit, Raudspepp and Liik, from the Tallinn Pedagogical Institute. It included an analysis of the projects presented for construction, a study of the opinions of the leaders to evaluate the success of this movement, and a study of the satisfaction of the residents in four of the existing complexes in the cities of Moscow, Saint Petersburg, Gatchina and Sverdiosk.

Now there are some small projects, projects by students on the preference of houses, perception of landscape, crowding in dormitories, colour preferences etc., directed from the Department of Psychology of the University of Tallinn.

Until the nineties the environmental research group of the Investigation Unit of the Tallinn Pedagogical Institute received research grants that came from the exterior. Mainly from Moscow, but also from other architectonic institutions which were interested in high-rise building and office buildings. Normally they were projects of two year's duration. There were seven people working in this unit during the eighties. Last year, the economic changes meant that there were no funds available, as the institutions which donated money for research have had problems themselves, on being divided into smaller units.

Now there is only one research project under development, which is an international project on environmental movements and housing changes in Estonia, Russia and Hungary, co-ordinated by the University of Canterbury in Kent. This is British money.

During the last few years four conferences have been organised, in 1981, 1983, 1985 with papers from the ex Soviet Union. In 1991 the last one was celebrated, with twelve people from western countries. It is interesting to verify through the proceedings that as well as assembling people from Russia, Kazakstan and other places in the east, the conference was attractive to authors from nearby Nordic countries, like Finland, not habitual in the congresses in the rest of Europe.

In the Russia of today there are some psychologists and architects interested in psychology in the Moscow Institute of Sociology, like Dridze (1991), who is working in social diagnostics for planning. They are also studying the individualisation and socialisation of relations between subject and environment in departments of large architectonic institutions like the State Committee for Architecture and Construction of Moscow. In Lviv, the Ukraine, with Dourmanov (1991) they are also working on the spatial image as a genetic basis in the development of physical environment.

In Russia there are not many different names from those mentioned in the previous sections. Niit only highlights the incorporation of some young people like Svetlana Gabidulina. Oleg Yanitsky has some influence. He is a sociologist who has organised some conferences within the MAB programme. He has some works on environment movements and the emergence of civil society in the Soviet Union (Yanitsky, 1991).

In Estonia, as well as Niit, the basic team of the environmental research unit of Tallinn is made up of Matic Heidmets and Jüori Kruusvall, who started it, Kadi Liik and Maaris Raudspepp, who have worked on environmental means for regulating some intergroup relations (Raudspepp, 1991).

As far as formation is concerned, in the whole area we have not heard of any programme of specialisation. Only in Tallinn, all the students of psychology have to go through a course on environmental psychology. They can have the degree of MA in the speciality that they desire, but this does not imply a programme of formal training, but rather a process of research.

The situation is not easy in these countries. Despite this, they hope to hold a conference in 1993, 'but I don't know how many people will come from Russia ...' said Niit.

8 Environmental psychology in Italy

The development of Environmental Psychology in Italy is still quite scanty, according to Felice Perussia's description (1983) and it has to be reduced to a growing handful of proper names, although with a punctual participation in some important international projects, as we will see, after Perussia's report, which will serve as a foundation (among others) for the development of this chapter.

As in the other European countries, the terminological problem of the non-establishment of a commonly accepted label is also registered. In such a way that we can find the indiscriminate use of the adjectives environmental, architectural, ecological and even geographical in the literature on the subject, accompanying the word psychology. All of this, in itself, is a symptom of exactly what the situation is. For Perussia it is foreseeable that the standard international label of "Environmental Psychology" will end up imposing itself. Even so, he points out, it is necessary not to forget the concept of "subjective ecology", as Environmental Psychology is one of the routes to the knowledge of "subjective human" ecology. His argument is the following:

> *The environment exists for the subject and acts on him through the same subject, so that what he does is not to interact with his environment "as if it were" the image that he has of it, but almost, as the environment (the ecological niche) is the internal representation that the individual has [Perussia, 1983: 264].*

8.1. The origin and roots

Faced with scarce autochthonous investigation, the sources of reference used habitually are English and American, and more occasionally French. There is not a sufficient Italian bibliographical foundation. The scarce materials available are translations and there are no manual written by Italian authors till 1992, except for a *reading* compiled by Bagnara and Misite, gathering articles by recognised authors from various countries, among them some Italian authors.

This situation can be extended, in general, to the whole of psychology and in particular scientific psychology. In fact, there is a great void which includes the fascist period to the sixties, where a strong interest in psychology was recorded.

The first text to be translated into Italian, and that in itself constitutes an exception for the impact that it caused, especially among architects, urban planners and designers in general, was the now classic *The Image Of The City*, by Lynch (1960), translated in 1964. It did not constitute, however, an immediate stimulus for empirical investigation, but in any case it did spur on theoretical and ideological approaches. It would be necessary to wait ten more years.

It would be necessary to wait until the end of the sixties to find any translations in the tradition of Environmental Psychology. In 1978 T. Lee's *Psychology and Environment* (1976), and the American Ittelson's (1973) *La Psychologia dell'Ambiente* (*Environment and Cognition*, in the original) were translated, the *reading* edited by Bagnara and Misite came out with American texts, and finally the German Kaminski's (1976) *Studi di Psicologia Ambientale* and the French Lévy-Leboyer's (1980) *Psicologia dell'Ambiente* were translated.

In the land of the Renaissance where in the 16th century programmatic approaches were already being made on what a city should be (like a big home) and the home (like a little city), the spaces and symbols to demonstrate the power of the prince, magnify the rituals, but also to promote communication, physical health, the exploitation of energy, how to weave stable relationships between the social strata through the structuration of urban space (Borsi, 1983), the notions of *Setting* and of "image" were perfectly defined. There was a long tradition of speculative approaches on parallel concepts to those that Environmental Psychology wished to develop, on the part of architects, artists, designers, urbanists, philosophers and politicians, but rarely psychologists. At the end of the sixties and the beginning of the seventies Italy lived through a series of participation experiences conducted by architects, among which we can highlight those by Aldo Rossi (1966) and some of De Carlo's theorisations (1971), close to the approaches in Environmental Psychology in other parts of the world.

8.2. **The fields of interest**

Within the strictly psychological field, the most prolific area has been that of the critical revisions, based essentially on frequently short English and American texts, published in general architecture, geography and psychology magazines (Bagnara, 1976; Oddono, 1979; Secchiaroli, 1979; Gasparini, 1980; Perussia, 1980*a*, 1980*b*, 1980*d*; Seassaro, 1980; Secchiaroli and Bonnes, 1980; Bianchi, 1982, to mention but a few).

The studies on cognition, perception and the image of the city have also had a certain development. Milan (Bianchi and Perussia, 1978; Bonnes and Secchiaroli, 1979), Bologna (Bonnes and Secchiaroli, 1982, 1984), Varese (Brusa, 1978) and Venice (Balboni *et al.*, 1978) have been the subjects of these studies. We should emphasize in particular the set of studies that are being developed under the common heading of "Urban ecology applied to the city of Rome" in the MAB programme of Unesco (Bonnes, 1987), among which figures an interesting international collaboration in the transcultural study of the representation of the urban environment, with Bonnes (Rome), Jodelet (Paris), Kruse (Hagen) and Stringer (Belfast). The categorisation and memory of the environment have been treated with a more theoretical approximation (Salmaso and Legrenzi, 1984; Peron, Baroni and Salmaso, 1984, of Padua) and efforts have been made of approximation to psychosocial models for a theory of aesthetics (Giulani, 1984, in Rome). Children have been the centre of interest in the studies of a cognitive theme, centred on the image of the city in the child (Perussia, 1982; Gaetti and Venini, 1982), of the home (Axia, Baroni and Peron, 1984, in Padua) and environmental education (Brusa, 1980). Another logical aspect attended to faced with the prosperous industry, is the influence of clothing in auto-image

(M. Bonaiuto and P. Bonaiuto, 1984, in Rome).

We should also mention the studies on the subjective perception of environmental quality and the impact of pollution, contaminants and nuclear energy (Perussia, 1980, De Carlo *et al.*, 1981), the problem of earthquakes (Catarinussi and Pelanda, 1981; Geipel, 1979) and the cognitive aspects of Italian emigrants (Gentileschi, 1980).

8.3. The institutional development

Independently from the limited resources for investigation, the institutional panorama is languid. There are no official courses on Environmental Psychology in the universities and the slow bureaucratic structure makes it difficult to foresee any change in the short term. We should also name the Institute of Psychology of the National Centre of Investigation CNR of Rome, the Faculty of Psychology of the University "La Sapeinza" in Rome where seminars are given in the framework of Social Psychology and The Institute of Psychology of the Faculty of Medicine in the State University, where isolated individuals work on this theme.

In the field of geography, paying special attention to environmental perception, it is necessary to mention the Institute of Human Geography in the University of Milan and the Institute of Geographical Sciences of the University of Parma.

There is not a tradition of meetings or symposia on the theme in Italy. In 1982 there was an attempt at organising the 7th Conference of the IAPS, which was not realised due to the lack of infrastructure and establishment of the domain. Its presence in the international conferences has been minimal, only constant through the architect C. Gavinnelli with approaches that are certainly not very close to psychology. It was not until the 8th Conference of the IAPS in 1984 in which there was a notable presence, as we will see later.

In Berlin, 1984, there was Italian participation in ten studies. P. and E. Bonaiuto and Romano on perceptive aspects of architectonic and environmental structures; Salmaso and Legrenzi on the categorisation of the environment in the examination of place; Peron, Baroni, Job and Salmaso on memory and cognitive strategies; on the same subject applied to children, Axia, Baroni and Peron; M. and P. Bonaiuto on the influence of clothing in the auto-image; Bonnes dealing with cognitive construction of the urban environment with a multidimensional and categorical perspective; Giuliani would take an interest in psychosocial models and aesthetic theories on home decoration; finally, in collaboration with Bernard in Paris, Bonnes, Amoni and Giuliani presented a paper on the structure, organisation and animation of domestic space. There are thirteen active authors in total, who make up a far from negligible balance.

8.4. Summary and conclusions till 1984

In spite of the institutional difficulties and the lack of tradition, Felice Perussia (1983) is very optimistic that Environmental Psychology will find its place in Italy, if it reaches a reconciliation with its dominant philosophical and epistemological tradition.

From our perspective we have evaluated the situation of EP in this country, as a development foreign to its background, forced by an outside influence that has revealed a certain "scientific" mysticism. The Italian tradition in the problem of the environment is rich in other perspectives, as we have seen, and the socio-environmental issue is in need of new contributions. Although the environmental psychologists initially have not been able to - or known how to - establish a bridge with their tradition, this is probably the influence of the dominant model in psychology. Opening out to other perspectives within this same discipline, they will be able to find the way to the reencounter with its own tradition and to have a greater impact on its theoretical approaches and in the resolution of problems in its cultural ambit.

Four years after Perussia's report this seems to have begun, in accordance with the themes of some of the acts and studies commented on. On the other hand, although there is a practically nonexistent academic development, the institutional development through investigation programmes seems to be getting stronger.

8.5. Epilogue: 1984-1992

As we saw throughout the chapter, six years ago in Italy there were practically only two teams working in a regular way, the groups in Rome and Padua. Moreover, they were working almost entirely alone, an isolated from each other. In 1992 it seems that more bonds have been established between them, and above all, that they have joined international projects.

The Rome team has Mirilia Bonnes of the University 'La Sapienza' as its most visible head. A collaborator since 1981 in the UNESCO's MAB-11 (man-biosphere) programme applied to Rome, she has been directing it since 1988. The team of psychologists is formed by Ardone, Bonaiauto, De Rosa and Ercolani. The team of urbanists is directed by Bagnasco and the biologists' group by Pignatti. They have published some reports (UNESCO MAB-11, 1981, 1987, 1991) on quality of life and applied ecology in the city of Rome. Its objectives are centred of the analysis of the M-E relationships and in the development of an integrated interdisciplinary methodology, through study and management (Bonnes, 1991). The programme seeks to make useful suggestions to those who have to make decisions. Moreover, they call thematic conferences of an international and interdisciplinary kind.

Apart from this programme, the Rome group has participated in precise projects on commission, like the revitalisation of a historic centre in the south of Italy, or the environmental requisites of an old people's centre, among others. For its part, also in the University of Rome, Paolo Bonaiuto and his team are working on the tradition of the psychology of perception, in the perception of buildings. From these perspectives they have participated in a national competition for the planning of a cemetery.

Also in Rome, the group associated with the National Council for Research with M. Vittoria Giuliani and Guisseppina Rullo is working on the significant, attachment (Giuliani, 1991), territorial behaviour (Giuliani, 1989, Giuliani, Roll and Bove, 1990) and Place identity (Giuliani and Rullo, 1986), a different construct from cognition which is gaining importance, and on which it seems there will be quite a lot of work carried out over the coming years, according to Bonnes (1992). Moreover, this group has a long tradition of work on use, behaviour and satisfaction with housing (Giuliani, Rullo and Caro, 1987, Giuliani and Bonnes, 1987, Giuliani and Rullo, 1988, Giuliani and Barbey,1991) and the aesthetic experience (Giuliani, 1986). It has also organised some international symposia on housing, like the one celebrated in Cortona (Giuliani Ed. 1992).

The nucleus of the University of Padua is especially active, with Maria Rosa Baroni, Erminielda Mainardi Peron, Giovanna Axia, Paola Salmaso and Remo Job. They are working from a cognitive perspective, following the Schema theories and the Scripts theory applied to the environmental field. They maintain a good connection with Tomy Gärling's Swedish team. Its interest is focused on basic research, centred on verbal exploration of the knowledge of the environment (Axia *et al.* 1988, Baroni and Peron, 1989, 1992), motivation in environmental psychology (Peron 1991), the effect of familiarity in environmental schemes (Peron *et al.* 1985, Baroni *et al.* 1985, Peron and Baroni 1990, Peron *et al.* 1990), memory of places (Salmaso and Legrenzi, 1984, Job, Peron, Baroni, 1986, Peron, Baroni and Job 1990) and life span (Axia, Peron, Baroni, 1991). In some cases they apply their basic research to some surroundings in particular like natural settings (Peron and Baroni, 1988), tourism (Peron, Baroni 1989), sports areas (Peron, Baroni, Falchero, 1991) or the classroom (Axia, Peron, 1990)

As well as the Rome and Padua nuclei, Bonnes (1992) mentions some other teams. The sociology groups in Bologna, with Secchiaroli, and psychology in Lecce with Mazzotta, collaborate in work on housing and the city (Bonnes, Secchiaroli, Mazzota, 1991; in Palermo, Sicily, on illegal building (Colaianni and Alaimo, 1992). Perussia is still active in Milan, basically working with geographers.

In October 1986 a congress in the Italian ambit was celebrated in Milan, organised by Perussia and Bianchi with the title "Subjective image and environment" in which forty-six papers were presented, organised in four sessions on "Personal space and built space", "Urban space", commonplace and habitual themes in this kind of meeting, but also two rather atypical subjects, or at least somewhat unusual, though no less attractive for that: "Theory and culture" and "Journeys and outings". At least the links of theory and culture seem to demonstrate a certain autochthonous specificity in the focus of the problem, in the same way as the theme as journeys and outings inasmuch as it connects with the characteristics of a touristic and Mediterranean country in which social life happens to a large extent in public spaces.

So the principal subjects of interest and research turn around three areas:

* Cognition and environmental perception, with relation to theories of cognitive schemes, environmental knowledge, script theories, etc., but also to the 'attachment' of significants of environment and 'Place identity'.
* Problems of environmental quality, environmental attitudes, ecological problems.
* Urban planning, revitalisation of old neighbourhoods, the design of institutions and services.

The work of consultants is basically carried out from the university and the principal client is the public administration. In some cases, however, there has been demand from private planning or organisational consultancy companies who are considering environmental problems. But there are some business consultants who are organised to carry out research of this kind.

In Italy there are no formal training programmes for the moment in Environmental Psychology. The only one there is the degree course in psychology. Within the social and organisational branch there are courses on environmental psychology, which can be taken as optional subjects. This is new since five years ago, thanks to the changes that have come about in the "Estatuto" which regulates the Italian universities. But for the moment, according to the information that Bonnes (1992) disposes of, they are only given in Padua and shortly in Rome. The problem is that there are not people adequately prepared to give the courses.

For the moment and in anticipation of the growth in education, as well as some translations of now classical texts, In 1992 Bonnes in Rome and Secchiaroli in Bologna have published *Psicologia ambientale. Introduzione alla psicologia sociale dell'ambiente*, the first manual by Italian authors. The book incorporates the ecological-naturalist ambit, making references to MAB. I also carries out an interesting revision of the traditional approaches of the psychology that have recognised some role in the environment, and values the plurality and convergence of paradigms within Environmental Psychology.

9 Environmental psychology in Spain

Environmental Psychology in Spain does not have a long tradition of its own, but since the end of the seventies an increasing interest has been revealed in this field, above all in some academic ambits, that could be fruitful in the middle or long term. In 1979 we carried out a survey (Pol, 1979) in the different departments of Psychology of the Universities to find out about the studies classifiable as Environmental Psychology, with or without being conscious of it, that were being realised. The results could not have been worse. The limited number of answers obtained had the sense of affirming that, in fact, all psychology was environmental in itself and therefore it was meaningless to speak about an Environmental Psychology. Three years later, in one of the first autochthonous publications (Jiménez-Burillo, 1982) articles appeared of the most diverse geographical precedence, some of them by authors directly linked to those who had transmitted the dissuasive answer to us. Many of them approaching the theme from a somewhat atypical bibliography (which indicates a certain opportunism, but on the other hand could be very creative). Now, the situation of Environmental Psychology is improving very much. I shall try to describe it.

9.1. Some antecedents

If we look back to the past we can find, during the first third of this century, a certain social dynamic and nuclei of thought that approached the environmental issue in specific surroundings, with parameters close to the actual ones in our field. One of the tangible ambits refers to school buildings. The deplorable state of some schools, especially state schools, would sensitise public opinion and cause debates on environmental conditions in town councils and professional meetings. The celebration of the "1st Spanish Congress of School Hygienists" in Barcelona in 1912, where doctors, architects and educators debated the state of the schools and the direction the solutions to be adopted should follow. Lighting, ventilation, toilets, furniture, distribution were themes brought into discussion. Equally the educational renovation movements described optimum conditions for the building and its situation.

In fact, in Catalonia, the whole of the *modern style movement*' of the end of the last century had a naturalist philosophy in the way of understanding the M-E relationship, as a reaction to the mechanisation of the industrialisation of the eighteen-thirties, which the architects took charge of capturing in their particular aesthetic. In part, a result of this same wave of thought was the placement sought for such teaching centres as the *Escola del Mar* (sea school) or the *Escola del Bosc* (forest school) in Barcelona, and the exploitation of natural resources as educational elements.

It is possible that we will find the clearest relationship between forms of thinking, educational approach, aesthetic and architectonic forms in the educational centres built in the twenties and thirties by the architect Goday, in the classicist style, in which there is a very clear and explicit conscience of the role that the environment plays in the formative process (Pol, 1982). Inscribed in the *'noucentist* movement', civic and regenerationalist, the whole reconstructive movement of the Catalan identity existed behind it, identified with a very particular aesthetic closely linked to the model of the ideal man that was desired to be reached.

On the level of thought, Eugeni d'Ors, the ideological inspiration of the Catalan *noucentisme*, and on a Spanish level Ortega y Gasset are historical figures who must be taken into account, but who have still not merited the consideration due to them from the present day social scientists who take an interest in the environment. As both of them were strongly influenced by the Germany of the twenties and thirties, it is not surprising that they should have considered the environment as an influential element in the formation of the individual. Julián Marías (1972), another philosopher disciple of Ortega, would later also take an interest in the subject of the city.

For d'Ors the environment of a determinate historical moment would mark the thinking of the epoch, and would even allow us to see its social organisation. His insistence in the peripheral and historical origins of many psychic processes, as Siguan (1981) points out, highlight the thinking of d'Ors as an antecedent *a posteriori* in the study of the M-E relationship, that would be very interesting to study in depth.

In a field strictly foreign to psychology, but profoundly impregnated with the social thinking of the period, it is necessary to mention some urbanistic approaches in the last century, such as the urban Plan of Cerdà (1860) in Barcelona or the Lineal City of Arturo Soria (1894) in Madrid. Later on, going through the *modernist* and *noucentist* movements, we come to the avant-garde of the thirties, which with the models of the Bauhaus and Le Corbusier's influence, were organised in the GATPAC (Artists and Technicians for the Progress of Contemporary Architecture).

We should also emphasise the treatment that the relationship of man with the environment received in the work of Ramon Turró (1854-1926). The relationship that he established between the origin of knowledge and alimentation, movement and knowledge of space and what we today call psychological processing of perceptive information, and his insistence in the mechanistic neutrality of the perceptive and sensitive processes (Siguán, 1981).

In the strictly psychological ambit we also find some preceding work in institutes of psychology created between 1910 and the thirties, especially dedicated to professional orientation, and where environmental aspects related with the workplace were considered (Kirchner, 1975).

9.2. The social conditions and the first steps

The development of the social sciences in Spain was interrupted by the civil war (1936-1939) and was marked by the marginalisation that the regime that emerged (1939-1975) submitted them to. Exile and repression meant that there was a radical break with the possible tradition of the country itself, and that new generations would have to look for their sources in the foreign bibliography that was most easily within reach. The French influence, due to its geographical and linguistic proximity, would find a fertile field in which to take root. Moreover, when the situation began to become a bit more relaxed, towards the end of the sixties, this would generate a dynamic of translations, strengthened by those that arrived from Latin America, not comparable with any other European country, at least in the field of Environmental Psychology.

At the beginning of the sixties, especially stimulated from architecture, urbanism and geography, very sensible in these moments to the environmental questions due, just as in the rest of Europe, to the urban concentration produced in the previous decades, the pioneering works in Environmental Psychology began to be translated. The publishing activity of the I.E.A.L (Institute of Studies of the Local Administration), dependent on the Ministry of Public Works in Madrid, and publishers like Gustavo Gili in Barcelona were to play an important part in this, in the broad perspective that their collections on architecture and urbanism considered. Even today it is necessary to resort to collections of this kind to find Environmental Psychology titles, rather than to collections on psychology, save a few exceptions. In 1972 the translation of *Psychology of Space* by A. Moles and E. Rohmer (1964) was published, which passed somewhat unnoticed. Hall in 1973 and Sommer in 1974, together with the belated arrival of the South-American edition of Lynch (1960) published in 1966 were landmarks, especially for the resonance they achieved among architects.

It would also be in 1973 when the Tomás Llorens, from Valencia, edited a pioneer *reading* called *Towards a Psychology of Architecture* through the Official College of Architects of Catalonia and the Balearics. A series of chosen articles by Canter, Stringer, Sommer and Lee were collected in this publication, with the conviction that (as Llorens explains) "with this book Architectural Psychology makes its entry in Spain, or Environmental Psychology, as it is also denominated on occasions [...]". After that, more or less specific translations and studies began to appear, especially connected with the professional practice of architecture and urbanism.

It would not be until 1978 that some now classic works appeared with the desire to approach Environmental Psychology as a global discipline. The traditional *reading* by Proshansky, Ittelson and Rivlin (1970), *Environmental Psychology. Man and his physical surroundings*, in a Mexican translation, the text by Canter, Stringer and coll. (1975) *Environmental Interaction*, the Rapoport's text (1977) *Human aspects of the urban form*, in spite of the fact that the same author had already published a selection of texts in 1974 edited in the original version as articles between 1967 and 1972, under the title of *Aspects of the quality of the environment*. A year later in 1979, the translations of Heimstra (1974) would arrive from Mexico, and of Kaminski (1976) from Argentina. This revision of the translations, that in no way seeks to be exhaustive, brings us to the appearance of the first autochthonous global studies.

In this first step, the text by Pinillos (1977) on *Psychopathology of urban life* merits a special mention. Pinillos was studying psychology in Germany before the Second World War and constitutes a link with the first Germanic Environmental Psychology.

9.3. The fields of interest

We will divide this section into the following subsections: in the first place we will deal with the work of a general perspective; then we will revise the basic investigation, considering the developmental studies, those centred on the perception and cognition of the environment, ecological psychology and the methodological concerns. Next we will revise the applied investigation on two sides, the urban problem and the studies on the school environment, to finalise with the work on environmental education.

a) General perspectives. In 1979 we find an initial revision in Catalan, that would be published later in Spanish (Pol, 1981) with the title *Psychology of the Environment*. The intention was to give a global vision of the state of the discipline, taking the proposal of K.H. Craik (1977) as an explicative model of six paradigms for the study of the relationship of man with his environment. Remesar and Hernández-Hernández (1980, 1982) approach the conditions that permit the emergence of the new domain, analyse the models used by Environmental Psychology and how they condition the methodological strategies adopted in research. Going more deeply into this theme has lead Hernández-Hernández (1985) to the study of diverse points of view of the ecological perspective in psychology. He make a critical revision and an epistemological reflection of the models used, as well as a comparison between the different disciplines that adopt an ecological perspective in the study of the M-E relationship, exemplifying on the trajectory followed by Roger Barker.

Some public organisms linked to the environmental issue such as MOPU (Ministry of Public Works) have made some punctual works of interest possible, like a monograph dedicated to the relationship between Psychology and Environment (Jiménez-Burillo [ed.], 1982). Some of the studies that had been realised on the subject up to this time in university media were published in this volume.

Some psychological magazines have also begun to open up to Environmental Psychology, as in the case of *Estudios de Psicología* (Studies on Psychology) which offered a useful documentary guide in 1982 prepared by Rodríguez-Sanabra and Fernández-Dols.

We should also mention the incorporation of Environmental Psychology themes in the manuals of social psychology that have been produced lately, as is the case of Jiménez-Burillo (1981), Blanc (1982) and Notó and Pañella (1986).

b) The basic research. In this section we should speak in the first place about the studies on the development of the notion of the city and the notion of place realised by an architect, Josep Muntañola (1973, 1978*a*, 1978*b*, 1979*a*, 1985; Muntañola, Hart, 1981; Muntañola, Morales, Pol 1981).

The studies centred on the perception and cognition of the environment constitute a wide field. In the first place we will comment of those that have concentrated on cognitive mapping and the image of the city. In 1970, following the work of Lynch, the architects Giménez and Llorens analysed an initial study on the image of the city of Valencia. Considering the city as a collective construction, they analysed its semantic content to classify how it entered modernity and how urban decay implies a loss of the collective image.

Juan Ignacio Aragonés (1983) and his collaborators from the Complutense University of Madrid prepared a theoretical and empirical study on cognitive mapping, applying it to Madrid. They sought to see how *Gestaltic* groups were formed around relevant elements of the city and linked the principal urban elements. Moreover they tried to ratify Lynch's categories in an empirical way with two data collection techniques, submitted to a multivariate analysis, only entering in doubt in the category of limit.

In the University of La Laguna, Bernardo Hernández-Ruiz (1983) elaborated a cognitive map of Santa Cruz de Tenerife, coinciding with the previous result concerning the secondary nature of the limits, and in this case of the neighbourhoods as well.

From a theoretical point of view we should also highlight the work on cognitive mapping and behavioural maps by two teams from the University of Barcelona. Carles

Riba, Fernando Hernández-Hernández and Antoni Remesar (1982*a*, 1982*b*, 1982*c*) tackle the problem of the relationships between mental representation and behaviour -between observables and indicators-, analysing the isomorphisms as formal and functional analogies demonstrated in the similarity of certain rules that organise both behaviour and its representations. The behavioural maps, they conclude, apart from their intrinsic value have a value as models and systems of indicators of cognitive maps.

The other team that we referred to basically demonstrated concern for the development of strategies of occupation and the use of the behavioural map as a basic technique for the analysis of the environment and its relationships with behavioural episodes (Anguera, 1980, Anguera and Blanco, 1982).

Since 1978 Rosa Gratacós, of the Autonomous University of Barcelona, has been making a research on perception of space in blind children. Starting from the proposal that it is through the action on the environment that it is possible to come to know it, Gratacós (1982, 1983, 1986) proposes a series of training activities to help blind people, through the other senses, to find points of reference for the structuration of space.

The incidence of cognitive and personality factors in the perception of the quality of the environment is an aspect that has generated numerous studies in the nucleus of the Autonomous University of Madrid. Some of them in collaboration with the Departments of Psychology and Ecology (González-Bernáldez and coll., 1979, 1982, 1984; Macià, 1979, 1980; Macià and Huici, 1981). Starting from the hypothesis that personality conditions the choice of landscapes, they conclude that there is a correspondence between the personality profile and the dichotomous choice between the humanised and natural landscape.

The last to be referred to in this section is the work of José C. Sanchez-Robles (1976) in Surrey. He tackled a critical study of the social conceptualisation of the home, applying Kelly's RGTs as elements of measurements. He concluded that middle class housewives disposed of a higher degree of conceptual structure relative to the evaluation of their homes than working class housewives. Moreover, the middle class housewives displayed a lower level of agreement among themselves inferior to those of the working class.

A second group of studies within this subsection refers to the effects of the built environment, social control and crowding. On the latter point we will only comment on the revision of studies made by Jiménez-Burillo (1981*b*, 1986), and concerning the effects of the built environment, Sangrador's revision (1981, 1986).

The built environment as a form of social control is one of the main centres of interest of the nucleus of the Autonomous University of Barcelona, either in the analysis of the distribution of school space (Ibáñez, Iñíguez, Elegebarrieta, 1982) or the field study and the theoretical revision from phenomenological perspectives adopted by Iñíguez (1983).

The third group of studies that we distinguish in this subsection is the one relative to Ecological Psychology, understanding this term in the sense that Barker uses and the notes made by Craik (1977) and Stokols (1977). Here we should highlight the work of theoretical conceptualisation realised by Hernández-Hernández (1982*a*, 1982*b*, 1985) and the reflections derived from the application of the notion of "behaviour setting" by Hernández-Hernández, Remesar and Riba (1982). It is also necessary to refer to Blanco's (1983) proposal of a model of quantitative analysis of behaviour in its natural contexts.

Finally we will mention the work of Ridruejo (1981, 1983), from the Autonomous University of Madrid, who, starting from Barker's proposals, takes the re-elaborations that R.H. Moos made on dealing with the environmental climate, analysing the implicative conditionals of the social climate in its syntactic and semantic aspects.

A fourth group of studies is that related to methodological concerns. Martínez-Arias (1981), of the Complutense University of Madrid, presents us with a systematic revision of the investigation designs followed by Environmental Psychology, observing a methodological pluralism and the tendency to use determinate designs according to the kind of problem to be studied. Fernández-Ballesteros (1980, 1982, 1983, 1986*b*) from the Autonomous University of Madrid, revised the different conceptualisations that "environment" has taken in psychology, and criticised the purely subjectivist vision of only taking the term into consideration from the perception of the individual. The authoress proposes a behavioural-ecological model of reciprocal determination in which behaviour, the person and the real environment are interactive.

Finally, the joining thread that permits us to interrelate all of Anguera's work (1974, 1980, 1982) from the University of Barcelona is the interest in different aspects of methodological character, especially related to the observation of spatial conduct, the patterns of registration, quantification and codification in natural studies, presenting various proposals.

c) The applied research. Applied research and especially its professional aspect or consulting has been scarce until 1987. There is only the exception of a few isolated cases of concrete investigations stimulated from academic activity, with scarce transcendence in social life at the moment and some assignments for the public administration, which seems to be beginning to become sensible to environmental matters from social sciences.

In the field of design, architecture and urbanism, we can find a psychological evaluation study of a project for children's playgrounds (Morales, 1981); some studies of design and the evaluation of old people's homes for the Barcelona City Council (Saura, 1984).

A specific field of application that has had a certain level of development is that of the school environment. The ICE, Institute of Sciences of Education of the University of Barcelona, together with the BCD Foundation (Barcelona Centre of Design) have shown a certain sensitivity to the theme for some years. Since 1979 they have organised four interdisciplinary conferences on the school environment (Pol and Morales eds., 1980, 1981, 1984) and have financed several basic research studies.

Sancho and Hernández-Hernández (1980) studied the environmental interaction in the classroom in the framework of the kindergarten. From 1980 to 1983 the SPIEE (Permanent Seminar on the School Environment) had concerned with the environmental interactions and behaviours of children in recreational space, (SPIEE, 1981, 1982, 1983). Both of these cases were supported by the ICE of the University of Barcelona.

We should refer to the study on the kind of schools desired by children as an element of reflection for the project of a new public centre for a small town in Girona (Pol, Morales, Ros, Presmanes, 1985), a commission for the Catalan Government. Through a series of tests, articles, interviews, drawings, visits to the new building site, constructive work with models, cut-outs and photomontages, they endeavoured to find out what the school desired by the children was like, evaluating both the functional and the formal levels, the symbolic and the environmental impact.

The evaluation of other institutional centres has been the object of several studies. Muntañola (1986) has analysed the penitentiary institutions of Barcelona from an architectonic and environmental point of view. Saura (1985) realised an inventory of resources in old people's homes for the Barcelona City Council.

Finally, within Barcelona City Council's programme for the improvement of the environment, we should mention the study on the social representation of environmental quality, mental health and the quality of life, by Domínguez and Pol (1986). Through different procedures of scales, questionnaires and free association of words, they evaluated the level of environmental satisfaction and tried to establish a standard of reference for the socio-cultural moment of the city, for the posterior evaluation of diverse aspects of the quality of life.

d) Environmental education. The last point of this revision of the fields of interest is also an application of the experience that Environmental Psychology can contribute. There is a lively controversy between what Tarrades (1983) calls environmentalists and educators in this field, according to whether you emphasise the qualifier (environmental) or the substantive (education). This discussion has been studied in depth, delimited and defined by the authors from the University of Mallorca, A. Colom and J. Sureda (1980*b*).

Another serious problem that is approached is whether the object of Environmental Education has to be the natural surroundings (create ecological conscience), the urban environment (knowing the city) or the environment as an integrated whole.

The reality of the knowledge of the environment is terribly low, above all in the urban area, according to Hernández-Hernández and Sancho's findings (1983). Lately the so called "Schools of Nature" have proliferated, the proposals of ecological itineraries, some urban itineraries, above all of a historicist character and some Centres of Environmental Education are being planned, like the Urban Environment Centre of the Barcelona's City Council. However, in these proposals the advances made in psychology on the interaction of the child (and the adult citizen) with his or her environment are rarely taken into consideration. These proposals, however, are explicitly gathered in the collection of books on Environmental Education edited by Muntañola, and in the videos that he has produced together with Pol and Morales on the subject (Videos, 1979, 1980, 1983, 1984).

9.4. The institutional development

The institutional development of Environmental Psychology in Spain is short. It seems, however, to obey a law -that can also be deduced from the revisions of other European countries-, indicating that before formally joining a field of knowledge the institutions promote the realisation of meetings, conferences, symposia and congresses on the theme. Or, seen from another point of view, those interested in a new disciplinary area will pressurise the institutions through the realisation of activities of this kind. So, in spite of the low level of development of academic training programmes and of proposals for investigation, the dynamic of meetings has been notable.

We have already referred to the four Conferences on the School Environment organised by the ICE of the University of Barcelona (1980, 1981, 1982, 1984). To these we should add those called on the same theme by the Catalan Government (1981) and the Spanish Ministry of Education (1983).

Otherwise, since 1981 all of biannual Spanish Congresses of Social Psychology held round tables and expositions of studies on our domain. There have been a great quantity of general and specific meetings, conferences and symposia on urban and environmental problems, but the participation of psychologists has been rare. The participation in the numerous conferences on Environmental Education has been equally scarce, although a little more abundant. As an example we will say that only in 1983 three general conferences were celebrated on the theme, with very few months difference.

The first event with the most transcendence and projection was the celebration of the 7th International Conference on Man and his Physical Surroundings (IAPS) in Barcelona in 1982. We will deal with this event more profoundly in the next section, on analysing the participation in the international conferences.

In November 1986, the First Conferences on Environmental Psychology were held, called through the initiative of the Department of Social Psychology of the Complutense and Autonomous Universities of Madrid. Considered as being a reduced symposium for specialists, the number of papers and applications for participation easily quadrupled the forecasts, the limit on inscription having to be set at 200, the majority being active participants. We mention this fact because we consider that it surpasses the anecdotal and is significative of the interest it aroused. The Organising Committee was formed by Jiménez-Burillo and Aragonés from the Complutense University of Madrid, Corraliza from the Autonomous University of Madrid, Hernández-Ruiz from La Laguna and Pol from Barcelona. The proceedings of the conferences gather a wide range of studies selected.

As for the academic programmes, after several years of regular optative seminars in Barcelona and Madrid, we should highlight the initiation of Environmental Psychology as a subject in the speciality of Social Psychology in the Complutense University of Madrid in 1986. One year later too in the Autonomous University of Madrid, the University of Barcelona, The Autonomous University of Barcelona, La Laguna, Seville and probably in others that we have not heard about, as a differentiated part of Applied Social Psychology.

Of the revision carried out until now four principal nuclei can be highlighted: La Laguna, Madrid and Barcelona. Valencia started with a study principally realised by architects linked to the School of Architecture, and in which the interest has basically been centred on the evaluation of built environments, but it decreases. La Laguna, centred on the Department of Psychology, with a reduced institutional development, but which has a certain dynamic of the production of studies, especially on cognitive mapping. Madrid, with a study centred on the methodological aspects of evaluation, on the perception of *crowding*, the perception of landscape, the evaluation of environmental climates and in cognitive maps, within the framework of the Faculties of Psychology of the Complutense University and the Autonomous University. In Barcelona we find different groups in which interdisciplinary collaboration occurs between architects and psychologists, working on methodological and theoretical problems of naturalist observation, behavioural and cognitive maps, the quality of life, the analysis of the school environment, developmental studies on the notion of place and the city in the child, Environmental Education, and the historical and epistemological analysis of Environmental Psychology. These studies have been developed in various centres linked to the University of Barcelona, the Faculty of Psychology, Faculty of Fine Art and the Institute of Sciences of Education, in the Autonomous University, in the Psychology Department; and in the Catalonia Polytechnic, in the School of Architecture.

9.5. The participation in the international conferences

The participation in the analysed conferences, between 1969 and 1984, is quite limited in comparison with the other works that we have reviewed. It is practically reduced to the Catalan participation and a single intervention by the ecologist González-Bernáldez from Madrid in Berlin, 1984. It began in Strasbourg with the architect Muntañola, and logically would explode in Barcelona with an important participation of psychologists, as well as architects and designers. In Berlin it went down again to four psychologists, one ecologists and an architect; in total, five studies.

The most productive Spanish authors in these conferences are Fernando Hernández-

Hernández, Josep Muntañola, Enric Pol and Montserrat Morales[1]. The impact and visibility were non-existent both in the conferences and the SSCI.

The most far reaching participation came about on a level of the organisation of the 7th International Conference on Man and his Physical Surroundings-IAPS, celebrated in July 1982. The committee was formed by Muntañola, Siguán, Morales and Pol (co-ordinator). There was no foreign participation in the local Committee, but there was the guidance of the executive of the recently formed IAPS, and especially by Simon, a member of the executive. The conference counted on the support of the three universities of Catalonia, the BCD (Barcelona Design Centre), a banking institution, the City Council and the Catalan Goverment. 250 attended from 21 countries and 157 papers were presented in the four official languages: Catalan, Spanish, English and French, for which simultaneous translation was provided.

The conference constituted a stimulus for progress in the consolidation of EP in our country.

9.6. Summary and conclusions till 1984

On the whole, we have found a belated but fairly prolific development. With a practically unexplored historical antecedents, EP began at the end of the sixties, influenced by the translations of texts that were then fashionable in the world of architecture and geography. Globally, there is not an openly dominant theoretical perspective, but rather there is a broader openness to the reception of new tendencies. Altogether we should say that a greater incidence of the cognitive and ecological perspective can be observed.

In general a panorama dominated by activity has appeared, linked to university life, in spite of the low level of formalisation of the academic problems. Moreover, there is still low social impact, and the interest and the expectations in complementary professional fields has diminished.

In spite of this last point, and according to what can be understood from that stated above, considering the development of the last five years, the expectations of formalisation in the new study plans and the multiplying effect that academic activity offers, together with the social concern for the state of the environment and quality of life, we can hope for a clear development and sedimentation of Environmental Psychology in Spain.

9.7. Epilogue: 1984-1992

This has been a notably rich period, especially since 1986. The seed of the first work, voluntaristic and isolated, began to germinate. There are five events which indicate a progressive consolidation.

1) The initial event to be described is the publication of the first 'orthodox' manual by a commercial publisher, entirely written by authors from the whole country who define themselves as environmental psychologists. A basic tool for teaching, it was compiled by Jiménez-Burillo and Aragonés (1986). Among other publications, we should highlight in

---

[1] *In the calculation of productivity the number of papers are considered with the number of congresses in which these have been exhibited. See 11.1.*

addition the book compiled by Fernández-Ballesteros (1987), with emphasis on methodological aspects. Corraliza (1987) proposed a procedure for the affective evaluation of the environment and Pol (1988) published an initial approximation to environmental psychology in Europe. Elsewhere, Sureda (1990) systematised the information on environmental education. In 1992 the first series of 'Psycho-Socio-Environmental Monographs' appeared (Pol, 1992) with an educational, informative and applied orientation, driven by the Master on Environmental Intervention in Barcelona.

2) The second event to be emphasised is the dynamic of conferences which articulates the informal constitution of an identified and recognised scientific community in the academic ambit. The regular celebration of conferences on Environmental Psychology of national scope in Madrid (Aragonés and Corraliza, 1986), Mallorca (Pol and Pich, 1989; Pol and Iñiguez, forthcoming), Seville (Castro, 1991) and La Laguna (1994) is most important. These conferences had wide resonance in university ambits, achieving an important effect of social impact where they were held. In addition to these, there have been some specialist conferences (Orellana 1992) or open and interdisciplinary events of incitement to collective reflection with respect to psychology (The lived city, Barcelona 1993). The invitation of environmental psychologists to participate in technical conferences on environmental engineering (Ambientalia '92, Reus) or the First National Congress on the Environment (Madrid, 1992) called by biologists could also be highlighted here.

3) In third place we should stress the beginning of regular teaching of environmental psychology on a pre-graduate level (degree courses) in many universities (Madrid 1986, Barcelona 1987, La Laguna 1988, Seville, Oviedo, Euskadi, Girona and with the new study plans of 1992 a few more).

4) The fourth event is the creation in 1988 of the MA in Environmental Intervention: Psychological, Social and Management Contexts, in Barcelona, with a founding team formed by Pol, Iñiguez, Morales (psychologists), Guardia (methodoligist), Serena (engineer), Muntañola (architect) and around twenty more lecturers. Its applied and interdisciplinary character in its conception and practice and the composition of its students has lead to the aperture of themes more closely related to the natural environment like the evaluation of environmental impact, audits or the working world, as well as of planning, management and quality of life. It has permitted us to attain a level of impact in the professional and political areas which direct these themes. Public and private assignments for applied investigation; the presence on environmental commissions in the university and public administration; professional practice of students in the public administration and environmental businesses; and the first posts are indicators of a certain consolidation. So the creation of the Master has been an element of articulation in research and professional and applied projection, as well as a formational instrument .

5) Psychologists are beginning to be incorporated into Environmental agencies in different autonomies and city councils. The first example was Andalusia, with Ricardo de Castro in Seville; after that Catalonia, with Salvador Rueda in Barcelona and several students in practice in various departments. Also in the local administration in Zaragoza and Barcelona.

6) The creation of the Commission on Environmental Psychology in the Official College of Psychologists in Madrid (1989) and of the Association of Environmental Masters and Consultants (AMCA), of interprofessional composition (1990), in Barcelona, have been

instruments for the creation of a professional and vehicular image with impact in society.

The dominant themes in this period have evolved from more cognitive considerations to more evaluative ones. The continuing work of Carreira on cognitive maps is outstanding, in close collaboration with Tomy Gärling (1984, 1985, 1986, 1989, 1990, 1991, 1992 a, b), Hernández-Ruiz (1986), Aragonés (1986) and Fernández-González (1985). Iñiguez (1988) approaches the dimension of time in the representation of space. Anguera (1986) and Anguera and Blanco (1986, 1987) consider methodological problems, along with Guardia et al. (1992). The aesthetic dimension in design is approached by Hernández-Hernández and Sancho (1989) and aesthetic preferences by Corraliza and coll. (1992).

Residential satisfaction has become a central theme (Aragonés and Amerigo, 1989, Amerigo, 1990, 1992, Galindo 1992), as well as quality of life (Dominguéz and Pol, 1986; Pol et al. 1990, 1992; Guardia et al. 1992). Place identity (Valera et al. 1988, 1990, 1992) is considered as a specific component but important for the quality of life, in the same way as the appropriation of space (Pol and Moreno, 1992, Pol 1993).

The evaluation of environmental impact from psychology was revised by Rubio in 1987. Some particular studies have been realised from the applied perspective: social rejection of a waste treatment plant (Moreno et al. 1988, 1990), of a motorway (Freixa et al. 1991), of a sports hall (Sierra et al. 1992), of noise (Lopez-Barrios, 1986, 1987, 1992; Herranz, 1992); on littering an ecological behaviour, Alcober et al. (1992), Hess and Hernandez-Ruiz (1992) and Suarez et al (1992). We need, also, mention the analysis of conceptual relationship between environmental psychology, environmental pedagogy and education by Romañà (1992).

Other studies on concrete spaces of the institutional kind have been centred on prisons (Muntañola 1986), youth custody centers (Llueca, Pol et al. 1992), resources for elderly people (Pol et al. 1989, 1991; Monreal, Martinez and Iñiguez, 1992); on deprived areas and shanty towns (Corraliza and Aragonés, 1990). Finally some studies on working environments (Aragall and Rovira 1991, 1992) and on commercial locations (Pol and Guardia, 1992) should also be emphasised.

The theoretical approaches are showing to be tremendously plural. At the beginning of the eighties the experimental and cognitive perspectives were dominant. At the beginning of the nineties the more positivist tendencies coexist with the more phenomenological perspectives. Constructionism, interactionism and the analysis of the discourse coexist with cognitivism, experimentalism and in a broad sense with empirism, according to the object and the object of the studies. In conclusion, we can affirm that Environmental Psychology in Spain is growing with strength.

Epilogue to part 2:
The Netherlands, Portugal, Greece and Turkey

As well as the large cultural areas revised in the previous chapters we should make mention of some others that for their relatively small geographical size and complicated language, like Holland, or for their scant development, like Portugal, Greece or Turkey, were not considered in the first versions of this study.

The Netherlands

Holland is a special case. Stringer and Kremer (1987) consider that there is scarce development of EP, according to their appraisal. But in view of their revision, it could be that it has a superior development to some of the areas that we have described. The first studies close to the ones we collected in our revision would be those of the sociologist Jonge (1962) on the image of the city; that of Wentholts (1968) on the perception of the reconstruction of the city of Rotterdam after the Second World War; those of Dijk and Steffen between 1966 and 1970 in Delft, gathering social aspects to which have to be answered by architecture; and the doctoral thesis of Ackermans on play behaviour.

However, there have been some congresses within the environmental area. The most important of these were the 1977 conference on perception of environmental quality, under the auspices of the UNESCO's MAB programme (Sanders, 1979) and the 10th conference of the IAPS in Delft in 1988.

In 1978 the Dutch Journal of Psychology dedicated an issue to environmental psychology. In 1979 the first manual appeared in Dutch (Smets, 1979) and in the same year Ester began to edit a collection on social aspects of the environment, going directly into the problems of natural resources, energetics, risk and environmental politics.

Stringer (1987) gathers ten centres where some kind of work linked to environmental psychology is being carried out, from architecture (Delft, Wageningen, Leiden, and now we should add Eindhoven), centred on environmental resources in traffic (Amsterdam, Delft, Haarlem) from sociology or social psychology (Tilburg, Eindhoven, Leiden, Nijmegen). In principal he underlines a limited academic development, but a relatively important professional development, although lately dissertations and the presence in international magazines have been developed more.

The Dutch presence in the IAPS conferences has been continuous and important, not so much for the number of participants as for their qualification and recognition. Prak, van Hoogdalam, van der Voordt, van Wegen, assiduous participants and organisers of the Delft conference (1988), Wurff from Psychology in Amsterdam and the Briton Stringer linked to psychology in Nijmegen and van Andel and Timmermans linked to the Technische Universiteit of Eindhoven are some noteworthy names.

Portugal

In the other areas that we collect in this epilogue, the situation is more diminished. Portugal has never participated in the conferences of the IAPS, but there are some nuclei to be described. In Lisbon we should mention Jorge Correia (1988), working on stress, Luis Soczka (1988) on social ecology of a neighbourhood and social identity, and José M. Palma Oliveira, working from a cognitive perspective of theoretical orientation on cognitive maps (Palma Oliveira, 1992). Palma Oliveira and his colleagues have participated in some of the Spanish congresses, and keep environmental psychology present in all the meetings of social psychology on a Portuguese level and some international conferences, like the workshop on Social and Environmental Psychology in the European Context (1986).

Greece

In this country there is only one notable group, in the architecture school of the Aristotil University in Thessalonica. Its members have been trained in postgraduate studies in England, like Aristides Mazis or Rena Papageorgiou. Others have received formation orientated towards sociology in Francophone areas, like Cleopatra Karaletsou in Belgium. From the beginnings of the group, around 1982, its activity has been centred more on academic teaching than on research. Moreover, they are beginning to do work in the Department of Education and in the Sociology department.

Research is orientated towards special groups and spaces, whether they be disabled people, old people or children, in the ambit of the city. The interactions and conflicts for space between different groups with contradictory interests are being studied. They employ a sociocultural perspective to interpret space, with a rather more qualitative and critical than quantitative orientation.

Money for research is very scarce and the investigators themselves have to go to look for it on their own initiative. The demand for research generated from outside the university is practically non-existent.

As the most notable event with the greatest projection internationally, we should mention the organisation of the 12th Conference of the IAPS in July 1992 in Thessalonica, organised by Mazis, Andreadou, Zafiroropoulos, Tsoukala, Papageorgiou and Karaletsou.

Turkey

In Turkey we should refer to the group in the Faculty of Architecture of the METU, Ankara, formed by Haluk Pamir and Vacit Imamoglu. These authors maintain a quite assiduous presence in the IAPS conferences. In 1980 they organised the 11th conference of the IAPS with the support of Necdet Teymur from London, who keeps up a close collaboration with his country of origin.

Like the majority of the other countries in which environmental psychology has incipient development, the celebration of an international conference supposes a certain consolidation for the organising team and stimulus and impact in the receiving society, at least in the university that supports it.

Part 3
EUROPE AS A WHOLE

In view of some common characteristics concerning the antecedents and first steps of Environmental Psychology, and how the dynamic of meetings, conferences and congresses crystallised in the formal constitution of the scientific community that expresses itself through the international conferences that are realised in Europe; in view of how the origin, development and sedimentation in each of the principal cultural areas that form it, with their own peculiarities, their own traditions, defined and at times differentiated mark a peculiar evolution but with not a few common characteristics; in view of how the different theoretical perspectives coexist in the early stages but fall out later when the level of development is higher; in view of, definitively, how a determinate scientific field advances along routes that if not different are at least slightly differentiated, because their social, institutional and individual context varies in spite of the internal logic of their ideas, we will now go on to attempt the global evaluation of Environmental Psychology in this melting pot that is Europe in its entirety.

In the light of the information exhibited in each cultural area and on the basis of the information that emerges from the bibliometric and sociometrical analysis of the conferences under consideration, we will tackle the substantivity or the pluridisciplinary subsidiarity of Environmental Psychology, from the comparative revision of its different origins, the analysis of its growth, consolidation and recognition, the terminological programme of its denomination and whether it is rooted in a common experience.

We shall analyse who is who concentrating on productivity, recognition received through the invitation to participate in relevant events, the scientific collaboration and the visibility and impact of the authors through the quotes they received.

Finally we will attempt to penetrate in the invisible structure of the social organisation of the science through the study of the networks of relationships, both of scientific collaboration expressed in the joint signing of the studies, and the social relationships expressed in the collaboration in various institutional and organisational applications, and the intellectual relationships expressed through debts, dependencies and authority shown in the quotes, co-quotes and mutual references. This will allow us to objectivise the hierarchical structuration that occurs in the whole scientific community, which has been given the name of "Invisible Colleges" and which in reduced and sufficiently compact communities can go beyond a theoretical intellectual leadership, acting as real nuclei of power.

All this will permit us to tackle the problem of whether or not there is a European Environmental Psychology.

10 Environmental psychology: Pluridisciplinary science versus applied psychology

Through the revision of each cultural area we have seen how a convergence of contextual factors occurs which facilitate the emergence of our field of study. The profound technological changes that were seen to be accelerated with the second world war and that were consolidated in the post-war period, which would profoundly affect psychology, being an important element in the crisis of the dominant paradigms of the moment (Caparrós, 1980), together with the expansion of the humanistic ideologies and a climate of social euphoria, in which the eschatological hopes of a better world were in full bloom, propitiated the first studies during this period, which in spite of everything, took on a technocratic orientation, converging with the spirit that reigned over the expansion and economic development of the period.

On the other hand, a strong process of degradation began to become evident in the environment. Despite the fact that the city had been an overcrowded and degraded setting since the industrial revolution, it has had the constant counterpoint of nature, distant at times but present and accessible, as a compensation, that would be diluted with time. It was in the moment in which the evolution of society and technology, of the reconstruction of Europe with a strong process of urban concentration and the migrations from the country to the city of the fifties and sixties, provoked worrying levels of degradation in the environment, with the distancing and technification of the more and more omnipotent and less accessible decision making centres to the citizen, regulating directly or indirectly the organisation of space and time, through urban planning and the effects of the "mass media". All this would lead to a situation of urban crisis, with an awareness of the necessity of a better adaptation of the environment to the individual and on the other hand to preserve the ecological balance and natural resources, leading to the emergence of the new urban movements of the last period.

These two historic moments, enjambed but slightly out of phase in time, generated what we could schematise in short into two contrary tendencies: the one corresponding to the second historical moment in its most extreme approaches, would try to act on, if not to change, the course of the technological evolution in believing that this goes against man himself and his environment; the other would propose that we are in an unstoppable but positive dynamic, and that in fact what is necessary is to advance even more in science and technology until their negative aspects disappear, as they are no more than momentary dysfunctions and are repairable with "progress". Obviously, behind them are two different ideological postures, that maintain themselves, coexist and fight.

Environmental Psychology has developed between both of them. After technocratic functionalist and positivist initial moments, it went on to cognitivist and interactionist approaches due to posterior social pressure and the changes in the framework of the reference. EP progressively integrated elements from the ecological issue, in principle quite foreign to its approaches. Due to the nominal coincidence, this gave it an important part
of the fame that it would attain, especially in the second half of the seventies.

10.1. Origins and interests

Of the eight cultural areas presented we can form two principal groups with respect to their origins. A first group made up of the countries with a long tradition and an emergence of the autochthonous and independent studies on H-E, which would be formed by Great Britain, Sweden, the Francophone area and the Germanophile area. The second group is formed by those that have a lesser tradition and in those that in fact the interest emerged in the second half of the seventies, basically through the influence of the previous ones, and above all of American EP; this is the case of the USSR, Italy and Spain. In these countries on the other hand, always with differing levels of tardy development in EP, a wide range of themes treated can be found, comparable with the most developed countries in this field.

In the countries with a long tradition the first studies can be found around the end of the fifties and the beginning of the sixties, with some peculiarities of the zone. On one side, the countries most severely castigated by the second world war had to confront the urban reconstruction and the global reconstruction of society, profoundly altered, moreover, as we have seen above, by the technological changes and those of productive systems. This would generate a notable development in urban studies, on housing, on the adaptation to new environments, the social production of space, etc., that in the first stage would generally be executed from sociology, with some punctual participations from psychology, especially in Great Britain, and social anthropology in France. Later on, with the economic progress it would be desired to tune in more on the problems of the built environment, and the architects would initiate the demand on psychologists for information on aspects of the individual's interaction with different surroundings.

Germany, in spite of the wealth of the first EP that developed in the first third of the century due to the interroption of the war and the dispersion or death of the authors, followed a similar development to that of other countries, not being able to speak of renewal. The case of Sweden is different as it did not suffer the hardships of the war, and due to its low density of population the issue of urban concentration played a secondary role. This explains why the interest of the first studies was concentrated more on questions related to the perception of the environment, light, colour, signification, and was questioned from the field of architecture to better resolve some problems of design that were presented.

Later on, now towards the end of the sixties and at the beginning of the following decade, once the work of international communication was established between those interested in this theme, under a strong American influence, the thematic range would open out and widen. Studies of functionalist analysis on concrete environments coexist with phenomenological, cognitivist, interactionist perspectives, centered on city or the home, taking an interest in the representation of space, of the appropriation of space, of proxemics, personality, adaptation to space, deviated behaviours, crowding, the urban movements, etc. The concern for the natural world within this scientific community described was practically nonexistent except recently in Germany, where the social movements have been stronger in this sense, and where in fact the first EP of the beginning of the century basically emerged from biology and meteorobiology, a tradition that they have attempted to recuperate over the last years.

In Great Britain, Sweden and Germany interdisciplinary collaboration occurred from the very beginning between architects and social scientists. In the Francophone area, on the other hand, this did not come about more than in a very sporadic way and quite rarely until after 1968. The interest in the themes, and their treatment, was taken in a parallel way from the different social disciplines, although frequently adopting a psychosocial perspective.

In Great Britain and France the first steps were taken in the basic concurrence of architecture and psychology, on the request of the interest of the architects. Engineers and the occasional geographer also participated, although to a lesser degree.

In the Francophone area the interest in space - space is the favourite term - came from the social sciences, especially urban sociology, anthropology and ethnology. When psychology intervened it was with a marked emphasis on social psychology. It crossed with architecture, also strong in the social tradition, at the end of the sixties. However, an area in itself like that of EP was not defined more than as a belated influence of the Anglophone tradition. Meanwhile, each one of the implicated disciplines revindicates the right to the consideration of space, the environment or the ecological, from its own identity.

Really it is not correct to speak of Environmental Psychology on referring to its origins. As we have seen, this denomination was adopted belatedly and with reservations (it is not like this in the USA, on the other hand). What is evident is the existence of a field of study, the relationship between man and the physical environment, which is the object of interest from the various perspectives (see 10.4.).

10.2. The growth of Environmental Psychology

The growth of the science in our century has been spectacular and has not escaped the interests of the scientists of science. Price, in 1951, formulated the so called Law of Exponential Growth of Science, according to which this grows more rapidly than social phenomena. Thus, while population doubles every fifty years, scientific literature does so every ten years and the scientific magazines and repertoires of abstracts every fifteen. The consequence of this is the high level of contemporaneity that characterises present day science. He considered, however, that this exponential growth corresponded to an abnormal global situation, as it is not possible to reach an infinite or absurd growth. A point of saturation would be reached - he says - that he attempts to define mathematically. We will not go into the details of his proposition, but we will take his declaration as a reference for the analysis of our ambit.

To carry out the evaluation of the growth of Environmental Psychology we have resorted to the analysis of the participation and productivity in the conferences referred to. We will synthesise some of their more relevant information.

In the fifteen years that cover the period studied, from 1969 to 1984, the participation of different authors in each conference goes from twelve in the first (1949) to 311 in the last (1984). This increase is more moderate if we concentrate on the number of signatures, of studies presented or of attendance in the conferences (see table 2, appendix). The growth fluctuates. This occurs for scientific reasons (the more or less rigorous selection of work, for example) and for power relationships within the scientific community (for example, double convocation of conferences in the same month).

In spite of these fluctuations the growth of Environmental Psychology in Europe starting from the conferences taken as an indicator complies with the exponential growth demonstrated by Price concerning authors, signatures and the number of studies (diagrams 1,2,3, appendix).

It is not fulfilled, however, for the growth of the global number of people attending the congresses. This is subject to fluctuations more dependent on external social factors than on the particular internal dynamic of the domain. The expectation that a new ambit can raise in fields more or less close to it , the disseminative function of the congresses, the kind of publicity channels chosen, the access to means and networks of communication, etc., have a strong impact on the response that each conference should have. Moreover, the differing levels of development of the disciplinary field in each country that hosted a conference and the impact on civil society would be seen to be reflected quantitively but also qualitatively in attendance. In any case, from 1979 onwards attendance seemed to stabilise in a ceiling that practically stopped growing, with an increase, however, in the rate of active participation (see table 2, appendix). In a positive reading, they go from disseminative conferences to conferences of informal exchange between active authors. In a negative reading, they could indicate a descent in the social expectations that they raise outwith the scientific community, or at least the academic community. We will concern ourselves with this and the impact of the conferences in the next section.

10.3. Consolidation and recognition

If in the previous section we saw the growth process, now we will analyse consolidation and impact in society. Although both concepts are closely related, they are not necessarily interdependent. There could be a strong consolidation and a low social impact and vice versa. This has already been picked up on in some way in previous chapters within the analysis of each cultural area. However, we believe that it is necessary to make a global and objective evaluation in its entirety.

Social recognition comes about in function of the social demand and the resources that the society itself offers. This means that both the level of institutional development in its most diverse forms, and the formal development in the academic structures will be valid indicators.

In previous sections we have reflected on how a progressive process of academicisation came about in EP, at least in the EP that expresses itself through the conferences analysed. The active participation of professionals who are not linked to academic centres in the specialised conferences could be taken as an indicator of the social impact and the expectations that EP raises. In absolute numbers, this participation reached its highest point in 1979, in the second Surrey Conference. In percentages of active participation the highpoint is situated in 1973 in Lund, falling off progressively after that (see table 9 and diagrams 7 and 8, appendix).

This marks out a period, between 1973 and 1979, which is notable as far as the social impact or expectations are concerned. This will be reaffirmed by other indicators.

As well as being a period of a strong social and creative dynamic concerning EP it was also the period in which the most quoted articles and books in the conferences were published (see tables 23 and 24, appendix). In 1979 the theoretical crisis that we have already mentioned became public knowledge, stimulated, moreover, by the crisis of relevance of the contributions of EP and the economic crisis itself that got progressively deeper after the so called oil war.

The aesthetic changes of fashion and thought in architecture, estranged from social approaches were not foreign from this crisis either.

In synthesis, there is empirical data that points to the period 1973-1979 as being a key period in the impact of EP. After this date the academic institutionalisation would continue to consolidate itself, but the direct impact in civil society at least became less evident, and was especially felt by those dedicated to the professional field applied to architecture, as we have seen in previous chapters.

However, due to the change in the social issues, the progressive reconciliation with the environmental problem from an "ecological" perspective (not necessarily "ecologist") lead to a greater participation in official and international programmes related to environmental planning and preservation, together with ecologists, biologists and health scientists. A sign of this are those linked to UNESCO's MAB programme (Man and Biosphere). This opened up a new field and new expectations during the eighties that had rarely been present in the previous period.

With respect to the investigation resources of official institutions for the study of urban problems, they are permitting us to transcend the limits of the strictly local framework to facilitate international collaboration in transcultural studies, as is the case of the programme on the psychological construction of the urban environment in cities of high cultural value "The Transcultural Study of the representation of the urban environment" (Bonnes, Italy; Jodelet, France; Kruse, Germany; Stringer, Great Britain, 1987) which could and is intended to explicitly promote and give cohesion to a European perspective on the psychological environmental treatment of the city.

On the other hand, the level of academic institutionalisation is very irregular. In Europe the most complete and widely disseminated M.A and doctorate programme in Environmental Psychology is the one in Guildford, Surrey. There are, however, many centres with partial programmes, investigation units, doctorate programmes or simply subjects. The highest level of institutionalisation is seen in Great Britain and Sweden. In Great Britain, as well as in Guildford, there are partial programmes in numerous centres, above all in architecture. Kingston, Cardiff, Strathclyde, Portsmouth, Birmingham, Oxford, London, to mention but a few, are all good examples. In Sweden there is a professorship in Stockholm, within the framework of a psychology centre, and various investigation and training units, the most powerful of which is the one in Lund Polytechnic.

In the Francophone area the panorama is more complex. As we have seen, according to Jodelet H-E relationships are being studied in six psychology centres and in total twenty-five in social science centres. The reality is wider, as under very diverse labels work is being carried out in numerous departments and laboratories of social psychology, integrated in their habitual parameters. The Institute of Social Psychology of the University of Strasbourg is an outstanding example, as are the School of Architecture of Louvain-la-Neuve in Belgium and the Federal Polytechnic in Lausanne in Switzerland.

In the rest of the areas the institutional development is less important. In Germany there are a certain number of investigation units and partial training programmes in Heidelberg, Tubinga and in Berlin. There is a similar situation in the USSR, as we can see from the corresponding section.

In Spain the Postgraduate Programme in Environmental Intervention in Barcelona has constituted a notable advance, as well as some other courses and projected courses in Madrid, Barcelona, Santiago, the Canary Islands, along with aid for investigation from the ICE, the IUCA of Madrid, and organisms of the Public Administration.

Another level of institutionalisation comes about due to the existing specialised magazines. In this way we find two in Great Britain: the old *Architectural Psychology*

Newsletter, now the *IAPS Newsletter*, the *Journal of Environmental Psychology* and a third which is bilingual in French and English, and is published in Switzerland under the title *Architecture et Comportement*. All three have an international voice and constitute a very positive balance. Also we have to take the general magazines on psychology and architecture into account which habitually publish articles related to EP.

One indication of the support that the field of study can receive from the institutions, aside from the formally established centres, are the meetings, symposia and congresses realised. Thus, we find a very long list in Great Britain, a goodly number in Sweden and the Francophone area, also in Germany, only two references in the USSR, one international conference in Barcelona, together with a considerable number of meetings and symposia in Barcelona and Madrid, and finally one conference in the Italian ambit in Milan.

Altogether there is a wide consensus - although not unanimous - in considering the congresses, which are the objects of this study, as points of interdisciplinary meeting and discussion that have set the standards of the development of H-E investigation. The fruit of these meetings, plural not only in disciplinary fields but also in theoretical perspectives, never free of controversy, has been the constitution of the IAPS Association, and specialised sections in European scientific societies, above all of social psychology and applied psychology.

10.4. Pluridisciplinarity and substantivity: professions and labels

On analysing the origins of our discipline we have seen as a constant the convergence of interests, fields of study and collaboration at least between psychology and architecture, taking on the denomination Psychology of Architecture.

This denomination would be the trade mark of five of the international conferences in Europe, between 1969 and 1979, those that were gathered under the IAPC logo (International Architectural Psychology Conference) and would constitute the label with which even today some authors identify themselves, especially in continental Europe. It would mark the title of one of the European magazines, *Architecture et Comportement* and the bulletin of the IAPS itself, which in its first phase, from 1970 to 1980 was called the *Architectural Psychology Newsletter*.

In Francophone ambits, as we have seen, and at the end of the sixties, without a full terminological agreement, the search for a pretendedly neutral term had greater weight (a disputed neutrality, however) (see chapter 5) in the so called "psychology of space", with influences more of sociological than architectonic origin. The titles of various pioneering books like those by Moles and Rohmer (1964) and Fisher (1967), or the selfsame minutes of the conferences that were held in Francophone areas, the 3rd and 5th IAPC, used the term psychology of space. Thus, if Psychology of Architecture suggests a "bi-disciplinary" relationship, Psychology of Space carries connotations of the use of the term in social psychology, anthropology and Marxist sociology. Both terms would be assessed as being restrictive to professional fields in the first case, and ideologically stamped or with too many theoretical connotations in the second.

In 1972, in the beginnings of the Surrey programme the title of "Environmental Psychology" was adopted for various reasons. In the first place for the recognised influence from the same ambit of study that was already strongly developed in the USA. Moreover, as its promoters recognise, the idea was not to restrict the interprofessional relationships only to architecture, as there were already contacts and collaborations of interest with other domains, such as geography for example, and they were interested in a wider field of study which would take in both the natural and built space. Also, the change of label allowed the differentiation from the dynamic that had developed in the previous years with certain levels of tensions and disputes (see 2.4 and 3.3).

In the Germanic area the discussion would be centred on the meeting in 1973 in Salzburg, agreeing to maintain the term *Umwelt Psychology* (Environmental Psychology) for the first Environmental Psychology of the beginning of the century (see 6.1 and 6.2), adopting Ecological Psychology or Ecopsychology (in a not necessarily Barkerian sense) for the present day science. Something similar would happen in Italy and in Spain, where even more variations coincided. The most generically accepted term, after about ten years of study, was Environmental Psychology, but in competition with Psychology of Surroundings, Psychology of Space, Psychology of the Setting, and even in Catalan we have to add "Psicologia de l'Environament". All of these infringed on differential variations that had their reasons to be and went beyond a merely semantic problem.

In any case the interdisciplinary nature of the study of environmental matters is what comes into play, what is discussed, what is collaborated on, but also what is controversial. The taxonomic activity as a form of appropriation of an ambit, even if it is a field of study, has lead to debate on the exclusivity, inclusion, margination or exclusion of determinate professional fields in environmental studies. The conscience of this fact meant that precisely the sibylline name of the International Association for the Study of People in their Physical Surroundings was adopted in the constitution of the IAPS, in a way that would recall the old logos of IAPC and ICEP, but would not mention or exclude any discipline which might be concerned with H-E from any perspective.

In spite of this, a heeling towards psychology has not been avoided, which leads to a frequent confusion that goes beyond the semantic to reach the epistemological, in as much as "Environmental Psychology" refers to both an applied field or a speciality of psychology, and to the environmental studies concerning the human being, but with different disciplinary and theoretical perspectives. We will deal with this later on, at the end of this chapter.

For the moment, to better illustrate this process we will revise the professional composition of the participants in the aforementioned conferences and the interprofessional scientific collaboration.

The professional profile of the active participants in the conferences varies with time. According to what can be seen from table 3 and diagrams 5 and 6 (appendix), the total number of psychologists and architects share a similar percentage. It is interesting, however, to see the fluctuation of this distribution, that outlines a mirror image diagram. This fluctuation could be explained by the kind of institution that welcomed and/or organised each conference. When it was a department of psychology the percentage of psychologists went up (with the exception of Kingston, 1970 and Surrey, 1974). When it was a department of architecture it is the other way around, the percentage of architects increased. Moreover, as we have already said, each host country contributed a higher number of studies than its usual standard and this has implications for the professional distribution.

The participation of other professions is very limited. We should emphasize, however, the presence of sociologists, quite notable in the conferences in the Francophone area and also the punctual presence of engineers. Also, the professional composition and its evolution bears a relationship to each country's tradition and the moment in which Environmental Psychology began to be developed in them. Thus, the countries in which there is a certain balance are those that most clearly began earlier in this field of study from the demands of architects, and those in which psychologists have taken on a preponderant role afterwards. This is the case of Great Britain, Sweden, The United States

105

and some other countries strongly influenced by these, such as Canada and the limited but productive Turkish nucleus. France and Belgium have a stronger tradition in social sciences and a belated homologation (though not with specific studies) in this scientific domain. Spain would join when the field was defined. In these three cases (see 5.7, 9.4 and 9.5) the development of EP emerged directly from psychology or sociology. The rest, given that their productivity is reduced to a limited spectrum of names, is not significant.

The general conclusion, as far as the dominant profession is concerned, is that in spite of the apparent balance between architecture and psychology, there has been a certain tendency to the predominance of the psychologists over the last few years. The participation of other professions is insignificant, except for a small percentage of sociologists proceeding above all from Francophone areas, and occasionally of engineers, geographers and anthropologists. We should emphasize the fact that despite the burgeoning of the problem of the environment that so profoundly occupies biologists and ecologists, they have almost never participated in the conferences analysed, with punctual exceptions in Berlin, 1984.

On the other hand, the institutional links are mainly with to centres of architecture (see table 9 and diagrams 7 and 8, respectively). Although the penetration of psychologists in these centres has been stable in time in absolute numbers, in percentages it becomes relative and loses specific weight.

Once this data has been analysed in the light of the qualitative evaluations of the analysis of each cultural area, seen in previous chapters, we can see how there was a change in the (numeric) leadership after the role played by architects and engineers in the emergence of EP as an applied field at the end of the seventies. Despite there being a good handful of "convinced" architects, there was an important drop in the expectations as to what the social sciences could contribute to architecture, with the applied work being affected above all. A necessity arose to establish or classify a theoretical framework for this dialogue, which lead to an "academicisation" expressed both in the emphasis of the role of psychology in the formation of the architect, and in the subjunction of the environmental or ecological paradigm by academic psychology.

There is another interesting aspect in our study that we would like to emphasize here, which is related to scientific collaboration. For the information that we have been able to deal with, in spite of the bias introduced by the number of authors of whom we have not been able to determine the formation or professional pertinence, the studies signed in collaboration by various disciplines is very low. Only 15 of the 538 studies analysed show collaboration between architects and psychologists, and 10 other professional combinations, in nine of which architects participated.

This demonstrates once again, despite the apparent desire for interdisciplinarity, the difficult coexistence and integration in one single domain of the different professions albeit to analyse the same object from different perspectives, in spite of the effort made to find an aseptic and comprehensive denomination. The polysemic qualifier that we gave earlier to the label of Environmental Psychology takes on its full meaning and needs to be clarified.

Duncan Joiner (1982) declared that each discipline in this broad field called Environmental Psychology, far from becoming integrated and reconciled, jealously guards its information and constructs, on the contrary to what some would like to believe and demonstrate (see the quote in 2.4).

However, if this is true, it has to be understood within the ambiguous context of the double use of the term EP, as a denomination of a pluridisciplinary field, or as a denomination of a specific branch of psychology.

In the time passed, psychology as a discipline has progressively integrated the environmental parameter in a fair number of its perspectives, at the same time as it has formed (or they have formed themselves) a handful of specialists in the environment. In this sense, to speak of a branch of psychology specialised in the environment is correct, given that its is defined and reconciled and has academic and institutional recognition.

In a broad sense, the denomination of EP in the H-E field of study, from the interdisciplinarity of its origin, is imprecise and Joiner's assertion proves to be true.

But as well as this, although EP is consolidated as a specialised branch of psychology that has developed parameters which traditionally have not been highly considered, it has also been capable, if not of being an agent or the cause, of acting as a catalyst in a reorientation (Altman, 1981 would speak of "revolution") of psychology from almost all of its perspectives. Psychology has discovered its surroundings. We will end this section with Altman's words:

> *Social psychology has begun to refer to the physical surroundings and the social context in relation to social behaviour; Psychology of development has discovered that it cannot understand this independently from the physical context; the psychologists of learning have begun to implant their learning focus points in the physical contexts on the grand scale; the experimentalists have incorporated aspects of the environment into their analysis of the perceptual phenomenon; the psychologists of personality continue struggling with the importance of the situation [...] [Altman, 1981: 5].*

10.5. The historical and functional roots

We will now concern ourselves with the antecedents on which EP grew and developed. We call them roots in as much as they are in the foundation -explicitly or implicitly- of today's production. Their approaches were not necessarily environmental, but their theoretical or empirical contributions inspired those of today. In many cases, however, Lévy-Leboyer's affirmation (1984) that they are roots sought *a posteriori* is fulfilled. The sources for this analysis will be, in the first place, the "Eminences" who recognise their debt to the authors interviewed. In the second place we will summarise the result of the corresponding bibliometrical analysis of the quotes made in the works presented in the conferences.

Starting from the interviews and documents consulted, the dispersion of sources is notable, as can be seen in table 10.1 (that without seeking to be exhaustive nor of statistical value, gathers the names of the recognised authors as points of reference or theoretical sources). The work of the first German EP sown with the dispersion of the war can be appreciated, although it is not recognized as such. Lewin was one of the most commonly recognised influences, together with Gestalt and the American Tolman. German phenomenology has a considerable influence in the Germanic and Francophone areas and in Sweden. Also Bachelard, Baudrillard and Foucault, who extended their influences to the British phenomenological group. And above all, the American pioneers, like Hall, Sommer, Lynch and Proshansky were those that have the most extensive influence.

Apart from these authors, each area presented sources more deeply rooted in their own traditions. If in Germany the first sources had been strongly related to biology and meteorobiology, in Great Britain what would bear more weight would be the anglo-american tradition, with outstanding authors recognised in Bartlett, such as Tolman, Kelly, Neisser, Festinger and Piaget (from American translations). In the Francophone area, on the other hand, what would have more influence would be the strong tradition in social sciences, from the classical sociology of Durkheim, Webber and Marx, to the urban

107

Table 10.1. *Authors with the function of classics and to whom the European "eminents" recognize their dept*

Great Britain	Sweden	French A.	Germany	USSR
Bachelard	Arnheim	Bachelard	Eco.Biology	Gibson
Bartlett	Berlyne	Barthes	Burgess	Hall
Festinger	Bird	Bastide	Compte	Leontiev
Foucault	Eysenk	Baudrillard	Durkheim	Sommer
Gestalt	Gibson	Blache	Chicago School	Stokols
Jahoda	Härd	Brunswik	Urban Studies	
Kelly	Herin	Castells	German	
Lewin	Katz	Duran	fenomenology	
Neisser	Köhler	Durheim	Haeckel	
Piaget	Langer	Freud	Hawley	
Marxitst Soc.	Ogden	Foucault	Hellpach	
	Osgood	Gestalt	Mckenzie	
	Richards	Greimas	Mukerjee	
		Halbwaks	Muchow	
		Heidegger	Park	
		Le Corbusier	Simmel	
		Ledrut	Sombart	
		Lefebvre	Thurawald	
		Lévy-Strauss	Uexküll	
		Lyotard		
		Margeraux		
		Marx		
		Sivadon		
		Tolman		
		Webber		
		Wallon		

Source: Personal interviews and reports of countries.

sociology of Lefebvre, Ledrut or Castells, going through the anthropology of Lévi-Strauss. The roots recognised by the Swedes were possibly the most universal and carefully related to their principal interests. In the rest of the cultural areas, they began from the inherited tradition transmitted by the American pioneers, even in the USSR where it would be necessary to add Leontiev to the list of acknowledgements.

The second source for the study of the antecedents in which the roots of present day EP penetrate comes to us through the bibliometric analysis of the references expressed in the conferences. For this we will differentiate between the so called "Historical Classics" and the "Functional Classics".

We understand "Historical Classics" as being those authors who receive quotations in publications at least 25 years before the beginning of the period studied, in our case before 1945. The result is a tremendously dispersed panorama. Altogether there are a total of 75 quotes from 78 authors and 74 different studies, 2.3% of the total of the quotes emitted in the conferences. There is no convergence in the few predominant authors. Neither do they attain an elevated specific weight as a whole. For this reason we can say that they fulfil the function of "Classics" rather than "Historical". We offer the overall list of these authors by cultural areas in table 10.2. It is made up of 25 studies published in the USA, 21 in Great Britain, 13 in France and 1 in the USSR and Sweden and 6 in which the source is not indicated. The place of publication indicates an American editorial supremacy rather than the in origin of the authors, as for example some French authors are quoted in American publications and vice versa.

Table 10.2. *Classic historical authors to cultural areas*

Name	Prof.	Language quotation	Year/Place publication.
French A.			
Anonymus	-	Fr.	1936 Fr.
Coppier, L.	-	Fr.	1932 Fr.
Dumezil, G.	Soc.	Fr.	1944 Fr.
Freud, S.	Psyc.	Fr.	1928 Fr.
Garnier, T.	Geogr.	Fr.	1918 Fr.
Hull, C.L.	Psyc.	Eng.	1943 EEUU
James, W.	Psyc.	Fr.	1909 Fr.
Pevner, N.	Arch.	Eng.	1936 Swit.
Piaget, J.	Psyc.	Fr.	1939 Fr.
Robert, J.	Arch.	Fr.	1939 Fr.
Vailland	Geogr.	Fr.	1955 Fr.
Wallon, H.	Psyc.	Fr.	1941 Fr.
Wallon, H.	Psyc.	Fr.	1949 Fr.
Germanic Area.			
Burgess, E.W.	Psyc. Soc.	Eng.	1925 USA
Gisbertz, W.	Arch.	Germ.	1936 Germ.
Howard, E.	Arch.	Germ.	1902 Germ.
Kampffmeyer, H.	Arch.	Germ.	1926 Aust.
Simmel, G.	Soc.	Germ.	1903 Germ.
Wirth, L.	Soc.	Eng.	1938 USA
Wulzinger,K.& Watzinger,C	Arch.	Germ.	1924 Germ.
USA Area.			
Allport, G.	Psyc.	Eng.	1937 USA
Anyal, A.	Psyc.	Germ.	1931 USA
Claparéde, E.	Psyc.	Fr.	1943 Fr.
Dewey, J.	Psyc.	Eng.	1916 USA
Murray, H.	Psyc.	Eng.	1938 USA
Partem, M.& Newhall, S.	Psyc.	Eng.	1943 USA
Updegraaf R.& Herbst, E.	Psyc.	Eng.	1933 -
Van Alstyhe, D. Psyc.	Eng.	1939	UK
Sweden			
Allesch, G.J.	Psyc.	Germ.	1925 Germ.
Eysenck, H.J.	Psyc.	Eng.	1941 UK
Gordon, K.	Psyc.	Eng.	1923 USA
Hering, E.	Psyc.	Germ.	1920 Germ.
Markelius, S.	-	Swed.	1930 Swed.
Ogden, CK.	Psyc.	Eng.	1943 UK
Trowbridge, CC.	Psyc.	Eng.	1913 -

(10.2. cont.)

Name	Prof.	Language quotation.	Year/Place publication	
Gread Britain				
Bartlett, FC.	Psyc.	Eng.	1932 UK	
Bedford, T.	Doc.	Eng.	1936 UK	
Beebe-Centre.	-	Eng.	1932 USA	
Bender, W.	Psyc.	Eng.	1933 UK	
Bridges, K.	Psyc.	Eng.	1929 UK	
Cohen & Nagel	Phil.	Eng.	1934 UK	
Collingoudod & Myres	Arch.	Eng.	1936 UK	
Chapin, P.S.	-	Eng.	1935 USA	
Dunker, K.	Psyc.	Eng.	1945 UK	
Eliot, T.S.	Phi.H.	Eng.	1932 UK	
Farwell, L.	Psyc.	Eng.	1932 -	
Firey, W.	Soc.	Eng.	1945 USA	
Freud, S.	Psyc.	Eng.	1922 -	
Gershun, A.	Eng.	Rus.	1939 USSR	
Houghten & Yaglou	Eng.	Eng.	1923 USA	
Mcdowell, M.S.	Psyc.	Eng.	1937 -	
Ogden & Richards	Psyc.	Eng.	1923 UK	
Piaget, J.	Psyc.	Eng.	1929 UK	
Plant, J.S.	Psyc.	Eng.	1937 USA	
Ruskin, J.	Arch.	Eng.	1907 UK	
Sargent, SS.	Psyc.	Eng.	1940 USA	
Sorokin, P.A.	Psyc.	Eng.	1937 UK	
Usher & Hunnybun	Psyc.	Eng.	1933 UK	
Wittenborn, J.R.	Psyc.	Eng.	1943 UK	
Australia				
Ahlschager, W.	Arch.	Germ.	1927 Germ.	
Cambria, F.	Arch.	Eng.	1927 USA	
Dorin, A.	Arch.	Eng.	1927 -	
Eberson, J.	Arch.	Eng.	1927 USA	
Henon, P.J.	Arch.	Eng.	1928 USA	
Lamb, T.	Arch.	Eng.	1928 USA	
Lee, S.C.	Arch.	Eng.	1929 USA	
M.P.N.	Arch.	Eng.	1929 USA	
Rapp, G. & Rapp, CW.	Arch.	Eng.	1923 USA	
Robertson, H.	Arch.	Eng.	1924 UK	
Other				
Birknoff G.D.	Psyc.	Eng.	1933 USA	
Campbell, N.R.	Psyc.	Eng.	1928 UK	
Maslow, A.N.	Psyc.	Eng.	1943 USA	
Ogden, CK.	Psyc.	Eng.	1943 USA	
Philipe, J.	Psyc.	Fr.	1904 Fr.	
Smuts, J.C.	Psyc.	Eng.	1926 UK	

With the "Functional Classics" the same thing happens as with the historical classics. For "Functional Classics" we understand them as being the authors who receive quotations to their works published between 10 and 25 years before the origin of the field of study (Tortosa, 1985). Although the number of quotations received by these authors is superior than the previous group, it is still a long way from what is necessary to fulfil the condition of "classic". In any case we will mention them as they are indicative of the sources of EP (see table 10.3).

Table: 10.3. *Classic functional authors according to cultural areas*

Great Britain.

Adorno	Gropius	Poincaré
Arnheim	Gullahorn	Popper
Barker	Hall	Riesman
Bartlett	Handel	2 Rossi
Bedford	Hess	Roy
Bennett	5 Kelly	Schmore
Blake	2 Kuper	Siegel
Brennan	Laing	Since Gray
Broadbent	Lee	Smith, J.G.
Bruner	Lewis	2 Sommer
Bursill	Lockwoodw	2 Spinley
Campbell	Loring	2 Sprott
Cattell	Lynch	Strauss
Chance	Mackworth	Teichner
4 Chapman	2 Maslow	Townsend
2 Chrenko	Mercer	Various
Dean	2 Merton	Vinacker
Duffy	2 Mintz	Waldram
Easterbrook	Moon	Walker
Edwards	Morris, C.	Webb
Euler, von	Morris, T.	Whorf
Eysenck	Neutra	2 Whyte
5 Festinger & col.	Orwell	Winch
Form	Osgood & col.	Wittgenstein
Gibson	2 Parsons	Young, J.Z.
2 Goffman	Pepler	Young, M.

French Area.

Anderson	Gurvith	Malraux
3 Bachelard	2 Hall	Mitrani
Bloch	Jones	Piaget
Chombart, P.H.	2 Kelly	Radoliffe
Dumezil	Klein	2 Spitz
Erikson	Lacan	Wallon
Festinger	2 Lagache	
Freud	Mancipoz	

Sweden.

Arnheim	Heyl	Moreno
2 Attneave	Hungerland	Ogden
3 Bruner	Johansson	5 Osgood
Cassirer	Kelly	Paulsson
Ekman	Lacey	Rauda
2 Gibson	Langer	Sandstrom
4 Guilford	Lévy-Strauss	Torgerson
4 Hesselgren	Long	

USA

Americ.Pub.Health Ass.	Hall	2 Osgood & col.
Attneave	Hinde	Osmond
Bruner	Hochberg	Piaget
Brunswik	Homans	Shils
Butler	Lynch	2 Simmel
Foote	Madge	Rossi
Gibson	Maslow	Zevi
2 Goffman	Merton	

(cont.)

Germanic Area		
Ashworth	KIose	Piaget
Hatt	Lindblom	Shannon
Hoffmeyer	Maslow	Siegel
Kelly	2 Osgood	

Other Areas.		
Allport	Fava	Maslow
Arnheim	2 Gibson	Oran
Ashby	Gorb	2 Osgood & col.
Barker	Guiroud	2 Piaget
Bernstein	Hall	Saussure
Blackshaw	Holling Shead	Schutz
Cassirer	Hopkinson	Sechehaye
Chaloner	Kafescioglu	Seyle
Chombart, M.J.	2 Kelly	Shannon y col.
Chomsky	Keuthe	Tomsu
Eldean	Langer	Tyrwhitt
Ergin bas	Lantz	Werner
Esser	Larson	
Eves	Lewin	

The number of quotations received by publications in the period 1945-1959 is still low. We should consider, however, that frequently authors who fulfil the conditions as "functional" receive the quotation in a reissue subsequent to their work, without recording the date of the original publication. Moreover, they are often precursors or pioneers in the field in which they remain active later than the period in question. For this reason, by continuing to name them in this way, in table 10.4. we record both the works referring to the specific period, which have served us for the preparation of the list of authors who fulfil this condition, and the references to the whole of the quotations that they receive albeit in publications with a later date. The result is a total of 287 that represent 8.9% of the conferences. 104 correspond strictly to the period defined, 3.2%.

This source of analysis, as well as the relationships of authors which are collected in the corresponding tables and the bibliographical references in both appendices, indicate the high level of contemporaneity of the sources used by EP and the low specific weight of the authors that could be called classics. An indication of this is the minimal explicit recognition of the German EP of the first third of the century. On the other hand, we should emphasize the historical role of phenomenology, especially in the Francophone and Germanic areas. Also the role facilitated for the development of the EP of the crisis of behaviourism and the emergence of cognitivism, which understood in broad sense (conceptually and historically) can define a good number of the authors quoted.

On the other hand, the influence of anthropology can be seen, from the Chicago School of Psychology and the pioneering work in proxemics. The aspects related to personality are concentrated in a small number of authors, while maintaining important differences in each cultural area.

Finally we would like to point out that although EP emerged as the fruit of a demand outside psychology itself, as we have repeatedly stated, it sought its references and theoretical frameworks inside the psychological tradition itself rather than in the architectonic tradition, even on the part of the architects themselves. This fact is significant in as much as it demonstrates a tendency to enter a disciplinary ambit different from its

Table 10.4. *Functional Classics*

(A)	(B)
Classified according to the quotations received.	*Classified according to the quotations received of publications between 1946-59*

Total	1946-59	Author	Subject quotation	
28	(2)	Sommer, R.	Prox.	12 Osgood, C.E.
27	(5)	Hall, E.F.	Prox.	11 Kelly, G.A.
23	(2)	Lynch, K.	Cog.	8 Gibson, J.J.
19	(5)	Piaget, J.	Cog.	6 Festinger, L.
13	(12)	Osgood, C.H.	Osg.	5 Hall, E.T.
12	(1)	Barker, R.	-	5 Piaget
12	(8)	Gibson, J.J.	Pers.	5 Maslow, A.H.
12	(4)	Goffman, E.	Pers.	4 Goffman, E.
12	(4)	Hesselgren, S.	Perc.	4 Hesselgren, S.
12	(11)	Kelly, G.A.	Pers.	4 Bruner, J.
9	(1)	Eysenck, H.J.	Pers.	4 Guilford, J.P.
9	(3)	Bachelard, G.	Fen.	4 Chapman
8	(1)	Bernstein	-	3 Bachelard, G.
8	(3)	Arnheim, R.	Gest.	3 Arnheim, R.
8	(5)	Maslow, A.H.	Pers.	3 Attneave
7	(6)	Festinger, L.	Grn.	3 Merton, R.K.
6	(4)	Bruner, J.	Cog.	3 Rossi, P.H.
5	(1)	Freud, S.	Pers.	2 Sommer, R.
5	(4)	Guilford, J.P.	Exp.	2 Lynch, K.
5	(1)	Lewin, K.	Prox.	2 Chombart de Lauuwe, P.H.
5	(1)	Lévy-Strauss, C.	Anthr.	1 Barker, R.
5	(1)	Simmel, G.	Soc.	1 Bernstein
4	(3)	Attneave, F.	Cog.	1 Chomsky
4	(4)	Chapman, D.	Soc.	1 Ekman, G.
4	(1)	Chomsky, N.	Cog.	1 Freud, S.
4	(1)	Ekmon, G.	Pers.	1 Gropius
4	(3)	Merton, R.K.	Soc.	1 Lewin, K.
4	(5)	Rossi, P.H.	Soc.	1 Lévy-Strauss, C.
4	(1)	Zevi, B.	Arch.	1 Ogden, C.K.
3	(2)	Chombart de Lauwe, P.H.	Anthr.	1 Simmel, G.
3	(1)	Ogden, Ck.	Cog.	1 Zevi, B.
3	(1)	Wallon, H.	Cog.	1 Wallon, H.

Total: 33 authors, 287 quotations: 8,9 % of all quotations.

own, in principle. To reason like a psychologist, to work like a psychologist or work in collaboration with a psychologist? This bibliometric skirmish permits the consideration of the three possibilities. In fact the first only implies formation and a capacity for dialogue, without necessarily supposing anything else. The second implies an appropriate formation in the best of cases and in the worst a simple infiltration, scorn and monopolizing of a space that the architect wants exclusively for himself. The third is the recognition that there is a joint space, the human being-environment relationship, that requires the concurrence, as a minimum, of both perspectives for its correct treatment. This is the approach that originated the whole of our story. But there is a fourth reading that would support the hypothesis of substantivity: in the course of the quarter of a century that we have been in this boat an integrated space has consolidated itself which creates a new professional, who requires a formation of pluridisciplinary origin, but which becomes specific and suitable for concrete necessities. This would imply having developed, or to be developing, its own ideas, theoretical and methodological perspectives. We will leave something for the conclusions...

10.6. Summary and conclusions

We have seen in the first section of this chapter how it was not correct to speak of a substantive Environmental Psychology on referring to its origins, both for the meagre tradition of studies, the ingenuity of the approaches and the non-existence of a theoretical body that would to support them. Moreover psychology entered this field of interest from the demand external to itself. This delineates an initial period in which the relationships between man and the physical environment complied as a pivot between at least two perfectly defined disciplines: architecture and plain psychology (or social psychology if you like).

In the second section we have seen how the development and growth of this "Environmental Psychology" as a pluridisciplinary field, open and indeterminate (the quantity of studies that were produced during the whole of the period studied on the architecture-psychology relationship are a demonstration of this indetermination), followed the laws of exponential growth of the disciplines that are considered to be consolidated in contemporary science. Is this an indication that it is being consolidated as a substantive discipline? It could be, but this alone is not a sufficient argument.

In the third section we have seen how there is a progressive consolidation and recognition, with a process of institutionalisation that came about from the creation of associations, periodical publications, scientific meetings, academic programmes and investigation funds. But we should add precision and read the fine print. The associations rarely maintain a neutral balance, as the American EDRA with its slant towards psychology or the selfsame IAPS heeled towards psychology demonstrate, at least until 1984, however many efforts to the contrary have been made. But also the same dominant theoretical perspectives within them demand to be taken into account, with the consequences of disintegration that we have seen. The same occurs with the periodical publications and the scientific meetings, with the honourable exceptions of the "convinced militants" of one or other disciplinary ambit.

We have seen how the active participation of professionals (especially non-psychologists) who were not linked to the academic world diminished from 1979 onwards, precisely when the crisis of theory and relevance exploded between conflicting psychological perspectives (I do not claim the existence of a causal relationship, I am only pointing out the coincidence). The process of academicisation was fortified, but precisely in the centres of psychology.

Finally, the investigation funds were maintained or increased, but we will have to see which direction they pointing in. It is true that the social problems have changed, as we have repeated opportunely. But for one reason or another, it seems that they are being orientated more towards questions that relate the environment with global social welfare or with problems of precise intervention in the neighbourhood or in population considered as a social organisation and the problems derived from it, from crime to traffic passing through unemployment, rather than as a constructive space before a social space. The selfsame MAB (Man and Biosphere) programmes of UNESCO give priority to the "ecologistical" perspective, but they are giving entry to environmental psychologists, etc. On the other hand, in the grand interventions of urban renovation design and monumentality count more than the impact and the services that they can offer to the affected social fabric.

From all this it does not seem to be clear that this space of interrogation between architecture is being precisely reinforced or consolidated as substantive, but quite the reverse. Moreover, although the congresses analysed have really been points of concurrence, dialogue and exchange between disciplines, maintaining a certain balance between the number of psychologists and architects, at the end of the seventies a move towards psychology came about. The dream of collaboration was perhaps no more than a dream. Of 538 studies analysed we have only been able to detect 25 interprofessional collaborations. More than collaboration it is a case of cohabitation.

The last aspect that underpins the indetermination of a substantive interdisciplinary field is the same dance of denominations and substantives that "cohabit", not only on a level of interdisciplinary relationships but also on the level of psychology itself.

Definitively, after twenty-five years of dialogue and cohabitation between architecture, it cannot be said that a common defined and substantive disciplinary ambit has been created. This does not mean that we cannot speak of Environmental Psychology, which moreover I believe enjoys good health. In any case the problem lies in the content of this term.

The confusion began in the initial denomination of "Psychology of Architecture" which defined and integrated both disciplines. By wanting to widen the ambit of the relationships and change architecture for "environmental", psychology was benefitted, and architecture was left out in the open. The consensus established in the origins was broken.

A specialised branch has been consolidated in psychology, which goes beyond a subsidiary applied psychology, and possesses its own ideas, theories at least of medium range, and its own specific methodological processes. This is being enriched with the assiduous contact with other disciplines, among which architecture is notable, but also ecology, sociology, medicine, anthropology and a long list of possible etceteras with which it is initiating and must deepen, a relationship which from the beginnings was prescribed.

The interdisciplinary meetings that have marked the milestones of its development should continue, in that it has been shown that they are productive, but they must deepen in their initial direction as a forum for dialogue and really open out to the other scientific domains.

11 Who is who: The most notable authors and nuclei

In this chapter we will attempt to sketch out the most notable names in the European environmental community. In a field whose main characteristic is to still have a short history and a high level of contemporaneity, the task which we propose looks extremely difficult. We will try to objectivise our process to the maximum by calling on indices and indicators that will allow us to refer to a social recognition sufficient to escape from subjectivisms.

For this reason we will refer exclusively to the authors who emerge from the analysis of the conferences. This introduces an important bias, but for the coherence of the argument we cannot avoid this. Some important nuclei do not appear because they have not actively participated in the conferences, which does not mean omission or contempt on our part. Insofar as we have been able to compile information on them, this has been offered in the analysis of each cultural area. In any case, for obvious reasons of the place of publication, we would like to mention our acknowledgement to our Spanish colleagues who, for their merits in this field, should appear in this chapter. This is the case of Jiménez-Burillo and Aragonés from the University of Alcalá de Henares, Fernández-Ballesteros, Blanco and Corraliza from the Autonomous University of Madrid, Bernardo Hernández from La Laguna, Anguera and Blanco of the University of Barcelona, Iñíguez of the Autonomous University of Barcelona and Sánchez-Robles from Valencia, among others.

So considering that in the revision of each cultural area some names with a wide consensus of recognition have already been stated, here we will limit ourselves to those that emerge from the conferences analysed, starting from four indicators: 1. the analysis of productivity; 2. notable participations; 3. the analysis of scientific collaboration and the nuclei that can be deduced from it, and 4. visibility measured through the quotations received.

11.1. The analysis of productivity

In previous chapters we have seen some initial information concerning exponential growth in science and how Price's law is obeyed in the EP that expresses itself through the conferences analysed. This growth is measured by the participation and productivity of the authors.

We understand active participation as being the attendance in congresses presenting papers, directing workshops and seminars or imparting magisterial conferences. We understand productivity as being the number of studies or active participations registered in a determinate unit of time (Carpintero and Peiró, 1979). A stratification comes about from productivity, as Cole and Cole have pointed out (1973), which means that those who have presented the most studies emerge and who in certain conditions can reach the category of "eminences".

For some, productivity bears a relationship to the scientific leadership and to the intellectual influence that it exercises (Dennis, 1954; Cole and Cole, 1973; Soler, 1984). For us it will have a basic use in two senses: 1) to detect who are the most prolific authors and who maintain a most assiduous presence in the conferences; 2) to detect the intellectual leadership through scientific and organisational collaboration, and the study of its specific weight in the block of references.

According to the Law formulated by Lotka in 1926, productivity in a scientific domain does not follow a "normal" random distribution (In the way of Gauss's curve). According to this law, 25% of the studies correspond to 75% of the authors, those of the lowest productivity, while those of the highest productivity accumulate another 25% and the average producers the remaining 50% (López Piñeiro, 1972).

Although this law would appear to be obeyed in the majority of the applications that have been made, in our field this does not seem to follow. We will attempt to analyse the reasons for this. This will allow us to highlight eminent authors and information on the state of consolidation in our field.

As we have seen in previous chapters (table 2, appendix), the total number of different authors who have participated in the conferences, according to the information to which we have had access, is 514. Logically many of them have participated presenting more than one study in some conferences, as well as having been able to do this individually or integrated in teams that sign the same paper together. Thus the number of signatures (independently of whether they have presented more than one study) is 741 and the total number of studies is 538.

We have defined criteria of classification on this total information on production and assiduity, which are resumed in four categories:

1) *High assiduous producers*, those who present a number of studies equal or superior to the number of the congresses (nine or more) distributed in more than half of these (five congresses). It is not the same for an author to present ten studies in two congresses as it is for one who presents them in nine congresses. Qualitatively we believe that this contributes different information on his personal course of development.
2) *Average assiduous producers*, the authors who present between five and nine papers in three or more different conferences.
3) *Small producers*, those that present two or more studies in two different conferences.
4) *Occasional or transient producers*, with one or more studies in a single conference.

From the resulting distribution (high producers, 0.4%; average, 3.5%; small, 11% and occasional or transient, 77%) (see tables 11 and 12, appendix) the high percentage of occasional authors and the low percentage of great assiduous producers is notable. In our case, the two most highly productive authors who exhaust the first category only gather 3.7% of the studies and not Lotka's 25%, while the 396 occasional producers accumulate

53.4% instead of 25%. The ten most productive and assiduous authors are gathered in table 11.1.

Table 11.1. *The most productive authors: number of works and conferences where they were presented*

Author	N. Works	Conferences
D. Canter	18:	0,0,1,2,SI,SI,SI,3,4,SII,SII,SII,SII,SII,SII,SI1,7,8
R. Küller	10:	1,1,2,3,3,4,SII,7,7,8
A. Lipman	7:	1,SI,3,4,SII,7,7
M. Krampen	7:	SI,SII,7,7,8,8,8
W.F. Preiser	7:	2,3,6,7,7,8,8
P. Korosec-Serf.	6:	2,3,7,7,8,8
P. Stringer	6:	0,1,3,5,6,8
R. Moore	6:	S1,7,7,7,8,8,
T. Lee	5:	0,1,6,6,8 (1 not published)
B. Mikellides	5:	2,4,7,7,8
J. Muntañola	6:	3,SII,7,7,7,7

(Min. 5 works in 3 different conferences).

Conferences code: 0 = Dalandhui, 1969; 1 = Kingston, 1970 (IAPC); 2 = Lund, 1973 (IAPC); S1 = Surrey, 1974 (ICEP); 3 = Strasbourg, 1976 (IAPC); 4 = Louvain-la-Neuve, 1979 (IAPC); SII = Surrey, 1979 (ICEP); 7 = Barcelona, 1982 (IAPS); 8 = Berlín, 1984 (IAPS).

This indicates a clearly established leadership for the period studied in the persons of Canter and Küller and on the other hand a low level of consolidation in this field of study. We will see later on to what point this leadership is intellectual, social or both at the same time.

Lotka's law requires a minimum period of a decade for its application. Although our study covers fifteen years, in reality if we take the nine conferences as a unit of time we would be below the minimum required. It is foreseeable, on the other hand, that high and average assiduous producers will continue with a presence within the same current tendency. If we consider that the growth is exponential and not linear, as we have explained, and probably with a wider perspective in time the resulting distribution would be more polarised at the extremes (table 11, appendix). On the other hand not all the EP that is produced in Europe has a sufficient level of presence in the conferences analysed.

Other explanations can be found in the rapid awareness and expansion of EP, in the nature of the conferences analysed and in the competitiveness of the social and especially the academic context in certain geographical ambits which stimulated the participation in the congresses of the new authors rather than the established ones, who accede more easily to other means of communication and social recognition. This has lead us to the need to consider the qualification of the presence of the most productive authors, to see if they play a notable role in the leading organs which have been given to the scientific community that we are analysing and in what type of scientific session they participate. First, however, we would like to complete the analysis of production with a brief comparative synthesis by cultural areas.

If the maximum absolute production corresponds to the authors that we have highlighted here, the knowledge of the specific weight that corresponds to each country in the European whole and the notable names in each area could be useful for the orientation of the evolution in each cultural area.

As can be seen from table 13 (appendix), three or four countries were incorporated in every new conference. However, the growth of participation is very different in each country. In this table the countries are grouped together by productivity, the first five being notable as the most prolific. Great Britain, the American participation, Germany, France and Sweden accumulate 68.5% of the total production between themselves. The eleven countries with average production gather 26%. These two categories accumulate altogether 94.6 % of all the studies.

These figures have to be differentiated according to assiduity. Among the high producers it is necessary to be more precise. The cases of Great Britain, Sweden and France present a constant production in almost all of the conferences, which increases relatively when the congress is in their country. The American participation has been moderate since the Lund Conference (1973) and multiplied by six in the Second Surrey conference (1979), remaining constant later. Germany, on the other hand, maintains an irregular presence, not particularly high, but comes into the group of maximum producers after a strong participation in Berlin (1984).

The constancy of the first two, as we have emphasized, is related to their function of leadership in the development of this field, making a strong impression on the specific orientation that it would take later on. France entered rather more belatedly, the social and sociological aspects mentioned in previous chapters being reflected in the fluctuation of its participation.

American participation shot up in 1979 in Surrey but not in Louvain-la-Neuve. In that year the USA was already beginning to overcome the western economic crisis of the previous years, which was to facilitate this arrival. Moreover, we have to understand its profusion in Surrey and not in Belgium for two other basic reasons. On one hand, the cultural proximity and language. On the other hand, partly related to the latter, the penetration and impact of British Environmental Psychology in the American community, especially by the Guildford group, which maintained strong links with it. This without a doubt facilitated the diffusion of the conference. The two following conferences also benefitted from this established relationship, after the integration of Guildford, Surrey and the original IAPC into the present day IAPS.

Among the average producers we find a similar case distribution, as far as assiduity is concerned. Greece, Turkey and Israel maintain an assiduous and constant presence, although it is limited. A second group comes into this category starting from a relatively strong participation in one or other of the conferences. This is clearly the case of Spain in the Barcelona Congress, or of Holland in the Berlin Conference. Others like Italy and Japan gradually increased their presence in the last conferences.

Elsewhere it is interesting to observe how Belgium or Switzerland have a greater presence in the conferences in which they have contributed in some way to the organisation and in which the French language was at least co-official. The same occurs in others such as Germany and Spain, respectively. On the other hand, countries like Canada and Australia display more important links to Great Britain. In general, there is a tendency towards a greater participation of the countries in congresses in nearby cultural areas as the examples mentioned of the Francophone area or the Commonwealth ambit demonstrate, both for the large producers and the small and occasional ones.

To complete this section we present (table 11.2) the list of the most prolific authors in the productive countries. We would like to insist that it is an evaluation of participation with respect to the conferences analysed, so its relevance is quite relative. We will leave the individual verification of the evaluation of each author with respect to the relevance in his or her cultural and scientific context to the interest of the reader.

11.2. Notable participations

Once the active participation has been quantitatively analysed, we will now go on to a qualitative analysis. A qualitative analysis in as much as we will concentrate on the relevance conceded to the author by the organisation in its notable situation in the programme and in the composition itself of the committees.

The participation in committees and the organisation of a conference is in itself a certain recognition of competence or at least of the work carried out. (We obviate to repeat here the list of organisation committees that can be consulted in table 1, appendix.)

Among the congresses that we are concerned with we can differentiate two typologies. A series of auto-convocated and in principle independent conferences, with a single nexus of union between them which is the presence of one same or the same organisers. This would be the case of Dalandhui (1969), Surrey 1 1974) and Surrey 2 (1979), in which the only common nexus is the function of David Canter, with different collaborators in each of the conferences. Starting from an institutional university structure and through the influence attained in formal and informal communicative channels of the scientific community, they achieved a notable success.

The second group, or the second typology, is composed of a group of congresses that while being formally independent, have a nexus of union between them through a linking committee named in each convocation, which had the function of organisation or at least of assessment of the local committee of the following conference.

In Dalandhui (1969) Honikman was commissioned to organise Kingston 1970; in Kingston Küller was commissioned to organise Lund 1973; there it was Korosec for Strasbourg 1976; Remy and Voyé would follow for Louvain-la-Neuve 1979, although the organisation would be in the charge of Simon.

After 1979, with the creation of the IAPS association, this would be in charge of commissioning the individual or group that was to organise the following conference, in close collaboration with the executive committee. Thus the team of Muntañola, Morales, Siguán and Pol were entrusted with the Barcelona conference in 1982 and Krampen with that of Berlin in 1984.

As can be seen in table 1 (appendix), D. Canter participated directly in the organisation of four of the nine conferences and indirectly through the IAPS executive in another. For their part, Stringer, Lee, Lipman, Küller and Simon participated in two committees. In total 43.2% of the organisational responsibilities were British, 13.6% Swedish and German, 9% Belgian and Catalan, 4.5% Franco-Swiss, 4.5% Dutch and 2.3% French. The British influence and especially that of the University of Surrey is overwhelming.

These proportions, on the other hand, were not maintained with relation to the invited conference members. The greatest percentages were of French and North-American authors. In the same table 1 (Appendix), where the names of the conferences and the percentages can be found, we can see how the magistral conference has not come about in any of the congresses celebrated in Great Britain. On the other hand this has happened in the continental ones, especially in Strasbourg. In this conference the participation of

120

Table 11.2. *The most productive authors in each cultural area*

Great Britain
D. Canter	18
A. Lipman	7
P. Stringer	6
T. Lee	5
B. Mikellides	5
P. Ellis	4
L. Smith	4
M. Symes	4
M. Edwards	3
I. Griffits	3
B.R. Lawson	3

French Area.
P.Korosec	6
G. Barbey	4
I. Bernard	3
A. Pezanou	3
J. Remy	3
L. Voyé	3

USA
W.F.E. Preiser	7
R. Moore	6
H. Sanoff	3

Sweden
R. Küller	10
T. Gärling	5
C.A. Acking	4
S. Hesselgren	3
J. Janssens	3

Germany
M. Krampen	7
H. Espe	4
A. Schmidt	4
J. Fritz	3
C.F. Graumann	3

Other
F. Henández	7 (Spain)
J. Muntañola	6 (Spain)
E. Pol	6 (Spain)
M. Mitropoulos	4 (Gree.)
M. Morales	4 (Spain)
A. Churchman	3 (Isr.)
J. Daish	3 (N.Z.)
D. Joiner	3 (N.Z.)
K. Lenartowicz	3 (Poland)
A. Peled	3 (Isr.)

recognised authors proceeding from different disciplinary fields who converged in their contributions to EP would be sought, as we have seen in previous chapters. Thus we find Canter, Moles Korosec and Proshansky as psychologists, Chombart de Lauwe as an anthropologist, Raymond and Sansot, sociologists, and the architect Peled. The tradition, initiated in Lund with Appleyard, Winkel and Canter, would continue in Louvain (1979) with Boudon, Longe and Rapoport; Barcelona (1982) with Muntañola, Lévy-Leboyer and Canter; and finally in Berlin (1984) (incorporating ecologists for the first time), with Galtung, Canter, Von Uexkül, Niit, Sánchez and Sommer.

France accumulates the maximum number of notable lecturers, although as we will see later on, it cannot be said that their impact is generalised. The second group is formed by the Americans with five conferences in four different congresses and a qualitatively important and prolonged impact, as they were prime American figures. The four magisterial British conferences were Canter's. The rest are occasional participations that do not always correspond to an assiduous presence in the congresses, nor to an important impact.

This list of lecturers and organisers, the monitoring of the active participation of the most eminent and the information on each cultural area, seen in previous chapters, shows us that there is not a bi-unanimous correspondence between recognition and productivity. Although high productivity can be a source of recognition, as this increases the active participation goes down in the conferences. In other words, those that have an elevated status within the scientific community or the academic institutional organisation tend to limit their participation to the notable sessions (invited lecturers, notable posts in the organogram) or at best, due to the situation of a discipline in formation, to the presentation of a single work. This is not the case, on the other hand, when in spite of having a certain recognition, the person is struggling to achieve a higher academic level or, inversely, he or she wishes to win a space or a recognition within the field of study (this is endorsed by the biographical path of some notable authors, which we have not published in this volume. This explains in part why Lotka's law is not fulfilled in the analysis of this manifestation of science). Later on we will deal with this information again in relation to the power structures and invisible colleges.

11.3. Nuclei of scientific collaboration

Another possible indicator of relevance (if not of eminence) from the perspective of the sociology of science, is the analysis of scientific collaboration through the joint signing of the studies. Collaboration expresses the network of relations existent in a scientific field and also its sedimentation or maturity, as this demands a certain theoretical agreement and the existence of a communications network (Price and Beaver, 1966; Morton, 1969; Crane, 1969, 1972). According to these proposals, highly consolidated scientific ambits demonstrate elevated collaboration indexes (CI). Natural sciences show a CI of 2.30 signatures per study (Clarke, 1964), physics 1.80 (Keenan, 1964) and sociology 1.60 (Hirch and Singleton, 1965). The International Psychology Congresses, according to Montoro's study (1982), place their index in 1.07 signatures per work. In our ambit the index is situated in 1.38 (see table 2, column 7, appendix).

In our study we have differentiated two basic types of collaboration, the collaboration between equals and the master-disciple collaboration.

The collaboration between equals, authors of the same rank or level of recognition or dedication, can occur between recognised, eminent or new authors. It can be punctual or stable. In our study it occurs predominantly in the countries that have a short tradition

The most productive nuclei of collaboration
(Criteria for their consideration: minimum of 3 works in 2 conferences)

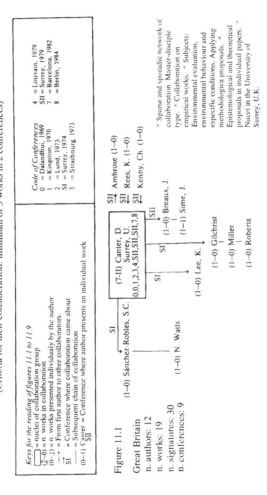

Figure 11.1

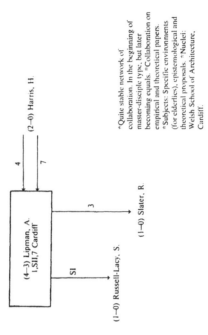

Figure 11.2
Great Britain
n. authors: 4
n. works: 7
n. signatures: 11
n. conferences: 5

Figure 11.3
Sweden
n. authors: 3
n. works: 13
n. signatures: 16
n. conferences: 6

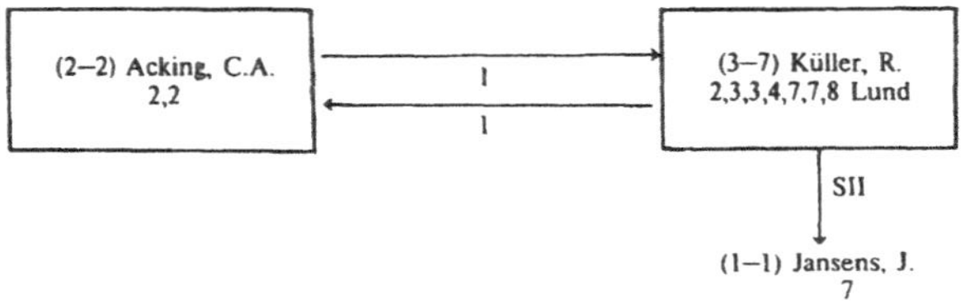

(2–2) Acking, C.A.
2,2

(3–7) Küller, R.
2,3,3,4,7,7,8 Lund

SI1

(1–1) Jansens, J.
7

*Sporadic network of collaboration. The main nuclei is between equals. The secondary nuclei is master-disciple type. *Collaboration on empirical and theoretical papers. *Subjects: Perception. Semantic differential applied to specific environments (for elderlies). *Nuclei: School of Architecture, Lund.

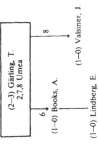

Figure 11.4
Sweden
n.authors: 4
n. works: 5
n. signatures: 6
n. conferences: 5

✻ Sporadic network of collaboration. ✻ Collaboration on empirical papers. Theoretical proposals on individual presentations. ✻ Subjects: Cognition. Behaviour of pedestrians in public spaces. ✻ Nuclei: Department of Psychology, Umea University.

Figure 11.5
France & Switzerland
n. authors: 7
n. works: 11
n. signatures: 17
n. conferences: 5

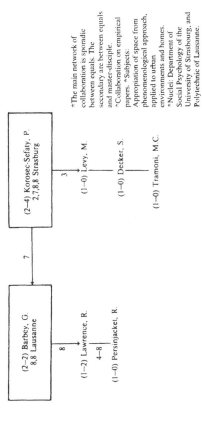

*The main network of collaboration is sporadic between equals. The secondary are between equals and master-disciple.
*Collaboration on empirical papers. *Subjects: Appropriation of space from phenomenological approach, applied to urban environments and homes.
*Nuclei: Department of Social Psychology of the University of Strasbourg, and Polytechnic of Lausanne.

127

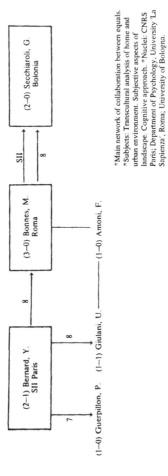

Figure: 11.6
France & Italy
n. authors: 6
n. works: 6
n. signatures: 12
n. conferences: 3

Figure 11.7
New Zealand
n. authors: 4
n. works: 5
n. signatures: 10
n. conferences: 3

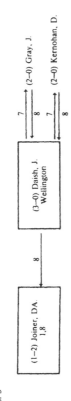

*Main network of collaboration between equals. *Collaboration on theoretical and applied papers. *Subjects: Post Occupancy Evaluation. Architecture and psychology relationships. *Nuclei: Department of Architecture of Victoria University. Ministery of Work and Development.

Figure 11.8
U.S.A
n. authors: 4
n. woks: 7
n. signatures: 10
n. conferences: 5

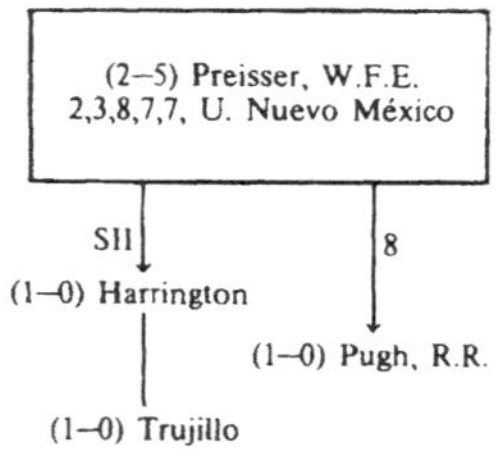

(2–5) Preisser, W.F.E.
2,3,8,7,7, U. Nuevo México

SI1
(1–0) Harrington

8
(1–0) Pugh, R.R.

(1–0) Trujillo

*Sporadic network of collaboration between master and disciples. *Collaboration on applied papers. *Subjects: Post Occupancy Evaluation. *University of New Mexico.

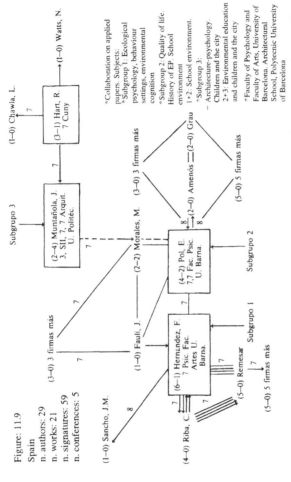

in the matter, without important figures who are outstanding, in which the more or less new authors choose team work among equals to enter what Crane (1969) calls the "Social circle of science", through the participation in congresses.

The master-disciple collaboration comes about through the composition of working teams in nuclei with a certain level of institutionalisation and/or sufficiently developed academic programmes. In these cases the teams are usually ephemeral and a trampoline for the protected promotion of new authors. At the same time they serve the "master" for the broadening of his productive capacity and his circle of influence (Carpintero and Peiró, 1981), creating a more powerful community of his own, and therefore gaining specific weight in the context of the general scientific community. In this case not only does it influence the theoretical quality of the nucleus of origin and its working dynamic, but also the resources that the institution puts at his disposal and the personality and abilities of the leader himself. In the countries with a longer tradition in EP this kind of collaboration lives in harmony with the previous kind.

In our discipline, as we can see in table 2 (appendix), the index of collaboration has grown in time as the field and applied work increases and is lesser in the theoretical and epistemological approaches and reflections. Therefore, it is lesser in the countries where this kind of work is dominant. The active presence of the countries with a longer tradition, such as Great Britain, Sweden and France, contribute an important percentage of work that mark the guidelines of the discipline from the authority reached by some authors. The countries of the most recent incorporation, like Germany, Italy or Spain, with the exception of some cases of recognised authors, contribute a relatively higher percentage of studies in collaboration, especially among new authors.

In figures 11.1 to 11.9 we schematise the nine most prolific nuclei of collaboration in the conferences studied. The criterion of selection has been the presentation of three or more papers in at least two different conferences. In each diagram a brief commentary is included on the nucleus.

The resulting nuclei are distributed in this way: two in Great Britain, two in Sweden, one Franco-Swiss, one Franco-Italian, one from New Zealand and one from Spain. Globally, of the 91 studies presented by the members of these nuclei, there are 39 in collaboration, 22 of which are of the master-disciple variety, 10 are among established equals and seven among new authors.

Globally, we could conclude that in the conferences studied large networks of scientific collaboration do not occur. This gives us an image of an unstructured scientific group. In any case, if we analyse the collaboration in the social organisation of the community we will see how the collaboration structure is more dense. We will deal with this later on analysing the nuclei of power and the Invisible colleges.

11.4. The visibility of the authors: the analysis of the impact

The ultimate significance of any study is its diffusion and its direct or indirect repercussion in the field of professional practice. The presentation of studies in congresses is a form of diffusion in the same way their its publication in book form or articles in magazines. Moreover, as Garfield has emphasized, the best way of protecting the authorship of an original work is its circulation (Carpintero and Peiró, 1981). But not everything that is published achieves the same influence in the scientific community or society. The quality of a study, the fashionable nature of a theme, the accessibility to the means of diffusion with the greatest implantation and the very position of an author in the power structures of the social organisation of the science have an influence in the echo that a study can find

Table 11.3. *The most visible authors with and without self-quotation. Percentage of self-quotation and country of origin*

Author		With self-quot.	%	Whithout self-quot.	%	Visibility Index **	% of self-quot.	Country
1	D. Canter	68	1,62	46	1,27	1,66	52,3	UK
2	A. Rapoport	39	0,93	27	0,74	1,43	30,7	USA
3	T. Lee	29	0,69	27	0,74	1,43	6,9	UK
4	R. Sommer	29	0,69	29	0,80	1,46	0	USA
5	H. Proshansky	29	0,69	25	0,69	1,39	13,8	USA
6	R. Küller	28	0,66	14	0,38	1,14	50	Swed.
7	*E.T. Hall	27	0,64	27	0,74	1,43	0	USA
8	*C. Alexander	24	0,57	24	0,66	1,38	0	USA
9	*K. Lynch	22	0,52	22	0,60	1,34	0	USA
1 0	D. Appleyard	20	0,47	17	0,47	1,23	15	USA
1 1	*J. Piaget	19	0,45	19	0,52	1,27	0	CH
1 2	S. Hesselgren	19	0,45	12	0,33	1,07	36,8	Swed.
1 3	*K.H. Craik	18	0,42	18	0,49	1,25	0	USA
1 4	*W.H. Ittelson	17	0,40	17	0,47	1,23	0	USA
1 5	M. Mitropoulos	16	0,38	0	0		100	Gree.
1 6	*R. Barker (86)	15	0,36	15	0,41	1,17	0	USA
1 6	*Inst. Soc. Urbaine	15	0,36	15	0,41	1,17	0	Fr.
1 6	*Ch.E. Osgood	15	0,36	15	0,41	1,17	0	USA
	Total: 18 authors	449	10,66	368	10,11		18,04	6 countries

* Authors who have not participated in any IAPS conference.
** Visibility index (Plata, 1965):Iogarithm of the number of received quotes.

in a given moment. This echo is what in sociology of science is denominated as *impact*, and is measured by a series of indexes, among them the number of quotes that a publication receives.

In this section we will summarise the result of the analysis of the impact of the studies presented in the conferences, through the study of the quotations that they receive and the global work of their authors, both within the organised community of the IAPS and the international community.

We will analyse who are the most quoted authors within the congresses, what impact the studies presented have within the subsequent conferences and in the international community through the Social Science Citation Index.

By applying Bradford's Law or Areas (see diagram 9 and tables 22 and 23, appendix) the total of references emitted, discounting auto-quotations to avoid the bias that their very high number would cause, three concentric circles are drawn with great dispersion. The very high percentage of authors who receive a unique quotation, that marks the number of areas that can be established, means that the nucleus gathers a very wide spectrum of different frequencies, which in the lowest are really irrelevant in spite of their nuclear position. That is to say, 101 quoted authors (4.08%) receive 28.8% of the quotations, while in the third area 1,494 authors (72.9%) receive a single quotation, displacing 41% of the total. In an initial reading this indicates to us the existence of an enormous dispersion with respect to the sources, and that the same impact that the authors who make up the nucleus can have will be quite relative.

In table 11.3 we can see the composition of the most visible group of authors.[1] Of the 18 authors, half of them have never participated in IAPS conferences. The six that head the list have, on the other hand.

The most visible author is Canter, who at the same time is the most productive, as we have seen above. Of the rest, only Küller and Lee are among the most productive. The others have had an average participation, like Hesselgren and Mitropoulos, or if not punctual but honorific, which is the case of Rapoport, Sommer, Proshansky and Appleyard, as invited lecturers.

If we subtract the auto-quotations from the quotations received, the group modifies its order. Sommer and Hall gain positions, Küller disappears from the list of the most visible authors; Rapoport, Lee and Appleyard go down places and Mitropoulos is left without any quotations at all.

Of the seventeen authors that are left, eleven are American and displace 6.48% of the quotes. Two are British, with 2.01% and two are Swedish with 0.71%. So we see how in spite of the fact that the most visible author, Canter, is British, within this nucleus the principal influence comes from the USA. On the other hand the first author from the Francophone area, the Swiss J. Piaget, is placed in eleventh place, and the first French author, the "Institut de Sociologie Urbaine" is in sixteenth place, with barely 0.5%. Together with Osgood these are the only representatives that are not directly linked to what we could call the "environmental community". In conclusion, there are no authors who exercise a determinant and agglutinating intellectual influence.

The same occurs on analysing the impact within the conferences of the 10 authors who are most productive in them. As a whole they displace only 3.23%. In tables 11.4 and

---

[1] *Index of visibility. Impact, just like productivity, according to Platz's (1965) proposition cannot be measured by the direct number of quotations received but rather by its logarithm, given that its growth is not lineal but exponential, and visibility grows much more slowly than the number of quotes.*

Table 11.4. *Impact of the 10 most productive authors in the conferences*

		Received quotation.				
	Number of works	With seif-quotations.	Without self-quotation.	%	Visibility index	Country
D. Canter	18	68 (1,62)	46 (1,27)	32,3	1,66	UK
R. Küller	10	28 (0,67)	14 (0,38)	50	1,14	Swed.
A. Lipman	7	8 (0,19)	7 (0,19)	12,5	0,84	UK
M. Krampen	7	10 (0,24)	5 (014)	50	0,69	RFA
W.F.E. Preiser	7	7 (0,17)	4 (11)	42,8	0,60	USA
P. Korosec-Ser.	6	12 (0,28)	6 (0,16)	50	0,77	Fr.
P. Stringer	6	13 (0,31)	5 (0,14)	61,5	0,69	UK
R. Moore	6	4 (0,09)	2 (0,05)	50	0,30	USA
T. Lee	5	29 (0,69)	27 (0,74)	6,9	1,43	UK
B. Mikellides	5	6 (0,14)	2 (0,05)	66,7	0,30	UK
Total:	77	185 (4,4)	118 (3,23)	36,2		6 countries

Table 11.5. *The most productive authors, arranged in order of quotes received*

1	D. Canter	46	
2	T. Lee	27	
3	R. Küller	14	
4	A. Lipman	7	Gather in the 1th. area of Bradford (nuclei)
5	P. Korosec	6	
6	P. Stringer	5	
7	M. Krampen	5	
8	W.F.E. Preiser	4	
9	B. Mikellides	2	Gather in the 2th. area
10	R. Moore	2	
	Total	118	

Note: As we can see in figure 9 (append.), these authors don't exhaust the total of the composition of the areas.

11.5 we see how these authors take on an order which reflects the order of importance of the nuclei of collaboration defined in the previous chapter.

The Surrey group, formed by Canter, Lee and Stringer during an important period, displaces 2.15% of the total of quotations, which represents 66.10% of the total of the references emitted in favour of the most productive. Küller's group is in second place with a limited 0.38% of the total, that is, 11.86% of the quotations received by the most productive. The second British group, lead by Lipman in Cardiff, and Korosec's French group in Strasbourg, are left in third and fourth places with a modest 0.19% and 0.16%, that is to say, 5.98% and 5.08% of the most prolific. So altogether, the most productive authors in the conferences receive exactly half the quotations of the 11 most visible Americans. This fact proves to be very illuminating for our initial objective of studying the social and intellectual leadership of European EP.

This group will see its impact reduced even more if we concentrate only on the studies that have deserved some kind of reference, having been presented in one of the conferences studied (tables 19 and 29, appendix).

The total number of references made to papers or studies presented in the conferences is 58, on a total of 538 studies presented. Moreover, considering that some papers receive mor than one quote, the number of works with any kind of impact is reduced even more. The comparative injustice is aggravated if we consider that the EDRA conferences have deserved 117 quotations. No comment.

The conferences in which the most references were emitted to others of the same series were those of Louvain-la-Neuve (1979), Strasbourg (1976) and Lund (1973). Those that received the most quotations were the first three: Dalandhui 1969, Kingston 1970 and Lund 1973. Globally, those that emit references in favour of studies presented in other conferences are authors highly integrated in the organisational structure of the congresses, average or high assiduous producers, and some of them occupy posts of responsibility in the organisational chart of the IAPS. Also, the conferences that received the most quotations were those that have generated a commercial publication through the publishers established in this ambit.

The last indicator of impact that we will consider is the visibility of the most productive authors in the international community, measured through the Social Science Citation Index. The only authors that appear are Lee, Canter and Lipman, and Küller and

Honikman with a noticeably lower impact. Krampen, Korosec and Mikellides appear in a testimonial way, but quoted by other studies that were not presented in the conferences (see table 21, appendix). Altogether, the quotations received by studies concerned with the conferences represent only 4.2% of the total of references emitted in favour of all these authors.

This scarce impact of the European conferences in the SSCI goes through the "minutes" of the first four and their compilers, whenever they have counted on sufficient diffusion. Without a doubt publication is an important factor in the possibilities of impact of a study. On the other hand, we should not forget that the SSCI introduces an important bias as a restraint to visibility and impact through the composition of the magazines and periodical publications that expounds on it, mainly American or Anglophone, which could tone down these results to some extent.

11.5. Summary and conclusions

In this chapter we have sketched out who are the most notable authors and nuclei starting from four indicators: the authors' productivity, the participation in notable sessions of the programme of the congresses, the most productive nuclei of scientific collaboration and the visibility of the authors measured by the quotations that they receive.

Table 11.6. *Outstanding authors according to the criteria of eminence (summary)*

Productivity and assiduousness	Visibility or received quotations	Outstanding lecturers
Canter (UK)	Canter (UK)	Canter (UK)
Küller (Swe.)	Rapoport (USA)	Chombart de L. (F)
Lipman (UK)	Lee (UK)	Raymond (F)
Krampen (G)	Sommer (USA)	Sansot (F)
Preiser (USA)	Proshansky (USA)	Korosec (F)
Korosec (Fr)	Küller (Swe)	Moles (F)
Moore (USA)	Appleyard (USA)	Boudon (F)
Lee (UK)	Hesselgren (Swe)	Levy-Leboyer (F)
		Appleyard (USA)
* Remy (B)	* Wools (UK)	Winkel (USA)
* Voyér (B)	* Raymond (F)	Proshansky (USA)
* Mikellides (UK)	* Haumon (F)	Rapoport (USA)
* Acking (Swe)	* Lugassy (F)	Sommer (USA)
* Hesselgren (Swe)	* Palmade (F)	Galtung (G)
* Janssens (Swe)	* Rivlin (USA)	Uexküll (G)
* Graumann (G)	* Sanoff (USA)	Peled (Isr)
* Espe (G)	* Krampen (G)	Jonge (NL)
* Schmidt (G)	* Seiwet (G)	Niit (USSR)
* Muntañola (S)	* Espe (G)	Sánchez (V)
* Fatouros (Gr)		

Code: B=Belgium F=France G=Germany Gr=Greece Isr=Israel NL=Netherland Swe=Sweden S=Spain UK=Great Britain USSR=Estonia and ex-USSR V=Venezuela.

The authors of the first and second column (superior part) are classified according to importance within each indicator. When preceeds a *, they enter in the list after the application of a corrective factor wich lists the outstanding authors within their cultural area.

We have seen how the EP that is produced in Europe presents a high level of contemporaneity, and does not follow the law of normality in scientific production which Lotka defined in 1926. We must point out that the results of these global analyses, both of productivity and visibility, overshadow the reality of the smallest or least productive cultural areas for the reasons that have already been explained. Given that a constant in this study, starting from our proposed objectives, was to consider the differential elements of each one of the cultural units that make up the European whole, we have applied the consideration of the most notable authors in relation to their area of origin as a criterium of correction. This is taken into account in the data which is presented in table 11.6, a summary of the partial results of the indicators of eminence used (see table 22, appendix).

In synthesis, in this chapter we have attempted to highlight the authors and nuclei of authors who play a notable role in the process of development of Environmental Psychology in Europe starting from the analysis of the participation, productivity and collaboration of the authors who participate in the European conferences. We have seen how from each one of these indicators names emerge that clearly endorse the qualitative analyses realised for each cultural area in the proceeding chapters and allow us to glimpse the organisational profiles that the scientific community will adopt which at the present is linked around the IAPS. The intellectual and theoretical leadership does not take shape so diaphanously, however. Starting from the bibliometric study of the references emitted in the conferences, we have confirmed the existence of a great dispersion of theoretical sources, both historical and contemporary, without the clearly established existence of any author or group that should agglutinate an important part of the quotations. We can affirm, on the other hand, the predominant Anglo-Saxon influence on the studies carried out, especially of North American authors and British authors from the Surrey group.

In spite of the theoretical dispersion of the studies among themselves and the low level of scientific collaboration, in comparison with other ambits of science there is a dense network of relationships and bonds which influence and mark the evolution of this ambit of study. This will be the object of our next chapter.

12 The invisible colleges

According to Price (1963), the studies on the exponential growth and the distributions of scientific production can lead to the idea that the studies are written only to be narrated by the deans, leaders and historians. However, this would be to exaggerate the weight of what we believe is one of the implicit functions of scientific production in the present day social and academic coordinates. In any case Price emphasized that each study represents at least a *quantum* of useful scientific information and some concrete contributions which mean that each author can be evaluated above the most prolific. This brings us to the necessity to penetrate into the knowledge of the social organisation of the science and the psychology of the scientist.

As we have mentioned before, Peiró (1981: 53) points out that the advance of science is not only the result of the internal logic of its ideas. The process of interaction between scientists contributes to and affects the development of science. This allows us to suppose that there is an interaction between cognitive and social eventualities (Crane, 1972). What is more, it has even been affirmed that a coherent structure of knowledge in a determinate field of science is inconceivable without being supported by an informal communication network among specialists (Garfield *et al.*, 1978).

As a contribution to the analysis of these interrelationships, in 1961 Price introduced the term "Invisible Colleges" to refer to groups of scientists who, while working in different places on similar themes, exchanged information through different means other than printed literature, in particular *pre-prints* (Peiró, 1981). He adopted the term "College" in the same sense as the group of scientists who had adopted it to promote the exchange of information, and who later would found the Royal Society. This permitted the conferral of a status and prestige onto each member which depended on the approval of his colleagues and facilitated communication by reducing a wide group to a select circle which permitted interpersonal relationships. These groups have permitted the consolidation of the status of the scientist without the need to increase the number of studies (which has been perfectly confirmed according to the data of our study in the community analysed). Also scientific exchange on a high level has become an important means of communication (Price, 1963). All this delineates the constitution of nuclei of power or pressure groups that orientate the direction of development of a field of science, which going to the ultimate consequences of Price's reasoning allows us to emphasize our affirmation, made in previous chapters, of the emergence of a double circuit of high science, very selective and of a minority, and of humble science-entertainment, low flying, to fill the time of the leisure society. Without specifically formalising nor defining these two circuits, they are implicit in Price's approach on differentiating Invisible Colleges from the "Great mass of scientists".

In his proposal, Price described and analysed in detail the forms of communication that underlie the Invisible colleges (IC) and which constitute the fundamental base of this level of organisation in science. His reasoning lead him to value the quantity of books and magazines an averagely prolific author should read, who at the end of his life, judging by the number of quotations, would have written around a hundred articles with original contributions. According to his calculation (we shall not go into detail on it here) it should be of 64 articles per day (!), which is obviously impossible. We will not discuss whether the calculation is correct or not and if the quantity is exaggerated, but of course even though it were much lower it seems to be clear that the information should be transmitted through other communication channels, which thus take on a clear caudal importance. "The authentic investigator," he says, "certainly does not read at all, but rather he acquires his information in other ways, through conversations and personal relationships" (Price, 1961: 121). From here the organisational methods and structures which permit the movement of this communication are established. He describes five:

> 1) The extra-official or extra-academical organisations, such as thematic organisations or professional associations.
> 2) Congresses and meetings are periodically called, which however turn out to be insufficient for direct or informal exchange.
> 3) Daily communication procedures are sought, such as correspondence, *pre-prints* and - we could add - computerised and telematic procedures.
> 4) These communication networks promote the realisation and signing of work in collaboration.
> 5) Finally, an ultimate manifestation of the communication among authors would be the influences shown through the mutual quotations in the blocks of references.

Price's proposal has not been free from criticism and alternative propositions (Mullins, 1968; Crane, 1969; Peiró, 1981), but in spite of this the concept has gradually been consolidated and has become quite common in studies on the social organisation of science.

The five points mentioned earlier set a standard that we shall follow in the development of this chapter. In our analysis we shall recuperate the information prepared in earlier chapters, integrating it in the data concerning institutional collaboration in the organisation of the conferences analysed, the nuclei of collaboration as a sign of a network of interrelationships and the mutual quotations as indicators of the direction of intellectual and theoretical interdependencies, both those manifested in interviews and those that emerged from the bibliometrical analysis. From here we will try to describe the Invisible colleges that occur in European Environmental Psychology, and we shall analyse their function as nuclei of power that direct the orientation that our discipline takes.

12.1. The nuclei of collaboration as indicators of the network of relationships

As we have seen in the introduction to this chapter, the analysis of collaboration is one of the principal indicators used for the detection of Invisible colleges (Price and Behaver, 1966; Peiró, 1981). In our study collaboration has already been analysed in chapter 11. In this chapter we shall recuperate some of its conclusive aspects, but we will not repeat ourselves unnecessarily.

As we have pointed out previously, in 1963 Price predicted that if the growing rhythm of collaboration in the signing of studies were to continue, by 1980 individual work would probably have disappeared or would be simply anecdotal. However, in EP, although there is a tendency to increase the index of collaborations, this is still notably low, displaying fluctuations in determinate moments which are related to the tradition and the productive style of each country. Precisely, this aspect appears as an authentic differential factor between the differential cultural areas.

On the other hand, we have confirmed that the countries with the longest tradition in EP are not those that present the highest indication of collaboration, but have a production based on individual contributions, from the authority acquired by the most recognised authors. The countries with a shorter tradition, on the other hand, except in the case of some recognised authors, show a greater incidence of studies in collaboration, especially among new authors.

From all this some interesting conclusions can be deduced as far as the structure of EP is concerned. Price (1963) - we have already picked up on this in chapter 11, but it deserves some modifications - said that the recognised and very prolific authors increase their productivity through being directors of teams that multiply their effects, as this implies that there is a connection with a powerful network of an Invisible college, and therefore with a powerful series of resource centres. This factor does not occur more than occasionally within the EP community which expresses itself through the conferences. Apart from the case of Canter's nucleus (master-disciple collaboration), the main contribution of the most productive and recognised authors is individual. It is the new authors and those of average production who present the greatest number of studies in collaboration.

Inverting the proposals we see that the presentation of studies alone, and even more if they are of the theoretical kind, constitute a sign of eminence, while the empirical or field studies are presented by the collaboration groups, above all among the new authors, although in some cases they are also presented by master-disciple groups. We have also seen the scarce collaboration between the most productive authors, practically nonexistent among the nuclei that we have described (see diagrams 11.1 to 11.9). Does this mean that there is not a network of relationships among the eminent authors? In short, no. Probably the fact that once the mechanisms of the hierarchical structuration of science are known, so repeatedly studied in the last decades, this has catalysed an alteration of these same mechanisms. The collaboration in the written whole of studies -at least in the group that we have analysed- has probably undergone a change of orientation. So it will be necessary to resort to other less transparent indicators to find these networks of connection. Institutional collaboration and mutual quotations will be our next indicators.

12.2. **Institutional collaboration**

If we have seen in the previous chapter that the collaboration in the signing of the studies does not show us the relationships that exist between the most eminent authors, these become clearer through the analysis of group participation in the institutional structures that the community studied is endowed with. Montoro (1982) points out how the act of participating in some organisational, structural petitions or in notable sessions within the conferences at least implies a certain recognition on the part of the community. Thus, in this chapter we shall analyse the implicit relationships in the group participations in the organisation committees of the conferences, in the executive and consultive (or full)

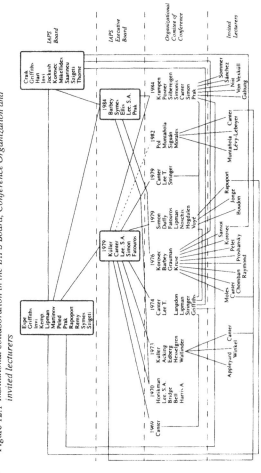

Figure 12.1 *Institutional collaboration in the IAPS Board, Conference Organization and invited lecturers*

committees of the IAPS since its constitution, and the recognition and relationships shown in the invitations that these give to certain authors to participate as lecturers or speakers in notable sessions.

Some of these affirmations can be deduced from the reading of diagram 12.1. The first executive committee maintained a much stronger presence or link than the new committee that emerged in 1984. Only one member was repeated, Sue Ann Lee, due to her connection with the *Newsletter* and with Kingston Polytechnic which helped in its financing. The outgoing executive committee had a direct participation in eight of the new organising committees of the conferences, through some of its members, while the incoming committee only had this in three. So the force of the new committee seemed to be inferior.

The strongest nexus of continuity between the various conferences passes through David Canter, present in four organising committees. The First Surrey Conference was the one that shared the most members of its organising committee with the other conferences and with the full committees of the IAPS, while on the other hand this was not the case in the Second Surrey Conference, where the team had been noticeably restricted. Moreover, as a whole the three central conferences were those that showed more links between the components of their committees. This network of relationships was diluted in the last three, going to the extreme case of Barcelona, where the only connection was the institutional contact between the organising committee and the executive of the IAPS. It was also diluted in another sense in Strasbourg 1976, which only had contact with the new leaders, and was left like an island, right in the middle.

All of this perfectly reflects the state of the power relationships that arise between these leading groups. For example, the disappearance of Lipman after his participation in the First Surrey and Louvain-la-Neuve committees is significant, in which he was the continuity link. It is also significant that after the isolation of the Strasbourg committee, with respect to the others and with respect to the first executive of the IAPS which emerged in 1980, one of its members, Barbey, went on to the presidency of the association in 1984, and another, Korosec, entered the full committee. Significant when in the three cases mentioned they have shown intellectual positions at least in disagreement with the dominant group of the first executive. All these aspects gave the new executive an image of "triumphant opposition", which despite never having been formally expressed, was implicitly shown on a level of the state of opinion of some circles of the association.

A third level of reading offers us the links that are demonstrated with the invited lecturers or notable sessions. We find Canter as the only person who repeated on three occasions, which gives him a clear position as leader in the relational structure, reinforced by being the only member of the committee who has developed this role. Korosec is the only lecturer who would later belong to the second consultive committee.

From this triple reading of figure 12.1 a fairly clearly defined dominant nucleus becomes clear, situated in Guildford, Surrey with David Canter as leader, which in the Berlin 1984 assembly seemed to be in a critical situation. A second incipient nucleus seems to emerge with Gilles Barbey and Perla Korosec as its leaders.

Figure 12.2 allows us to lower the level of reading from the institutional level to the individual level. The collaborations in some institutional instances (group participation in some committee) have been taken into account in its preparation.

As we can see, in spite of the dense network of interconnections, a central nucleus is formed around Canter, and six peripheral nuclei are connected to him. What changes is the intensity of the connection to the central nucleus. It is noticeable that Canter's central position is previous to the constitution of the IAPS as an association. Within this, the links that are reinforced are those more related to the figure of the General Secretary, Canter,

Figure 12.2 *Individual relationship network through institutional participation in the IAPS activities (in conferences and association chart)*

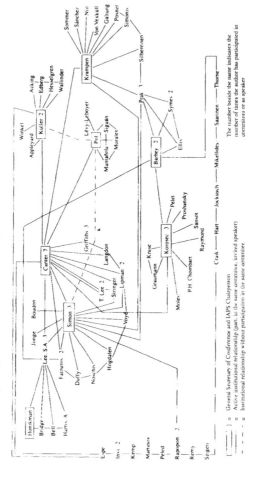

than to the President, Küller, who only appears in two organisational instances, logically with links with all the members of the executive.

The most dense network of relationships comes about between the organising nuclei headed by Canter, Simon, Küller and Honikman, who at the same time lead the formationof the IAPS. In Honikman's case, the visible head of the Kingston conference, his maximum connection is seen through Sue Ann Lee and not directly. Sue Ann Lee is the only participant in the two executives, as we have pointed out previously, and so one of the few links with the Barbey-Korosec nucleus. In the case of Simon, his centrality in the subgroup which he defines maintains a surprising and elevated connection with Canter's central nucleus. Surprising in the light of the events before the 1979 crisis, but which it is necessary to understand in the light of the desire to maintain an active presence of the francophone and southern European area, a situation reinforced by the participation of Fatourus. The most important connection in the nucleus appears through three names that repeatedly participated until the date indicated with the nucleus lead by Canter. The direct link comes about through Simon's participation in the first executive of consensus of the IAPS, and is repeatedly critical, especially in matters related to the question of language, not inasmuch as to the official policy as to real practice. One of the names that drives the contact between the two nuclei, and that in its moment was able to reinforce the connections, but which after 1979 would be diminished in power, is that of Lipman (see section 3.7 and foll.)

The nucleus centred around Krampen emerged solely and exclusively in Berlin. So it is a young connection, though powerful. The specific weight of Canter's participation in the Berlin organisation was very intense.

The team most clearly connected with Canter is the one from Guildford, Surrey itself. Lee and Stringer, without institutional links with other authors or nuclei, and also Griffiths, a close collaborator in various institutional cases, who as figures of prestige reinforce Canter's central position.

The organising nucleus of the Barcelona Conference was the one that proved to be the least integrated in the network of interrelationships. In fact, it only maintained institutional contact -though not formal collaborations in a single committee- with the IAPS executive, especially with four of its members, Canter, Simon, Küller and S.A. Lee.

The last nucleus that we shall analyse is the Barbey-Korosec tandem. As we have seen previously in the analysis of diagram 12.1, the organising team of the Strasbourg conference lead by Korosec did not display links with more than the new executive and full committee of the IAPS. Moreover, as we have already pointed out, the mew executive shows very few links with the previous one. This configured a quite autonomous and alternative nucleus.

The only direct connection with Canter is its participation in a notable plenary meeting in Strasbourg, but it does not openly present any institutional relationship. The other three points of contact with Canter's large nucleus are through names that act as a pivot: S.A. Lee, Niels Prak and the members of the full committee, renovated in the Haifa, Israel conference in July 1986. So the connection is very weak.

In synthesis, on an organisational and institutional level, in the EP that expresses itself through the conferences analysed we see how two principal nuclei are formed: the most powerful is lead by Canter, who displays a strong network of relationships that includes 41 names, plus the members of the consultive committee or full committee; the second nucleus is centred on the Barbey-Korosec tandem and includes 14 names, plus the corresponding full committee, with a notable central European specific weight.

12.3. Mutual quotations as a manifestation of intellectual relationships

Until now we have seen the relationships shown in scientific and organisational or social collaboration. We will now tackle the relationships of convergence and/or intellectual dependency among the notable authors in each cultural area who participate in the conferences, the last link that serves to complete the relationships that allow us to realise the analysis of the invisible colleges according to what we saw at the beginning of this chapter.

The thirty notable authors, who constitute the base that we shall study here, emerge from the conjugation of the criteria related to production and visibility. In the first place we have taken the globally most productive authors in the whole of the conferences. In second place, the most visible or the most quoted of those that participate. In both cases we have completed the direct list of global consideration with those that emerged from the application of the corrective factor of the cultural areas, that is to say, the most productive and visible authors who participated in each one of these zones, although their specific weight with respect to the whole of Europe is limited, according to what has already been defined in the previous chapters.

Within the group that would be formed we shall analyse which nuclei of dependency or intellectual relationships appear, and in what direction. Of the 30 authors, some integrated intellectually into one or another of these nuclei, others to a wider community of European or western scope and others showed that they did not belong to either. We have defined four kinds of intellectual links, always referring to the mutual quotations between the 30 notable authors. We consider that they are intellectually integrated in the European EP community when, apart from the cultural area itself, they quote authors from other European areas, as well as the possible reference to the American participants. Due to the important specific weight of the American participation and the influence that it exercises, we have defined a second category composed of the authors who as well as their own area only quote American participants as outsiders. The third will be formed by those who only quote the notable authors from their own area. Finally, a fourth category includes the authors who do not demonstrate any intellectual links with the group analysed.

From the result of its application in table 12.1, we can see how there is a dominant tendency towards references from the same area or not to show any link with the group analysed. The formation of a network of European intellectual interdependencies within this group only passes through these authors: Korosec, Küller, Mikellides, Graumann and Muntañola, while such important authors as Canter and Lipman, and on another level Krampen and Voyé, show a clear connection with the American participants and with their own area. It will be necessary, however, to go into the relationships existent within each area more deeply.

In Great Britain there is a nucleus of clear intellectual leadership formed by Lee and Canter. This nucleus reveals a link with the Americans, and no intellectual debt to the rest of Europe. On the other hand it exercises influence that reaches Küller and Muntañola, but does not affect Francophones or Germans.

Lee and Stringer are outstanding as leaders of the chain, who receive quotations from the eminences, but do not emit them. What has repeatedly been seen in this study is reflected here: the influence of the Guildford group.

Mikellides, who appears in a position of intellectual dependence, is the one who would seem to be the most open to influences. In a way this author appears as a prototype of what is an emergent generation within the IAPC-ICEP-IAPS tradition, with the conscience and desire to build a European EP.

Table 12.1. *Intellectual links of outstanding authors in the different cultural areas*

Intellectual links	British Area	Swedish Area	French Area	USA participation	Germanic Area	Other Areas	Total
1) Integrated European Community*	Mikellides	Küller	Korosec		Graumann	Muntañola	5
2) Own area and americans	Canter Lipman		Voyé		Krampen		4
3) Only own area		Acking Janssens	Haumon Lugassy Raymond	Appleyard Moore Preiser Proshansky Rapoport Rivlin Sanoff Sommer	Espe		12
4) No links with analysed groups	Stringer Wools Lee	Hesselgren	Remy Palmade			Fatouros	9
Total	6	4	7	8	3	2	30

* Integrated intellectually in the European Community: authors that quotes author from other European areas, in addition to them of their own area.

The Swedish nucleus proves once again to be the most integrated in itself and at the same time well balanced in the international intellectual dependencies. The four authors who make up this group quote each other and demonstrate a generational structure. While Hesselgren, as a pioneer, exercises a certain intellectual leadership, Acking and Küller, of the same generation, quote each other, and Janssens, as a disciple, quotes Küller. This constitutes the nexus with the rest of the eminences, with intellectual debts to Canter and Sommer and influence from Korosec.

The Germanic area does not form any integrated nucleus, excepting a mutual quotation between Espe and Krampen.

The Francophone area appears to be noticeably closed in on itself. Basically formed by sociologists, geographers and architects, it receives an important influence from Rapoport, who has a perspective clearly adjacent to the dominant parameters in this area, at the same time as he is practically outside the dominant nucleus of his country.

The opening out to other common influences, especially American but also including Swedish and Germanic influences passes through Korosec (the only author who fully identifies with EP as a domain, as we have seen), but who at the same time has a very low influence on this nucleus, only linking herself intellectually with two authors from the Francophone nucleus. Here it is necessary to argue for a generational justification of pertinence. However, she constitutes the weak nexus with the rest of the European community. The influence of this nucleus hardly surpasses the frontiers of its cultural area, except in the case of Mikellides (GB).

The American group present in Europe clearly exercises a function of intellectual leadership which extends its influences over all the cultural areas under consideration.

The Rivlin-Proshansky nucleus is paradigmatic, gathering nine of the notable European authors around itself, from all the theoretical tendencies. Internally they do not receive any influence from authors outside their area. It clearly exercises a function of intellectual leadership, fruit of its pioneering work and of the enormously influential publications that they have generated, which have constituted one of the liberating factors of EP in Europe. We should also mention Robert Sommer, with similar characteristics to those of this nucleus.

Rapoport shares leadership with a perspective centred more on the analysis of architecture and urbanism from an anthropological point of view which correlates well with the tradition of social sciences in the francophone area, but also extends his influence to the other cultural areas, except Sweden. On the other hand, he does not display any intellectual links with the group of Americans present in Europe.

The third nucleus, that of Sanoff, is very small. It is practically a mini-link with a single connection with Europe through Canter.

Altogether five nuclei are formed in which the specific weight of each cultural area is very important. The Francophone and Swedish nuclei reveal powerful interdependencies that go beyond the influences from outside the area.

The British nucleus exercises a punctual leadership on authors from all of the other European areas, but shows itself to be completely integrated intellectually with the Americans.

The Americans exercise full influence through their principal nuclei in all of the European areas. We can affirm that the intellectual leadership of EP in the strict sense corresponds to them.

The integration of the different communities comes about through the most active members of the IAPS, like Canter, Küller, Korosec and Mikellides. On the other hand, the situation of the organisation of conferences and the promoters of the association confirms once again the truth of the hypothesis that these activities imply an improvement of the status of their promoters in the whole of the scientific community or communities of membership.

In synthesis, all this shows us a clear dependency on the theoretical constructs developed by American EP, which confirms what we have seen in previous chapters, and makes it difficult to speak of a European EP. However, it allows us to detect the still strong identity of each cultural area, as well as the aspects of professional composition, thematics, theoretical orientation and interest seen before, as well as that concerning the relationships, links and/or mutual intellectual dependencies. A group that we could call generational is delineated, and which could reach the point of forming this cultural integration (category 1, table 12.1), dominated by contents of phenomenological, cognitive and epistemological-genetical orientation.

12.4. **The Invisible Colleges as nuclei of power**

Following the logic defined in the introduction to this chapter, and once we have seen the relationships that emerge in the three levels established to define the possible Invisible colleges, next we will go on to express them in concrete terms. Once they are established, we will analyse their relationship with the power structures that come about within the community studied.

On considering the relationships established as a whole in the three previous chapters, figure 12.3 emerges, which will serve as a foundation for the global consideration of the Invisible colleges. To facilitate clarity in its representation we have restricted the relational lines to the American authors who have a punctual presence as lecturers or who, moreover, are repeatedly quoted but do not participate in the organisational structure, to a single connection to the head of the Invisible college or to the focal points or nuclei, through which we would like to synthesise the link or the influence exercised on the nucleus. On the other hand, the quotations that the secondary authors make only appear when they refer to other authors from the group of 30 celebrities. Also, in the representation we have graphically differentiated the institutional collaboration, the collaboration in the joint signing of the works and the intellectual links expressed by the quotations.

In a strict sense, a single Invisible college is formed, starting from the dense network of relationships headed by Canter, with whom all the focal points or nuclei have some direct or indirect relationship, at least through some of their members. In spite of this, there are some nuclei that have a very weak connection and that can be prefigured as Invisible or potential colleges. Also, due to the short period studied, we have not reflected the qualitative incidences in the diagram that come about over time. On the level of evaluation we all see how there are important differential variations.

To systematise the qualitative analysis of these relationships here we swill apply the Spectral Model of Group Analysis (Munné, 1985). In the analysis of the relationships of a group four factors or levels come into play: the thematic, the functional, the cognitive and the affective.

Applying this model to the analysis of a scientific community, we consider that in order to be able to define a group as an Invisible college, the relationships between its members have to cover at least two of these levels. The first or thematic level is obligatory, as it is co-substantial with the existence of the same possible College. The second level, which we consider to be of obligatory fulfilment, can vary between the functional or the cognitive or both at the same time. In the Price's definition of an Invisible college (1961), he referred to authors that while working in different places on

Figure 12.3 *Synthesis of quotations, co-quotations and mutual influences among outstanding authors in the Conferences*

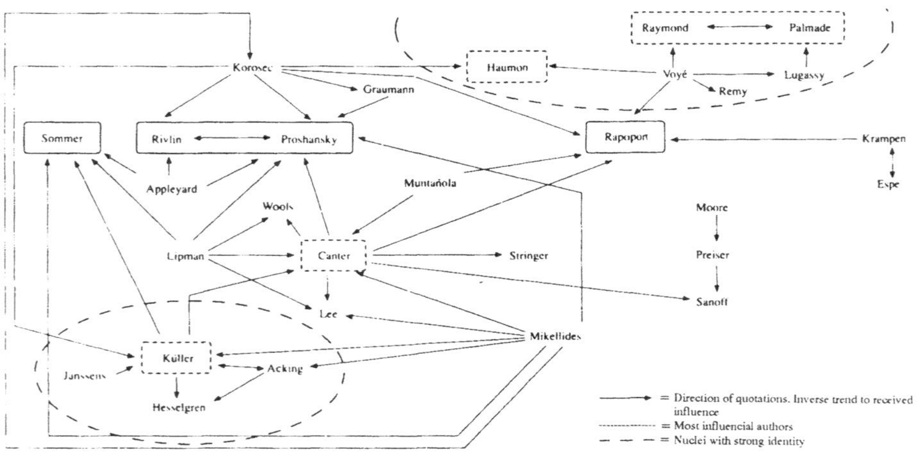

150

similar themes (thematic level) exchange information (functional level) and logically have mutual influence over each other (cognitive level). Given that in some cases the intellectual dependencies or links between a group of authors are considered as an indicator, independently of the channel through which the information on an author has arrived to others (that is to say, it can be a formal route, not necessarily through a direct contact or exchange of *pre-prints*), we consider the third or cognitive level as being sufficient, without the functional level. The reverse is also true. There can also be an exchange or participation in some organisational instances (functional level) and on the other hand the maintenance of different positions. This case (thethematic level plus the functional level, but not the cognitive level) can result in (but not necessarily) an open controversy, and probably a breakdown of relations.

Within the collective of authors that we are analysing, a thematic unit occurs, the H-E relationships, which guarantees the first level. The second, or functional level, is present in the majority of the relationships that are established, through a convergence in the IAPS conferences. However, this link can on occasions be so weak that it can virtually be considered accidental or nonexistent. This is the case of some of the Francophone lecturers, totally outside this community. The third level, the cognitive, intellectual or theoretical, does not always transpire. In the collective studied there are differentiated theoretical positions that act as diversifiers of the possible Invisible colleges that could emerge. The fourth, or affective level is very difficult to detect. It is only clearly seen in some of the natural groups in concrete cultural areas.

Canter's large Invisible college develops on the four levels mentioned, although with a different specific weight in each one of the connections expressed. Following an anti-clockwise direction, with the centre on Canter, we shall analyse the kind of link with each one of the focal points that appear in diagram 12.4.

Starting with the group of names that only collaborate in the signing of works, in a collaboration that we defined in chapter 11 as being of the "master-disciple" variety, this clearly covers the first, third and fourth levels and to a lesser degree the second (functional) as it has a low participation in the organisational structure of the IAPS. The integration of this group in the IC is clear, at least at the moment of collaboration.

The Barcelona focal point is situated on the first, second and third levels. The thematic link is evident. The functional level comes about through the organisation of the conference in 1982. The intellectual level appears through the quotations emitted and Canter's invitation to participate as a lecturer in the Barcelona Congress. In any case, this link appears punctually on all its levels, especially the functional, as the organisation of the conference did not involve any connection with the organisational structure. On the other hand, the links with other members of the formally analysed group are reduced to the members of the executive committee of the moment (1982).

The Swedish focal point, centred on Küller, maintains a strong network of internal connections, with a radial structure that develops on the four levels. In this sense, we could say that it has a strong degree of internal cohesion which together with the data we have seen in other chapters on themes and sources, gives it a clear autonomous identity. However, the strong institutional connection (the second level) of its head with Canter's nucleus, also its intellectual link (third level) and the influence exercised over other members mean that its link with this IC is sufficiently strong for it to be considered as integrated, in spite of its identity. The relationships of the fourth level, on the other hand, are not clear. Starting from a certain reiterated conflict due to the parallel conferences of 1973, 1974 and 1979, from the beginning it maintained itself apart from the constitution of the IAPS. Later the relations would be normalised.

Following the order in which we are carrying out this revision, we now go into

Figure 12.4 *Invisible Colleges in European Environmental Psychology that express themselves through the IAPS Conferences*

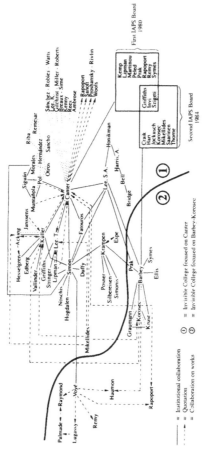

NOTE: 1) To simplify the diagram, quotations made by secondary authors appear only when they refer to other authors different from the 30 most quoted. 2) Due to the brief term analysed, we don't consider temporal evolution. 3) To simplify the diagram, we have removed dependencies from American or other authors who are only quoted.

a nebula of names that form the basic nucleus of Canter's Invisible college. Anyway, it is possible to distinguish sub-groups as far as the levels of connection are concerned.

In the first place the sub-group of the authors closest to Canter on all levels is notable: T. Lee, Stringer and Griffiths. These four formed the basic Guildford nucleus and remained together until Stringer left for Holland. This is a very powerful nucleus, profoundly integrated on the four levels defined, despite the fact that at times they have shown some intellectual divergences. T. Lee is the only European author of the group under study to whom Canter recognises an intellectual debt. A travelling companion, integrated on the four levels but with whom the connection would break quite radically in 1979 is Lipman (seechapter 3). Fatourus, for his part, maintains an isolated and personal connection as the sole Greek representative. His connection is thematic, functional and affective, but we do not have explicit evidence of the intellectual link (although we do not believe we would be wrong to confirm it). In spite of its internal cohesion, this sub-group does not manifest influence in third parties of the group studied.

The case of the Belgian author Simon is somewhat unusual. He and the nucleus that is centred around him, as well as the thematic connection, maintains an noticeably conflictive institutional link with the IAPS (the second level, functional). In 1979 the parallel conferences directed respectively by Simon and Canter were organised. After this date Simon joined the executive on the constitution of the IAPS, a committee in which the role of the non-Anglophone cultural areas would be constantly and staunchly defended, especially the French speaking areas. For its sources and perspective, the fact that the diagram does not register intellectual links with any member of the community is the result of a theoretical perspective and position different from Voyé's. Obviously, the fourth level link (affective) is nonexistent. In spite of this, he is connected with this IC by the strong institutional (functional) relationships exercised.

Krampen forms the link with a limited German group which connects with Canter's IC solely on a thematic and functional level, starting from the organisation of the Berlin conference in 1984. The functional links with Simon and Prak also come from this point. In fact, excepting the intellectual link with Espe, relationships of another kind do not appear inside or outside the sub-group, and for this reason it is not possible to speak of a German focal point with identity connected to this IC.

Simon is only linked in a functional way to Voyé; and through them the connection occurs with a nucleus which Simon could belong to given his link on the first, third and fourth levels. It is a characteristic nucleus of the Francophone area, basically adjusted to the first and third levels. The intellectual link between them is strong, in spite of their diverse professional profiles (sociologists, geographers, architects and psychologists), but there is no collaboration in studies and nor are there institutional links. Their only intellectual link with outside authors belonging to the group studied is with Rapoport. By rights, they do not belong to Canter's IC, but their internal links are not strong enough to define it as an Invisible college. In fact, this group could be linked to the one lead by Barbey and Korosec, given that their participation in Strasbourg came about as a result of their invitation and coincides in authors that have influence over them.

The Barbey-Korosec nucleus presents notable differential characteristics that give it an unusual identity. Its complete composition is a consolidated Central European network with Graumann and Kruse, which clearly develops on the four levels. This group, as well as having a close formal and informal relationship in a regular series of public and private study meetings (see chapter 5), counts the organisation of the Strasbourg conference as one of its assets, through which it connected with the Francophone nucleus mentioned previously. It shares intellectual references with this nucleus, Rapoport being the most influential American. Until 1984, except for the membership of the IAPS, it did not maintain any connection with Canter's nucleus, with which it maintains a rather mutually critical relationship. Both Korosec and Kruse have explained their consciousness and desire to promote the phenomenological perspective as a differential characteristic, or the most notable contribution, of European EP (see points 5 and 6).

In 1984, this nucleus widened its relational network when Barbey became the president of the IAPS, with members of the executive committee and the full consultive committee of the association. In this way, on a basically thematic and functional level, and to a lesser extent cognitive and intellectual, the network embraced Switzerland, France, Belgium and Germany, as well as Holland, Great Britain and to a lesser degree the USA, Canada and Japan. Among the basically functional links we find Prak and above all S.A. Lee, who constitute the only functional connection of importance with Canter's nucleus. The functional link with the council or full committee has been very weak and of little influence, both for the first and the second executive. On an intellectual level, there is a strong link through references, due to institutional questions, from Korosec to Küller and Canter, as well as the sole quotation made to Korosec from this group by Mikellides, whose situation is not clear in Canter's IC.

Globally, then, in spite of a certain punctual connection of this focal point with Canter's IC, the strong integration on the four levels of the base group and the links with different specific weights of each level with the sub-groups described, as well as the openly assumed theoretical identity and positions, permit the identification of this nucleus, not just as one more focal point of the principal IC, but rather as an authentic Invisible college with pretensions and a will to be an alternative to the previous one.

So altogether, we can define the existence in their own right of two unique Invisible colleges: the broadest one centred on Canter, which extends its connections to all the cultural areas, and the second, more incipient, which we could denominate as being eminently Central European, based on the stronger tradition in the Francophone area, and in part on the similar characteristics of the German phenomenological tradition.

They are both integrated into the same organisational structure, but as we have seen, the links between them are weak. Moreover, the authors that lead them have declared profound mutual theoretical discrepancies to us.

The participation in the factitious power structures is also very different. While the Central European IC participated more sparingly in these structures until 1984, not having any organ of institutional expression at its disposal (magazines, etc.), and only counting on a part of the academic programmes within psychology (Heidelberg), within social psychology (Strasbourg) and within architecture (Lausanne). The IC headed by Canter -and in the background, but with a strong influence, by T. Lee- starts from a powerful academic programme on the economic and institutional level of Environmental Psychology in Guildford, Surrey. Moreover, from the possibilities given to it by this, it counts on a strong dynamic of local, national and international meetings (conferences and congresses), substantial projection and contacts within the American community, and an important participation in formal channels of expression, publishers, magazines, etc., especially in the *Journal of Environmental Psychology*, Academic Press and other important international Anglophone publishers. Also, as we have seen throughout this study, it played a very important role in the constitution of the IAPS itself and the orientation it followed in its earliest moments.

This detection of key points in the power structures does not only result in the promotion of the lines of theory and application of their proposals, but it also leads to some very delicate moments in the identity of the field of study itself, as is quite clearly demonstrated in the controversy analysed in section 3.7. In spite of the theoretical and disciplinary openness that, by definition, the IAPS wishes to give to itself, what is true is that in its first years of formal existence it has taken on an orientation marked to a large extent by empiricism, which would be answered but at the cost the explicit abandonment of some professionals and the bid for isolation from the "Environmental Psychology" label on the part of others. The association, however, has not broken, and those who have remained, conscious of the necessity for the definitive configuration of this field of study, seek new orientations in the perceptive renovation of its governing bodies.

In any case, apart from what refers strictly to the IAPS as an organisation, the "invisible" power structure in EP has been created, and the IC focused on Guildford through the disposition of other institutional channels will follow its own dynamic and influences.

12.5. Summary and conclusions

In this chapter we have revised the notions that have served as a foundation for its preparation and the three indicators used for the analysis and configuration of Invisible colleges have been defined.

In the first place, we have considered the nuclei of collaboration, which were already studied in chapter 11, as indicators of a network of relationships that assert a direct link between authors.

Next the mutual quotations have been analysed as demonstrations of intellectual relationships and dependencies, concentrating on the 30 most notable authors, starting from their productivity and visibility.

In third place we have studied institutional collaboration in the structures of the social organisation of EP, particularly within the two traditions of conferences, the IAPC and ICEP, which converged in the IAPS.

Finally, from the global consideration of the relationships expressed in the previous paragraphs, we have tackled the configuration of the Invisible colleges, taking Munné's proposal of spectral analysis of group relationships (1985) into consideration for the qualitative study of the links established between the authors involved, both on a thematic, functional, intellectual and affective level.

The result of this analysis is the configuration of two Invisible colleges headed by Canter-Lee and Barbey-Korosec, of differing theoretical orientations, with geographical and cultural areas of defined implantation and a differentiated participation in the power structures.

13 Prospective and general conclusion

After two long decades of the burgeoning of environmental studies, modern Environmental Psychology has gradually formed its own space, but at the same time undergoing a profound change in its theoretical approaches.

EP is the fruit of the social context of the postwar western world. The technological revolution facilitated new forms of production, of urban concentration in a Europe under reconstruction, that together with the new methods of mass communication would change cultural habits and rituals. The consolidation of the nuclear family took place and the transmission of social values went from the family tradition to the confirmation of states of opinion through the *mass media*, with the emergence of the generational conflict through the adaptation to the new reality. This was accompanied by the bid for the functional specialisation of the city's spaces, in a climate of developmentalist euphoria, along with the possibility that for the first time the human being could over-exploit natural resources on a world scale. All this projected the environmental factor to the front line of the social issue, accompanied by the increase in the possibilities of social control and the withdrawal of the citizen from the organs of decision and power.

The consequences of this new structure in society awakened the interest of the social sciences and especially psychology in the effects of the new habitat on individuals. It was born from positivist and empiricist approaches in the beginning, just when on the other hand neopositivism and behaviourism initiated their profound paradigmatic crisis. With a predominance of *Case Studies* and some theoretical efforts seeking antecedents in periods in the past, EP began its voyage at the end of the fifties, to settle a decade later.

It would not be until the crisis and the paradigmatic change of psychology were consolidated that EP could free itself of the dead weight of behaviourism that had caused the miserable results of its first steps and the first disenchantment of the hopes raised among the environmental designers and managers.

It would not take long, however, to go into a new crisis, this time induced by an important change in the position taken before the environment. As a result of the energy crisis of the early seventies and the accusation of the new urban movements which arose after 1968, especially the ecological movements, what Aranguren (1984) called the "step from Modernity to Post-modernity" took place, which was marked by a change in attitude that went from the transformation into a "product" of the immense and inherent "store" of raw materials, store and at the same time "workshop", workshop and at the same time "supermarket" in which to consume the environment, to a consideration of the surroundings and the human being as a undefinable unit in which the environment goes from being an outside object, external and consumable to forming an integral part of ourselves. EP slowly but progressively registered this change in its perspectives of analysis and in the themes that it tackled.

Indirectly and unconsciously the change in the attitude towards the environment on a conceptual level brought the approaches of EP closer to those of the notion of "vital space" that Lewin had formulated, imbued in the context of the old German Environmental Psychology of Hellman and Muschow at the beginning of the century, and to the phenomenological approaches, which seem to have shown a recuperation in the last few years within this domain, judging from the debates and conflicts that have emerged in the heart of the scientific community, and the selfsame configuration of an Invisible college with this tendency that has recently gained positions in the European formal power structures. However, empiricism and positivism are still dominant, and along with them the matter still to be resolved -a constant preoccupation throughout the period- of a theoretical framework, the elaboration of which has been desired in determinate moments starting from the already notable accumulation of empirical data and at other times has been sought in the adaptation or reference to external propositions in principle, such as the case of George Kelly's Theory of personal constructs, which has had a certain response especially in some moments in the Guildford group.

The reorientation of some powerful American authors (like Altman, little quoted directly by the European authors but with an important medium of diffusion, such as his series *Environment and Human Behavior* which analysed and divulged not always convergent studies on orientation with those dominant in those states, as well as Stokols and Altman's [eds.] "Handbook", which we have already mentioned), the opening and interest in European phenomenology, together with a generational disassociation from the old approaches of the pioneers (Altman, 1981) in a transposition of convergent positions with which on a philosophical level we mentioned Aranguren, allow us to glimpse what could be the foreseeable development of this disciplinary field in a near future.

A future that is marked by the professional evolution of the authors implicated towards a prevalence of psychologists - with the participation of sociologists especially in the Francophone area -, with presence in centres of architecture and a marked process of academicisation in the domain. Acadamicisation accompanied by a certain loss of impact in society, but with a maintenance or even an increase in indirect influence.

An academicisation that on the other hand concentrates our attention on the reflection on the significance or not of Environmental Psychology as an interdisciplinary ambit in which there is still neither agreement on the denomination nor on the integration of the theoretical constructs from which each of the interested disciplines should depart, in spite of the desire for survival and the more and more formalised social organisation of the community involved. On the other hand the pertinence or not of speaking of Environmental Psychology as a specialised branch of psychology.

In the corresponding section we have concluded on the necessity to preserve the term EP for the specialisation of psychology that has developed traditionally disconsidered parameters, elaborating its own constructs, middle distance theories and methodological processes, at the same time as it has acted as a catalyst for a reorientation (Altman, 1981, would speak of revolution) of psychology in general, integrating the environmental aspects in all of its approaches. It will be necessary, however, to preserve the space of dialogue and collaboration which in its moment stimulated the origin of EP itself with other disciplines, especially architecture and urbanism, but opening itself to other ambits like ecology, biology, health sciences, etc. A tendency that can already be confirmed in the latest proposals for investigation that are being prepared, which is coherent with the evolution of the philosophical substrate and the new social problems that have to be tackled, as we have mentioned earlier.

At this point, with a consolidated institutional and academic recognition, this EP has taken on a marked psychological and social modulation since the meeting that Moscovici called in Paris in 1981 marking theoretical correlations in its various perspectives, which brought it closer to social psychology, something on the other hand widely shared by other authors from the different cultural areas, it could be illuminating to call it "Environmental Psychology".

Finally, once the qualitative and quantitative data expounded on have been analysed, we can describe the Invisible colleges that in some ways act as nuclei of power: those of Surrey and Central Europe.

a) The Surrey college, with David Canter at its head, appears to be powerful and consolidated. It holds abundant power resources, it has consolidated training programmes, communication media such as books and magazines, and has been present (generally in the organisation) in more than half of the IAPS conferences. Identified with the Anglo-Saxon empirical tradition, it enjoys an intense network of influences.

b) The Central European College (Strasbourg, Lausanne, Heidelberg) is the youngest. Emerging from the phenomenological tradition it has disposed of scarce power resources until now, although from 1984 onwards it has held the presidency of the IAPS in the person of G. Barbey, and has less institutionally powerful formation programmes. Together with Barbey, Korosec, Graumann and Kruse have been the most visible heads up to now.

In conclusion, these two nuclei are defined as power structures within the organisation that the scientific community which expresses itself through the IAPS conferences takes on, demonstrating two differentiated theoretical perspectives that seem to structure their areas of influence in a proportional way.

Is there an European Environmental Psychology?

In this context, in view of some of the partial conclusions concerning the emergence and expansion of EP in Europe, its impact on society, the most notable authors and their professional profile, the institutional framework within which it develops, the intellectual dependencies and the social, organisational and scientific structure that it is endowed with, we shall now try to answer the million dollar question implicitly formulated throughout the book: Is there a European Environmental Psychology?

In the light of the data from the different sources analysed, we have to conclude with an announced answer: NO. But a no with modulations.

The Environmental Psychology that is developing in Europe has:

- its own infrastructure of scientific organisation:
- formal and informal communication channels, and
- even its own history.

In the sense of the social organisation of the science, we could say that the Environmental Psychology that is carried out in Europe has its own identity, which even goes as far as the formation of autochthonous Invisible colleges.

However, the intellectual dependency on the American pioneers and constructs, probably due to a force outside the very dynamic of the ideas connected above all with the context of the relationships of social forces and power structures, above all concerning the communication channels, which dilute this potential identity, integrating it in a differentiated way if you like, in a western community, with a "traction engine" centred on the United States. On the other hand, the plurality of perspectives, sources and the selfsame denominations that this field takes constitute the wealth of its approaches, but at the same time drown this possible identity. However, this constitutes a promise for the future that time will take charge of clarifying.

POSTSCRIPT: 1984–1992

14 From architectural psychology to green environmental psychology

On renewing the work on this text, and attempting to complete a vision for the development of Environmental Psychology in the period between 1984 and 1992, I have made a series of reflections that can not follow the same structure as the rest of the book. Some years have passed from its first edition and you see things in a different way, whether you like it or not. I still endorse the preceding text, although now I would have organised it in another way and I would have emphasised other aspects. We are in times of very rapid change. In 1986 it was difficult to foresee the great transformation that Europe and the world has suffered since then. This transformation, in all senses adopted from the logic of the perspective of analysis makes it evident that it is influencing Environmental Psychology. I have already commented on some aspects in the epilogues of the cultural areas, in the second part of the book, and now these will help me to organize this postscript.

At this time, with eight years of perspective, one can now speak of three births or three stages in Environmental Psychology and not two as we had done before. The new world of social, political and ecological co-ordinates presents us with new themes and a reorientation of issues and objectives. The dominant theoretical paradigms are also in crisis and we seem to be seeking new perspectives. The profession of the environmental psychologist still appears to be the great pending issue, despite some advances made and the new expectations for the future which seem to be opening out in the field of social impact and the management of natural resources. This implies taking on new travelling companions with whom we have had little relations until now, like biologists, engineers, economists and lawyers, as well as others. Altogether this brings us to the need to consider a new prospective on Environmental Psychology.

1. Three stages

Throughout the book, by adopting the approaches of Kruse and Graumann (1987), I have repeatedly spoken about the double birth of Environmental Psychology in Europe. The first Environmental Psychology was that of the first third of the century, basically Germanic in nature. The second was the one which emerged at the end of the fifties and the beginning of the sixties, simultaneously in diverse parts of the continent. Two births that in principle were unconnected, independent and centred on one similar object of study: the behaviour in space of individuals, the social production of space, the effects of urbanism and the effect of the built surroundings in general. In the second half of the eighties, however, there began to be indications of the beginning of a new stage: a transition from the Architectural Psychology to a "green" Environmental Psychology.

1.1. *From the first to the second Environmental Psychology: the American transition*

The "environmentalist" spirit that impregnated psychology, sociology, architecture, biology and other branches of knowledge and social philosophy during the first third of the century, on the contrary to what has been confirmed at times, did not disappear with the second world war. At most, it changed form or name. The existence of diffuse or little explored nexus (to the best of my knowledge) has already been pointed out in chapters 2 and 3. It would be a good object of study to go into these nexus more deeply. In any case I would like to emphasise their existence here to show how in fact what we called a break is not a "break" as such, but rather displacement and a transition through American psychology, which returned its influence to Europe.

Throughout the book, especially in chapter 12, the real and recognised influence of American Environmental Psychology on the European has been demonstrated. This American Environmental Psychology receives the European influence without any kind of break. So we find ourselves before a double nexus between the first and the second European Environmental Psychology:

a) A first nexus by direct influence, even if it is through what Lévy-Leboyer (1980) qualifies as "origins a posteriori," when the present day authors seek antecedents in the best known classics. Of course, Lewin is the first in ranking and passes as a 'founding father', when in fact -without wishing to wrest from him the importance and leadership that he deserves- he does no more than transmit a part of the influence he received in his formation. On the other hand, Hellpach's books can still be found in some European countries, and Simmel's are still republished. What is more, it so happens that some of the authors of the period were present in the earliest meetings. This is the case of Jahoda in the first meetings on the theme held by the British Psychological society at the beginning of the sixties, as Terrence Lee told us (1984). Other unknown nexus occurred. We could mention José Luis Pinillos (1977), the first Spanish author to publish a paper on the theme, with an impact on the whole of the first generation of environmental psychologists in this country. A work of maturity, the author himself recognises it as an influence of the formation received in his youth in pre-war Germany (Pinillos, 1990, personal communication). Investigating in this direction the list could be longer.

b) A second nexus by indirect influence through American Environmental Psychology and sociology, that had then picked up the European influence through the authors that emigrated to the United States.

The Chicago School owes a great deal to Simmel, especially through Park, who had been a pupil of his in Berlin. Burgess, Park and Wirth encouraged the translation of Simmel's work into English. What is more, the links of the School with the origins of Symbolic Interactionism in psychology, and Proshansky's links with interactionism, allows us to follow the trail of German social and environmental thought of the first third of the century by one of the founders of modern American Environmental Psychology. All this without belittling other intellectual debts that Proshansky could recognise.

164

Of course, Lewin left a very strong impression on American psychology in general. In part, the auto-criticism of Tolman's behaviourism is due to his contact with Lewin. And as it is known, the first modern approach to the notion of 'mental maps' emerged from this auto-criticism (Tolman, 1948), which would lead to all the studies of cognitive maps, without wishing to wrest merit from Lynch.

A second direct and transcendent impact of Lewin for American psychology, which returned its influence on the European, especially on Germany through Kaminski, is Barker's Ecological Psychology, as Hernández has demonstrated (1985).

In fact, the very change of name from psychology of architecture to environmental psychology, forced by Canter in 1972 from the Surrey programme, was no more than another American influence, just as the generalisation of this name in the Francophone area clashes with the psychology of space. We should remember that the first recorded time (for the moment) that the label environmental psychology was used was by Hellpach in Germany.

I do not pretend to be exhaustive in this description, but I do want to point out, as I have said, that the twin births of environmental psychology in Europe have more links than those directly visible. In fact they form two stages linked by a transition period that passed the United States.

1.2. *From the Psychology of Architecture to 'Green' Psychology*

What characterised what we will continue to call the second birth of Environmental Psychology in Europe -with an end to maintaining unity-, concerned itself with problems of design, construction and planning in an answer to the social demands of the moment. As it has been explained in chapters 2 and 3, the contexts were slightly different in each cultural area, and therefore the results were also slightly different approaches. The improvement in the design of housing, of neighbourhoods or workplaces was spurred on by economic circumstances, by a social philosophy, but above all by a new and common situation to all the areas: what has been called the 'Technological Revolution'.

The Technological Revolution implied an important change in the forms of production, an important concentration of the population in the cities -under reconstruction through the effects of the war or not- and the beginning of enormous growth in the tertiary or service sector. First, the concern for the new housing for the migrant population that came from the country concentrated the problem on construction and the Architectural Psychology began to be discussed.

Later, with a certain settling and consolidation of the social fabric, which allowed it to acquire group conscience and capacity for revindication, together with a social discontent for unsatisfactory living conditions, the urban social movements and the alternative movements would begin. The nucleus of Environmental Psychology began to evolve towards more social aspects, related to residential satisfaction and the quality of life. It would continue, however, to be centred on urbanistic, architectonic and to a lesser degree organisational or labour aspects.

Both in the first case and in the second, the reaction was delayed with respect to necessity. The response was given with and approximate delay of five or ten years. The same delay with which it reacted to a new challenge: the ecological problem.

Under the label of the ecological problem, I will cover everything that has to do with the conservation of nature, of energy, of resources, pollution, industrial and technological risks, ecologically 'responsible' behaviour of the population and everything that has been called a 'green' movement.

The first 'green' movements appeared among intellectuals, marginal or alternative groups at the end of the sixties. They really blossomed in the mid-seventies, spurred on by the oil crisis of 1973 and the economic and social crisis that it carried with it. They took on social and political power at the beginning of the eighties. An important change in social and environmental philosophy began to be detected, even on a level of the European governments, in the middle of the eighties. 'Green Capitalism' began to be mentioned at the beginning of the nineties, with the environmental commitments of some leading companies, which demonstrate that investing in the preservation of the environment could even be good business. On the other hand, the European Community legislation, that contemplated the evaluation of environmental impact (1985) was completed with new regulations (1992) of obligatory fulfilment from 1993 onwards.

Until the middle of the eighties, environmental psychology in Europe did not enter into the theme more than in an anecdotal way. All this in spite of the fact that on creating the Surrey programme in 1972, the reason given by Canter for adopting the name of Environmental Psychology and not Psychology of Architecture was to open out to these themes. In fact, it is not until 1984 that we find the first 'green' report in an IAPS conference. It is not a coincidence that it was in Berlin, West Germany, birthplace of the green political movements. Before, such papers only appeared in an isolated and sporadic way.

In my interviews with the outstanding authors of all the cultural areas in the summer of 1992, the majority expressed that they had recorded a certain social demand in this sense. In some cases with small but important advances. The participation of a team of psychologists formed by people from France, Italy, England, Holland, Switzerland and Germany in the UNESCO's MAB (Man and Biosphere) programme is notable, along with some actions in England, the participation of psychologists in technological and environmental governmental commissions in Germany and Catalonia. What is more, we should add the influence of the American texts and a few of the European ones, that as early as the eighties incorporated chapters dedicated to energy conservation, reduction of residues and ecologically responsible behaviour.

This little tour around the ecological aspects initiated at the end of the eighties, reinforced by what was derived from the coming into force of the new community environmental regulations -which obliges the analysis of the social impact of any impact or environmental audit- together with changing world co-ordinates that propose a new dimension in the social issue, is marking what in Europe has begun to be a third stage -a stage of a more social, more organisational and greener Environmental Psychology- but once again with a delay.

2. New world co-ordinates

If the environmental psychology of the second stage was spurred on by constructive, urban and social problems due to the important country - city migrations, the green Environmental Psychology of the third stage would be marked by a change in world co-ordinates, both as far as the ecological problems and the generalisation of south - north migratory processes are concerned, a result of the ecological, economic, technological and demographic imbalances.

Great technological advances have placed the human being in a position of being able to over-exploit natural resources well beyond the capacity for regeneration that the environment possesses for the first time in history. This was already the case in the fifties, but it would not be until the eighties that, encouraged by some years of the accusations of the more or less marginal naturalist movements, western society took note that this was the case in a fairly generalised way. Western society took note that each time that an activity is exercised a potentially irreversible impact in the environment is being caused, and that in the final analysis we too are part of this environment.

2.1. *A reflection on some contextual aspects*

Sociologists, economists and politicians has described many years ago how the great problem of world development and economic growth would be displaced towards north-south relationships. Often they described it from economic, political or even racial parameters, but seldom as an environmental and/or ecological issue, apart from some references to deforestation to extend the area of cultivation driven by hunger and the problems derived from the demographic explosion, or for the need to import wood to the west. Frequently treated as the breaking of local and isolated balances.

In the eighties the voices that considered the world dimension of environmental problems became more and more numerous. In some cases they found a real popular echo, such as the shrinking of the ozone layer by CFC's. An important reduction in the consumption of aerosol sprays with CFC's occurred spontaneously, forcing the industries to change the composition of their products.

However, it would not be until 1987 that the Bruntland report appeared ('Our Common Future') promoted by the UN, on the environment and development. In the spring of 1992, the Rio de Janeiro World Summit took place, also promoted by the UN, which demonstrated the difficulties for the states to come to agreement on the solutions to adopt, in spite of recognising the problems linked to the over-exploitation of natural resources, the unsuitable nature of production processes, the reduction in the ozone layer, the overheating of the earth and the need to maintain what has been called biodiversity.

2.2 *Socio-economic conditioners*

In the nineties, and also in the eighties, one of the contextual issues of the fifties which spurred on the birth of Environmental Psychology (the country-city emigration) became universal. It is no longer country - city, but south - north, third world - first world, poor countries - rich countries, with different factors propitiating it. Factors that are not new, but which the world has become conscious of and is seeing the reach of their effects.

1) Over-exploitation of the environment and natural resources of the countries of the south on the part of the north rich countries, without the north having contributed technology and resources for a certain rebalancing and their its self-management.
2) The contribution of the north obsolete technology or only partial technologies which make them totally dependent.
3) Processes of apparent decolonisation that have left countries in the hands of dictators, when they are not personalist regimes, which are not capable of carrying out a balanced development between the population, resources and exterior relations.
4) The growing influence of the mass media that is constantly projecting standards of living, models of ways of life, different and attractive social and cultural values that generate disconformity with their reality and so break their more ecologically more balanced ways of being. That is to say, it generates licit aspirations but that are dysfunctional in their context.
5) At the same time, this influence of the mass media generates another contradictory tendency: it is experienced as an external aggression against the very identity of the

people, culture or ethnic group, and generates a withdrawal of the group into itself in defence of their values, more and more eroded and concentrated in a few emblematic differential elements, which can lead to attitudes of self-defence which to a certain point are legitimate or to not easily acceptable aggressive and expansive self-defence.

6) Existence of different demographic tendencies potentially conflictive. On one side birth control provokes a deviation in the developed countries which, convinced that it is necessary to reduce the alarming growth of world population, lowers their demographic growth rate to 0. On the other hand, they have basically exported technology for the control of infant mortality to the poor countries, but not for the solution of global problems of development. This solves a part of the problem (more than anything else the ethical problem of the western world!) but causes a grave new ecological imbalance, the demographic boom.

The increase of a population in the poor countries, to which we cannot give support for work and subsistence unleashes serious ecological and social consequences:

1) Increases in deforestation takes place to gain more cultivable area, to face the increase in population. Frequently, however, this implies the beginning of a desertification process.

2) Increases occure in conflicts among peoples, tribes and ethnic groups who need to widen their domains to survive.

3) Geographic mobility is facilitated through means of transport, which permit the covering of much greater distances with less effort, danger and time.

4) New waves of massive migration occure for populations expelled from their lands by hunger. This migrations in part have been stimulated by the 'advanced' countries need for cheap labour (now that slaves are no longer bought and sold...). Their need to correct the inversion of their age pyramid and the problems that it generates to maintain the highly expensive social security and service systems that they have provided themselves.

5) In situations of economic recession (like that of the nineties) the tendency emerges to stop or expel this migrant population.

All this generates the emergence of the present day racial and ethnic conflicts in the old Europe. Frequently they are no more than class conflicts tinged by differential elements excessively divergent or encysted as being emblematic (race, culture, religion, ways of life) for each group, that they give them the appearance off racism. The appearances and behaviours which in principal are not racism, end up by generating consolidated attitudes of racism that are unscrupulously exploited by ideological groups who use racism as their banner.

Although this marks an important change in the parameters within which Environmental Psychology has to move, it is not our objective to suggest solutions to these problems.

When we approached 'simple' problems like the adaptation of new residential suburbs of people who came in from the country, we were dealing with questions linked to different ways of life within similar or very close cultural, religious and ethnic parameters. Now things are no longer like that.

The big differences between groups tend to close themselves off as a form of protection of their own identity. At the same time this closing terndency works also as a positive way of facilitating people adaptation to a new social, cultural and economic environment. For us this marks the importance of taking into consideration socio-cultural and economic parameters in the analysis of situations in order to understand the behaviour of individuals and groups, and for the proposal of solutions.

The 'old' parameters -still prevalent in the majority of cases in Environmental Psychology- could become a luxury offered to an elite that rarely ask for them and frequently reject them, when they are not merely an academic exercise.

The tendency in the last years in Environmental Psychology of being more and more socially orientated has to be assumed and reinforced in all of its consequences. The individual psychological processes can be explained and understood less and less outside their social context and this context is changing rapidly and profoundly. The new challenge for Environmental Psychology is not in abandoning its domaine and the knowledge attained in the previous stages but in knowing how to incorporate the new reference parameters, ecological, social and economic, into its reflection and analysis of reality. Things seem to be going in this direction, as we will see later on.

2.3. *The idyllic 'past' and the evil 'present'*

It is very common to hear the use of the expression 'that didn't happen before' or 'things were different before'. An old Spanish saying says that 'all time past was better'. Psychologically this 'attachment' to the past experience which confirms and justifies our present identity can be explained . However, what does it mean when we speak of a hetairist and at times idyllic past and a present that is different and at least more dangerous? Where is the dividing line?

I will attempt to establish some of the significant differences as far as aspects are related to the environment by contrasting some of the emblematic characteristics of the 'past' with the 'present'.

a) 'Past' and 'Present' from an economic, productive and demographic point of view.

Past	*Present*
1 - Balanced forms of production in restricted geographical areas. Local, regional and at the most national or statal economic units.	1 - Forms of production orientated to markets that are becoming more and more universal. This implies passing from a reduced and tolerable impact of the productive activity to its gigantic multiplication to limits indigestible for nature (e.g.. quarries and cement production, car production, the chemical industry...).
2 - Relatively low necessary energy resources in function of the possibilities of the geographic zone	2 - Necessary energy resources well above those available to the geographic environment, which implies the need to import them from other zones. This importation increases in itself the necessary resources.

(cont.)

(cont.)

Past	Present

3 - Rudimentary and expensive transport in time and money, therefore short distances were covered and the centres of production were diversified. That is, the work was distributed geographically. This implied a necessary tendency towards autarchy.

3 - The ease and generalisation of transport relatively reduces the dimensions of the world, concentrates the work places and leads to the internationalisation of the demographic, productive and economic issues.

4 - Diversified production in each geographic zone, with a tendency towards self-sufficiency.

4 - A tendency towards specialisation in each zone and the increase in world interdependencies.

5 - A limited and relatively balanced technological capacity for the exploitation of resources.

5 - An unlimited technological capacity for the exploitation of resources, which could lead to their depletion (exhaustion) or destruction.

6 - Social and political issues of local, regional and national or statal reach.

6 - Social and political issues of universal consequences and dimensions.

b) 'Past' and 'Present' from the point of view of Social attitudes, social customs and the 'Culture of recycling':

Past	Present

1 - More contact of the individual with natural surroundings. Knowledge and personal experience of the cycle of nature and of the vital animal and human cycle (birth, growth, maturity and death). This facilitated respectful attitudes towards the environment.

1 - 80% of the western population lives in a technological environment (generally urban). The cycle of nature and the vital cycle are unknown by a large part of the population. Neither births nor deaths occur in everyday surroundings. We appear or disappear miraculously in a hospital or an undertaker's. Difficulties and artificialities for the creation or promotion of pro-environmental attitudes.

(cont.)

(cont.)

Past	Present
2 - The majority of products had a known cycle. From the origin of the raw materials, their manufacture and use, to their reuse for a secondary function. If this was not possible one had to try to get rid of them through a rag and bone man, spread them on the fields or burn them. People knew very well what could be used in the fields and what could not	2 - The cycle of the product is totally unknown, its nature, composition, its possibilities of reuse. It is simply abandoned in the street or a container and it disappears. We do not know what happens afterwards
3 - Few services. Everybody was responsible for covering their own necessities. Those that functioned were organised directly by the community or managed by people in charge very close to the user. The use of the service was known, along with its necessity and function. Liability and appropriation (attachment) to services.	3 - Services exist 'in their own right'. They have to exist and that is all. we do not know how they originate nor their working processes. We only know that when we need them they have to answer our expectations. Services are responsible for covering our needs. They are magic. People no longer feel responsable, they ignore them. But what is more, services can become oppressive. They end up by saying how and what we can do and what we cannot do. People give up on their sense of liability, get into a feeling of being disenfranchised and alienated.
4 - Culture and popular aesthetics looked favourably on products manufactured by the user himself, in spite of imperfections and a possibly 'vulgar' appearance.	4 - The culture and aesthetics of design do not admit the manufactured product with a 'vulgar' appearance.
5 - Labour was cheaper than raw materials. But even in situations that were not so, there was a culture of recycling. Everything that was finished for its main use had solutions for secondary uses or was burnt. A low volume of waste production.	5 - Labour is more expensive than raw materials. Things are not recycled, they are thrown away. What is more, there are new materials which are difficult to recycle (plastics, etc.). An enormous volume of production of urban and industrial waste. For years the volume of waste produced has been considered as an indicator of social development, with positive connotations.

This contrast between past and present clearly indicates the need to work on the complicated task of creating new attitudes and a new culture of recycling and of modification of the processes and philosophies of production. As Boulding pointed out as early as 1966 it was necessary to move from the culture and 'economy of the cowboy' (he who always has enormous spaces at his disposal and does not have to take care of any kind of recycling nor of improving his use of resources nor of reducing waste) to a culture and an 'economy of an astronaut' (he who manages a closed system in which all materials are recycled, space is a scarce commodity and where the conservation of the capital stock has priority) (Coll. 1991,57).

Although it is necessary to make people conscious or produce conditions for making citizens responsible, it is necessary also to regulate production and actuation processes via rules and regulations. For example: the prohibition of non-returnable bottles in those products that traditionally have been feasible; stimulate the incentives to the collection of bottles; restrict, reuse or penalise disposable packing, etc. Fortunately, the new community rules seem to be going in this direction.

As psychologists we have an important task to carry out in the direct presentation of new attitudes and behaviours, and of social representations and states of opinion with respect to the environment that reinforce them. We should stimulate new pro-environmental behaviour from techniques of social marketing and modelling. We must not, however, forget that under certain circumstances it is the change of conduct which modifies attitudes and values, as the theories of cognitive dissonances and social comparison demonstrate. So, the regulatory route is essential. As well we should enter into new relationships with new travelling companions, like legislators and managers, whilie mantaining our previous ones.

3. Reconsidering the objective

To develop this section I will concentrate on some reflections abouth the thematic evolution -with strong epistemological repercussions- of Environmental Psychology which expresses itself through the IAPS conferences has undergone and is undergoing, the opinions expressed by notable authors in the different European cultural areas and the analysis of the European legislation which will come into force in 1993.

3.1 *The thematic evolution*

The thematic evolution of the IAPS Conferences between 1969 and 1984 is collected in tables 14 to 18 (appendix). This evolution could be representative of the dominant subject in Environmental Psychology in this period. During this period the most frequent themes have been centred on: the evaluation of environments in general and creating methodological proposals; the relationships between architecture and social sciences, the image and significance of places from a cognitive point of view; territorial behaviour; the evaluation and experience of the dwelling, the uses of space and the perception of the environment. In this order, the categories mentioned represent 63.2% of the works presented.

Independently of the fact that I now consider the system of thematic categories that I employed to be excessively broad and that it disperses the studies too much, I believe that it reflects the experience of these years quite well. With the perspective that the passing of time gives, I should like to make a few comments.

a) In the first place I highlight the great methodological concern that can be inferred as a reflection of the dominant scientific dogmas in this period, but without it being possible to affirm that definitive advances have been made by now. In spite of the recurrence of the papers analysed, there is still a certain disorientation and minimally standardised generally accepted procedures and techniques have not been established by the scientific community. This is probably because, as the epistemological discussions repeatedly demonstrate, these tasks are tremendously difficult -or even impossible for some- for the social sciences to achieve.

Within the third stage of Environmental Psychology, some authors closely linked to the most experimental approaches recognise that they are coming out of the laboratories, becoming more holistic in their approaches and more committed to the social reality in their analyses, while maintaining a strict methodology.

This methodological evolution has been forced -and at the same time has permitted- by an evolution of the object of study. From the earliest molecular analyses things have gone on to more molar approaches. In Environmental Psychology this has meant going from the study of the evaluation of a concrete space as a 'stimulus', to analysing the experience of this same place as one more component in the system it is framed in. A system which comprises other spaces, the whole of values and cultural and social reference points of the group or subject. That is to say, there has been an inevitable change of object. Now satisfaction with design is not talked about as much as residential satisfaction; the aesthetic agreeability of the office or its primary functionality is not talked about as much as what the surroundings contribute to or hinder in the satisfaction or quality in work.

Quality of life has become central. This implies the consideration of environmental aspects as elements of a dynamic and changing system by definition, which reaches a wider and wider range. Wider and open to culture, forms and ways of life, coexistence among races and ethnic groups with very differentiated ways of being who have to share the same space, the forms of production, economic and political interests and comportment with respect to natural resources etc. All of these themes emerging in the four IAPS conferences held in the latest period (Haifa, Israel, [1986], Delft, Holland, [1988], Ankara, Turkey [1990] and Thessalonica, Greece, [1992]), but especially in the 1992 conference. What is more, in the four conferences their central themes have reflected the idea of change or evolution in relation to culture and history: 'Environments in Transition', 'Looking Back to the Future', 'Culture, Space, History' and 'Socio-Environmental Metamorphoses: Builtscape, Landscape, Ethnoscape, Euroscape'.

b) A second comment to be made is the important specific weight of the works dedicated to analysing, annotating, justifying or revindicating the role of the social sciences in design and architecture. They reflect a search for meaning and identity, that after the Barcelona conference (1982) seemed to be beginning to decline. In Delft there was a symposium on a similar theme, but more centred on the theory of architecture than on the justification of the social sciences. In Thessalonica there were none. However, a novelty was proposed: Lenelis Kruse requested to the plenary committee of the association (IAPS) that the organisation of the Manchester conference in 1994 should include the theme of the world ecological issue in its initial programme, to stimulate the reflection on the theme from Environmental Psychology. It was no longer necessary to demonstrate the meaning of Environmental Psychology for design, but it was necessary to go more deeply into the consideration of another field which has still not been penetrated to any great degree.

In fact the theme is not a new one. In Strasbourg (1976), Surrey (1979), Barcelona (1982), Berlin (1984) and Ankara (1988) more or less isolated works had been presented on ecological aspects. Moreover, in Berlin it had been the theme of an invited conference, and in Haifa (1986) a theme proposed for a round table. What is new is the explicit desire to stimulate reflection and work as a theme of the present and the future in which

Environmental Psychology has to become more explicitly involved. It is not a coincidence that once again it was a German who proposed this.

c) A third comment stems from another of the dominant themes in the previous account: the cognitive aspects of the image and significance of the environment. In the period 1986-92 this theme continued to have an important presence. However, an important change was registered in its treatment. From an approach dominated by proposals more or less close to the theories of cognitive maps, basically centred on structural aspects (maps as cognitive schemes and their working) the point of view has moved on to the study of significants, the formation of significants and their relationship with the identity of the Self and Place Identity.

It is a similar evolution to the tendency of the present day crisis of cognitivism, in general psychology. The self-criticisms of emblematic authors like Bruner (1990), who revindicates the return to a psychology of the significants and not only of the processes, open to popular psychology and culture, to the analysis of the construction that subjects carry out of reality through their own discourse, close to the approaches of Gergen and the constructionists, are not alien to this transition. The latest studies presented on the signification of space are beginning to owe more - at times without knowing it - to these approaches and to symbolic interactionism than to 'classic' cognitivism.

To conclude this subheading, I would like to point out that if although it still cannot be affirmed that we are going into a new paradigm (in Kuhnian terms), what I have tried to reflect in these observations is that we are not in a period of 'normal science' but rather that everything seems to be pointing to the fact that we are going into a situation of paradigmatic crisis. The changes of attitude and the adaptation of the rigidity of theoretical postures which had characterised some periods of Environmental Psychology, above all between 1979 and 1982, which is reflected in the two 'invisible colleges' described in chapter 12, seem to confirm this as we will see later on.

3.2. *Themes by areas, as seen by the eminents*

During the summer of 1992 I was able to interview a series of notable and recognised authors from each one of the European cultural areas. I tried to be sure that I spoke to the same authors as I had in the 1986 study. I was not always able to achieve this. This time it was David Canter, Jonathan Sime and Sue Ann Lee in Great Britain, Perla Korosec-Serfaty from France (at present in Quebec), Lenelis Kruse from Germany, Thomas Niit from Estonia (Ex-USSR), Mirilia Bonnes from Italy, Rikart Küller from Sweden and Cleopatra Karaletsou from Greece. They have been a great help in the writing of the epilogues for each cultural area referring to the period 1984-1992 in the second part of the book. I will now try to synthesise the most relevant opinions of the conversations, referring to the thematic aspects. I hope I have done this without betraying the meaning of their considerations.

In these interviews, together with what resulted from the analysis of publications produced in the period 84-92, it can be seen that the dominant themes answer the evolution which I have pointed out in the previous paragraph. At the moment in all of the European cultural areas themes related to the quality of urban life are dominant. Whether it be from the consideration of the dwelling as a part of the system of resources necessary for social welfare, or whether it be from the consideration of the environmental quality of the neighbourhood or public spaces, as components of this same system. All this concerned with planning or evaluation of constructed environments.

A transition can be observed -and the eminents recognise this- from the most structural approaches (functionality, cognition etc.) to more experience orientated and symbolic considerations (satisfaction, place identity, etc.), in all the areas. In the same way the basic studies seem to lose specific weight, to give way to theoretical proposals from eminently social applied works. Traffic, elderly people, institutions for minors could be good examples of the Lund, Barcelona and Tubingen groups. Citizen's safety, vandalism and urbanity for the Paris groups.

In Great Britain, especially in Surrey, the interest for environmental aspects from what Canter calls 'more commercial disciplines' is revealed, such as marketing projects, organisation and management of services, organisational aspects of security at work and of the perception of risk. Themes in which Lévy-Leboyer's Paris group also has a tradition.

Aspects linked to health and interdisciplinary work with doctors have begun to appear, especially in Lund (Küller, 1992). Frequently analysing aspects that are not new in themselves, like the effects of noise, the quality of air, exterior and interior pollution of buildings, especially in the Stockholm groups and the new Strasbourg laboratory.

In this line a growth of interest in the field of the natural resources, the ecological issue and pro-environmental behaviour can also be detected. Kruse argues strongly for the need for more implication of environmental psychologists in these themes, on which important institutional steps have begun to be taken in Germany. It is necessary to advance more in this, especially after the reunification and the grave problems of preservation of resources and of nature that the Lands and the ex east have.

Canter also describes an increase in the interest of the Surrey students in ecological themes. In the M.A. program in Barcelona they have become a central theme in the formation and in the applied studies that are developed in it. They also have an important weight in Seville, Madrid and La Laguna. In France, Italy, Greece or the ex-USSR, this subject is still very marginal, being centred mainly on urban aspects.

Probably, the application of new European Community environmental regulations will give new opportunities - if we know how to make the most of them - to psychologists, in areas that are not completely strange to us, such as the evaluation of social impact on studies on environmental impact, the organisational analysis in environmental audits. But we will deal with this in the next chapter.

3.3. *The environmental regulations*

The environmental regulations are different in each European country, but they show some common points. There are two large blocs that should be differentiated: on the one side what affects the urban environments and on the other what is derived from the large infrastructures and industries with respect to the environment. The first, an old and long tradition in each country, is highly diversified. The second, much more recent, emanates from general community directives which each country has to develop.

As far as the urban environments are concerned, the diverse legislations habitually regulate planning and management processes of activities, especially those considered a nuisance or dangerous. More or less explicitly, they contemplate the consideration of the necessities and the social effects of planning. At times they contemplate direct routes for the participation of the citizens. This defines a field of activity for social scientists, in which the environmental psychologist can participate, and is already participating.

On the other hand, there are European Community programmes and from the Council of Europe which stimulates work on the evaluation of the quality of life in the cities, considering the environmental aspects from the perspective of health, like that of the 'Healthy Cities' or what is directed towards the regeneration of the old city centres and the preservation of the historic heritage, like the programme of 'New life for the cities'.

As for the natural resources and the ecological problem, there is the basic regulation of the European Community (Directive 85/337/EEC) on environmental impact and that which regulates the Environmental Audits and Ecological Labelling (Official Bulletin E.C. C76/2 of 23.3.92).

The first contemplates that any evaluation previous to the declaration of ecological impact has to consider the effects on the population and patrimony, as well as foreseeing the acceptance or rejection that it could generate in the affected population. Effects to be measured through the socio-economic modifications that the intervention would provoke, demographic changes, modifications in the ways of life and the well-being of the people.

The second, as well as the social impacts on the context of the industry or service, contemplates possible organisational transformations to adapt to the new forms of production and new technologies which pollute less, the establishment of an environmental policy to the organisation of security programmes which could lead to changes in its organisational culture, training, adaptation and awareness programmes for members of the company. As well as the concession and authorisation of the use of the ecological label to distinguish the clean forms of manufacture as a way of giving prestige to products, this generates a new aspect related to marketing, promotion and the pro-environmental attitudes of consumers.

Both cases demonstrate a behavioural and attitudinal dimension contemplated by the law which the social scientists have to develop, among whom environmental psychologists can make important contributions.

To sum up, in concluding this chapter, we can see how (through the evolution itself the philosophy of science and its theoretical and methodological implications) there has been a gradual transition from the molecular to the molar, from the reactions of individuals to the environment conceptualised as a stimulus to the analysis of experiences, the significants and the well being of the subjects in their social context. A change of interest almost uniquely focused on design, architecture and certain aspects of the city to urbanistic considerations more concentrated on the quality of life, history and culture. As well as a growing interest in the ecological aspects which are beginning to be promoted by the development and application of the regulations which refer to them.

4. On the Invisible Colleges (I.C.).

There are reasonable indications that allow us to think that some things have changed between 1984 and 1992, with respect to the invisible colleges defined in chapter 12. I have not redone the calculations, nor have I re-elaborated the indicators developed from the objectivable data that I then made use of, but it could be that it is not indispensable to do this, nor might it have a great deal of sense. What I do wish to remember, however, is that the colleges established themselves from a series of links on four levels: thematic, functional, cognitive and affective.

On contrasting evaluations with the protagonists themselves of the colleges described, there is practical unanimity in that the situation has changed, that some of the most determinant conditioners have vanished. I will avoid the temptation (and the embarrassment) of pointing out who wins and who loses, nor of saying that there are no longer any theoretical controversies nor differentiated groups. History - also in science - is dialectic and in contrast to fairy tales never has an ending, not even a happy ending. We build it daily and its episodes surpass its protagonists.

The first confirmation that I would like to make is that the thematic unity has been maintained, in spite of the different theoretical and epistemological perspectives which seem to characterise the two I.C's, and the implications on the possible displacement or change of object of study (in the sense dealt with in the previous chapter), which break the scientific community (as can be interpreted from some of the abandonments rooted in the tensions of 1979) (see chapters 2 and 3). Despite this, as I have described, the object of study has been displaced and widened, but in a similar way.

On a functional level, there are unitary conferences without more conflict than the logical theoretical discussions. Canter, Lee, Küller, Kruse, Barbey, Koresec, Churchman, Teymur, among others, usually play an important role. There are four or five research projects being carried out by international teams, financed by the European Community, with a well distributed participation. (The links between the people is becoming institutional, says Canter (1992*b*)). Canter still disposes of the maximum accessibility to the publication media, but as Korosec has pointed out (1992), admitting works carried out from the most diverse perspectives.

On a cognitive level (now I believe it would be better to call it theoretical, despite that at the right moment the term cognitive is justified (see 12.4)) the general evolution of psychology has facilitated the opening out to the typically antagonistic approaches. The displacement of the object of study towards more complex phenomena and considerations has meant that -in the words of Küller (1992)- we 'are more holistic in essence, but still maintaining a strict methodology, working in a systematic way.' The coexistence of empirical and phenomenological focal points , of quantitative and qualitative, observational correlative and hermeneutic methodologies, although with an important drop in the experimental methodologies, leads Kruse (1992) to point out that 'In fact, we could say that the dominant current is that of the multi-method approach.' This evolution does not respond as much to the fact that the specific weight of the name of a member of one I.C. or another has been balanced, or that one has lost force with respect to another, but rather to the evolutions of theoretical postures of the authors themselves.

We are left with the last of the analytical levels, the affective. I commented in section 12.4 on the difficulty in detecting objective indicators for their consideration. This difficulty is still obvious. What we seem to be able to affirm is that, as a human collective, the scientific community that has agglutinated around the IAPS goes well beyond intellectual or thematic communion. Even more so, it does when it is formed by a relatively reduced nucleus of assiduous members (around 300, and many of these occasional or transitory participants). Ten years of existence and formal association, but with a common history for more than twenty years, having some very critical moments, represents practically the totality of the professional life of the majority of its members. Without a doubt it is an element that has ended up having its influence on possible dogmatisms.

(We are getting close to 'soap opera', in which everything is due to apparently simple causes, but is complicated by an accumulation of absurdities, misunderstandings and coincidences in time. I would not like these four levels of analysis to be interpreted as a reductionist analysis, but as 'some' elements that help us to reflect on reality. The reality is very complex and the construction the we inevitably make around reality could be even more so.).

There are other factors which have helped alongside these to prevent exploring the discrepancies of this scientific community more deeply. On the one side, we must insist - once again- in the paradigmatic crisis of present day psychology, referred to previously, which has sweetened the tone of the respective discourses a great deal. On the other hand, it is important to point out that despite the years of history, we have still not managed to obtain some elements that seem to be fundamental - and which we are still engaged in -

like defining a recognised professional profile, finding a place in society (attain the fashioning of a normalised social demand) beyond what we have tried to attribute to ourselves but that is not always recognised from the outside; wishing to go into new ambits of reflection and analysis in which without a doubt contributions can be made, but arriving with a certain delay, which hinders penetration. These have been - and are still - two elements of group cohesion that can be summed up in just one; ' the common enemy'. All this, despite some more or less successful models like that of American environmental psychology, which has reached some milestones more than the European in this field.

Looking to the future inevitably leads us to approach a duality of theoretical strategies, fields, objects, and perspectives. On one side, the possibility of taking the technological option - but technocratic - of being able to anticipate the appearance of social and environmental imbalances, and to elaborate the instruments necessary for the implementation of the correct pertinent measures, according to an optimum model of reference which still has to be defined and that of course must be plural. I am referring to the model of an ideal world that we aspire to. The analysis of the European legislation with respect to urbanism, the environment and social rights could be useful to establish those 'standards', if it is possible.

On the other hand, and it could be in a contradictory way -or could it be complementary?- to contribute elements of reconsideration of society which allow us to leave the circle of technocracy, 'marked by the only criterion of the operativeness, of efficiency, of prevision, of the anticipation of the ends through the "exact" provision of the means . . . and in this way is only possible through culture, which is the only thing that escapes from its power, and that technical thought tries to domesticate and dominate.' (Argullol i Trias, 1992, 110). Faced with this affirmation, it could be necessary to consider whether we domesticate progress from culture or whether we domesticate culture from technology.

To sum up, rather than two objectives, we are faced with two cosmologies that go a long way beyond the methodological discussions and traditional epistemologies.

5. The professional problem

I have already recorded (epilogue to chapter 3) the evaluation that Canter made (1992*b*) referring to the fact that there are few professional posts with a profile clearly of environmental psychologists, but there is a wide spectrum of professionals who are developing programmes, themes, or evaluations that are environmental in themselves. Both referring to aspects of design and planning, residential satisfaction, institutional and organisational analysis, risks, dangers and safety in businesses, environmental attitudes, evaluation of landscapes, or in commercial disciplines. Lee (1992), however, said that it is not clear that there are created posts, although some have begun to work as consultants.

If things are like this in Great Britain, one of the countries with the longest traditions and the clearest and most consolidated training programmes, in the rest of Europe the situation is equally or even more meagre. The majority of projects are carried out from academic requests or official research centres, financed by public budgeting. The demands from private companies are still only testimonial, although there are some things from commercial and organisational builders. This is worrying, but only relatively.

The problem could be considered as merely a problem of strategy. Canter, in an interview for the professional magazine of the psychologists of Catalonia, manifested that 'for the good development of Environmental Psychology it is necessary to have a generation of our students in posts of responsibility, of management with broad responsibility, because then they will give posts to people with this curriculum. This is

Figure 14.1. *Conceptualisation of environment and social roles in environmental intervention*

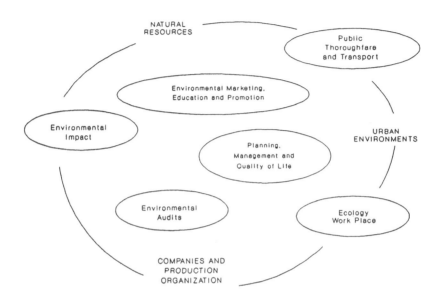

Source: Pol, E. (1992) Professional social profiles in environmental intervention. Barcelona, Psycho-social-environmental monographs. University of Barcelona.

already happening. We have students from fifteen years ago who are giving work to our present students. It is a slow process, but it is already being noted.' (Canter 1992a:26). With less tradition, we have already experienced this in Barcelona.

As we saw previously, the participation of environmental psychologists in international, state or local committees, is extended over the majority of European countries. At times mainly orientated towards urban and social issues, as in France or Italy, other times more concentrated on architecture and health, as in Sweden, or also incorporating interiors into the ecological ambit, as in Germany and Spain or the organisational as in Great Britain, France, Spain and Holland. (I apologise to the small countries, for not having the information to frame them correctly). This permits expectations for the future.

A different question is whether we have a clearly defined 'product'. That is to say, what does an environmental psychologist contribute that is different from any other professional? How does he work and who is interested or who do we think he has to interest? Here the situation is more confused. There are some respectable examples in Canada, as Korosec showed (1992), and in the United States (Guifford 1987 made some useful contributions).

In Barcelona we are working on basic professional profiles, with distinct objectives, focuses, procedures, roles and clients (Pol, 1992), and from the beginning of the work with an interdisciplinary team. I will try to synthesise them.

In the figure 14.1, the idea is to represent the conception of the environment with which we are working in the Barcelona M.A. program. The environment is a unit in which the human being has to form a part, as an individual and as a participant in social and productive organisations. There are three sides to this (the text in capitals) with different expressions and forms: a misnamed 'natural environment' (that here we call 'natural resources'): an urban environment and special environments of an organisational-productive kind, which can be placed in both rural and urban ambits.

In small type and encircled appear the professional profiles in which the social scientists can develop their activity, and therefore the environmental psychologist too. Although all six profiles affect the three kinds of environments, in each one we give more weight to some than to others. The relative situation of the profiles with respect to the three sides that characterise the environment indicate their relative weight as a field of study, the nature of the object of study and the kind of environment most affected. I will describe briefly their contents.

5.1 *Planning, management and quality of life*

This profile agglutinates a goodly part of what has traditionally been the field of study of environmental psychologists. It reaches the planning and management of natural resources and the landscape, but is basically centred on urban environments. Housing, neighbourhoods, towns and cities are their nuclei, and in them, the prevision, analysis and evaluation for the achievement of a greater social welfare and a better quality of life.

Its necessity and social function has the origin in the presumed right of individuals and collectives to preserve their own individual and group identity, the cultural forms, shared symbolic values and forms of relation, without them being altered by alien interventions beyond what is desired.

The intervention of the environmental psychologist is orientated towards seeking the balance between the always necessary adaptation of individuals or groups to the conditions of life that are given to them, the uncontrollable modifications that these conditions of life always suffer and the adaptation of the environment to their particular necessities.

This role takes on meaning when large scale planning and construction for a mass of anonymous populations provokes the physical, relational, cultural and economic distancing between the designer-planner and the future user. The designer-planner has the necessity for a programme to develop in his work, which gathers the customs, habitual behaviours, relational styles, space, symbolic references and values of the future users, not always standardisable. The designer-planner himself frequently attempts to get closer to this. However, frequently he also does not have the instruments, time nor the training to be able to understand them sufficiently.

The social system is not static, but dynamic, that is to say changing. What is valid in a determinate moment, was not so before and will not be afterwards. In the same way, the standards that we can denominate quality of life are also changeable. Therefore the designer -planner and the manager have to find the balance between the introduction of elements of renovation-innovation and the maintenance of the identity of the place, collectives and individuals.

The contribution of the social scientist can occur in three moments of the process, which mark three types of different relationships with the architects or planners and the client:

a) preparatory phase to the design, in the detection and establishment of the programme of necessities;
b) the design phase itself;
c) post-occupational evaluation.

The fields of intervention, as a consultant, as a member of staff of a design company or as a public civil servant can be synthesised in:

* Architectural design consultancies
* Urban planning consultancies
* Social management institutions
* Public administration
* Assessment and defence of neighbours.

In each country of the Community there are laws that regulate processes of intervention of the citizens in the planning and management of the urban spaces and resources. As far as the construction of buildings for housing or institutions is concerned, it is not generally accustomed to contemplate any obligatory participation route. At least, for the elaboration of plans and projects participation is regulated through periods of public exhibition in which the citizens can present allegations.

In any case, as experience shows, the development of a professional profile of the social scientist in this field depends on the goodwill or the professionalism of the designers, planners and managers on one side and of the neighbourhood collectives on the other.

5.2 *Public thoroughfares and transport*

This second profile is, in fact, a particular case of the previous one. It has, however, some specificities that give it entity. Centred of everything that has to do with the mobility of people, it participates fully in the issue of the urban environment at the same time as it affects the resources and the 'natural surroundings'.

It takes on its own identity as mobility has become universal, the car and collective means of transport have become generalised and the distances have been reduced by speed. Passive security of the streets and roads, the risk and the perception of risk of drivers and other users which affects their behaviour, the clarity of signposting, the perceptive habits and distortion of perception due to speed, to fatigue or the consumption of determinate products, are central and specific themes that require a psycho-social consideration. The car as the main cause of the atmospheric and sonic pollution of our cities and therefore, on the rebound, pollution as an effect of individual decisions on the choice of the means of transport, and how this effects the planning of services and/or of the adoption of opinion campaigns or dissuasory measures, complete the object of study.

Just as in the previous profile, the environmental psychologist can act as a consultant or as a member of staff in:

* Traffic planning and management centres.
* Transport planning and management.
* Industries related to ergonomics and automotion.
* Architectural design offices.
* Urban planning offices.

181

* Public administration.
* Assessment and defence of neighbourhoods and consumers.

There is already quite a lot of experience in this field, both from the Surrey groups, and those of Tubingen and other places in Germany, in Lund and Barcelona. The list could certainly have been much longer, but frequently, although they are dealing with these themes, the professionals do not identify themselves as environmental psychologists. (it is worth asking whether it is necessary!)

5.3. *Evaluation of the Environmental Impact*

In 1969 the Studies of Evaluation of Environmental Impact began to become widely accepted starting from the National Environmental Policy Act in the USA. In Europe, this profile was derived from the actual legislation and the environmental issue, in the dimensions we have seen before, related to the forms of production, cultural changes and the capacity for over-exploitation of resources as a result of human activity. Therefore, this profile is quite concentrated on resources and the 'natural environment', although it participates on the organisational-productive side.

The present Community legislation (Directive 85/337/EEC), contemplates both the ecological and social aspects in the evaluation of Environmental Impact. The legislation itself describes extensively the objectives and procedures, defining what kinds of activities have to be carried out and how.

The last objective of the studies on environmental impact is to evaluate the positive and negative effects that an intervention will have (large industry, roads and motorways, railways, ports, etc.), to take the correct pertinent measures, in the elaboration phase of the project. Moreover, to give elements to the competent public administration so that it can make a **declaration of impact** (compatible, moderate, severe or critical effects), and by consequence refuse or authorise the execution of the project, imposing, if necessary, more corrective measures.

In the E.I.A. the social scientists habitually tackle variables which could be foreseeable, like for example socio-demographic variables (sex, age, active population, ageing), institutional demands (services, infrastructures, etc.), voluntary or forced migrations, socio-economic factors, cohesion of the community, health, style and quality of life, symbolic places, cultural and historic patrimony in danger etc. Each case, however, has its peculiarities and the approach has to be specific.

Just as in the evaluation of the ecological impact, it is a case of trying to be able to define the present situation and the social effects of an environmental intervention. This requires the consideration, description and elaboration of a prospective in three situations.

* The place as it is socially in the moment that the intervention is considered
* The evolution that the place(s) will undergo if the planned intervention is carried out.
* The evolution of the place(s) if the intervention is not carried out.

By making contrasts between these three descriptions, we can arrive at the extent of the social impact the intervention will have, to be considered together with the ecological impact and so define the global environmental impact.

The studies of environmental impact can be made by the same company or organism promoting the project, if they dispose of a sufficient technical team on their staff. It is, however, frequent that they commission an external specialised company or consultant.

There is a great variety among consulting companies, for their size and organisational structure. We can distinguish three basic types:

a) The largest companies offer a complete service - a 'full service' - from the organisation itself, of an interdisciplinary nature, which can go from the realisation of the EIA to the execution of the project itself.
b) The medium sized companies offer the same 'full service', but not from within their organisational structure, but through co-ordinated specialised companies.
c) The small companies formed by a restricted personnel team, interprofessional or not, who punctually contract persons to execute works, and concrete parts of the studies to other specialised assessors.

The first and second are normally managed by professional businessmen or by engineers, incorporating other specialised professionals, like lawyers, biologists, geologists, economists and occasionally, architects too. Among the small companies there are many formed by small groups of engineers, biologists or geologists, and at times by the possible combination between two or three of these professionals.

The role of the social scientist is seldom contemplated in any of them, especially for the frequently partial application that is used of the legislation in force. When it is executed, in the social aspects it is frequent that this is left to some 'appraisals' by engineers or biologists themselves, or in some cases in statistical socio-economic and demographic valuations made by an economist. When social scientists do intervene, normally it is by contracting an external team to realise this part of the study. The perspective of the toughening of the levels of fulfilment of the legislation means that an important change in this situation can be foreseen. Due to the very nature of the problem, it is fundamental that environmental psychologists should be prepared to contribute their knowledge from specialised consultancies, integrating themselves into the staff of 'full service' companies, or participating in the creation of new interdisciplinary companies.

5.4 Environmental Audits

The profile related to the Environmental Audits comes wholly into the ambit of the company, but orientated towards natural resources. The Official Bulletin of the European Communities No. C76/2 of 23.3.92) records the now approved proposal of the regulation of the community system of Environmental Audits and eco-labelling.

The objective of the system of environmental audits, which establishes the regulation, is: to promote the improvement of the results of the industrial activities in relation to the environment through the establishment of protection mechanisms, the systematic, objective, and periodical evaluation of the results obtained and the provision of appropriate information to the public with respect to behaviour in environmental matters.

It implies the formal establishment of a policy by companies, objectives, a programme of action and a system of environmental management defined from the highest management, periodically evaluated, ensuring the appropriate participation of the workers or their representatives in the process.

In the evaluation (auditing) the control and prevention of the repercussions of the activity on the various components of the environment are considered; the management, saving and choice of energy; the reduction, recycling and reuse, transport and elimination of residues; the selection of production processes; product planning; the prevention of accidents; information, formation and participation of the personnel in environmental themes; the external information and public participation, as well as the response to their complaints.

The responsibility for the establishment of the plan of the company lies with the upper management, but counting on the collaboration of the workers representatives and fomenting the sense of responsibility with relation to the environment among the employees.

In this way, the environmental audit evaluates the systems of management undertaken, the fulfilment of the community, state and local regulations, and the coherence of the environmental policy of the company. A special emphasis is made in the regulations on management systems, the rules of environmental behaviour and their fulfilment. Once the evaluation has been carried out, the next step is to proceed to make an 'Environmental Declaration' which has to be ratified by an environmental inspector. Once the process has been successfully concluded, the company can 'give prestige' to its products with the logo that accredits its participation in the system of Eco-audits.

The audits have to be carried out by persons, teams or companies independent from the activity that they are evaluating, but with a 'sufficient knowledge of the sector (...) and the sufficient training and expertise as auditors to reach the objectives set' (O.B. 27.3.92,9). On the other hand, it does not specify a unique formational profile of the inspectors, who are accredited by the communities according to criteria established in collaboration with, among others, the professional associations.

If we analyse what we have described up to now, we can see how an important part of the aspects to be evaluated have to do with organisation, management, risk and the perception of risk, the behaviour of personnel, the formation of the workers and internal and external information. All of these aspects have been traditionally linked to the field of the psychology of organisations, but which require in this case a strong formation in environmental themes. Therefore, environmental psychologists with a good training in organisational aspects is a serious natural candidate to work in this emerging field.

5.5 *Ecology of the workplace*

In the previous profile we mentioned the environmental effects of the business in its environment and of the minimisation of the expenditure of natural resources in their forms of production. But this does not exhaust the environmental issue of the working world. To the labour organisations, their physical structure plays an important part for their good working, both from the functional and the aesthetic and symbolic points of view, in production and labour satisfaction of the workers. The building is not a simple neutral container. It affects the structure of the organisation, it permits, facilitates or inhibits ways of doing things or interactions between its members. It constitutes a good part of its 'organisational culture' and represents the public image of the company. Moreover, its situation generates - or alters - the ebb and flow of exchange with society and of circulation within its context. It configures a landscape. The choice of its situation goes beyond covering functional necessities. The ecology of the workplace is obviously centred on the business side, but orientated towards built environments and to urban surroundings.

The working world is one of the oldest subjects in which environmental psychology has intervened. Frequently, however, only from the functionality of design or the mere adaptation of the structure of the space to the organisational necessities, starting from behavioural observation, as is usually the case in the majority of the P.O.E. (Post-Occupancy Evaluation). Our approach to an 'Ecology of the workplace' seeks to open the ambit of intervention to organisational, symbolic, ergonomic and work health aspects. The contribution of the prestigious Valencian professor of psychology of organisations, Josep Maria Peiró (1991), in a symposium on 'Formation and Profession of the Environmental Psychologist', brought us the parameters from outside Environmental Psychology that help us to define this profile.

Peiró describes two perspectives of intervention of the environmental psychologist in organisations: the organisation as an environment and the environments in which the organisation act.

With relation to the first perspective, he highlights five especially relevant aspects: the physical conditions of work, the tasks to be carried out, the technology, the social surroundings of work and the organisational structure.

The physical conditions of work have frequently been approached from a perspective of safety and hygiene, for which the ambit of working medicine and ergonomics have been circumscribed. However, as the quality of working life is emphasised and not only in the correction of dysfunctions, the importance of the contributions of the environmental psychologist is becoming more and more evident in aspects related to the working environment. For the traditional valuations of luminosity, noise, temperature, design of spaces, level of saturation of a space, etc. it is necessary to add aspects related to the capacity of management of the worker over his space, the possibilities of appropriation and personalisation, of privacy, the use of status symbols, etc.

The tasks to be realised and their grouping in work places, to a large extent delimitate the spaces and instruments to use, which would have to have acceptable ergonomic qualities. Technology is becoming more and more a critical component in the organisational surroundings. Its importance does not stem so much from its determining character of human, social, and organisational processes (as technological determinism has claimed), but because a same technology can design itself and implant itself in multiple forms. The choice that is made will have to take the social system of the organisation into account as a socio-technical system. The process of accelerated change is having strong implications for the working environment on multiple levels, but especially in the adaptation of workers to new situations, new organisational, physical and aesthetic structures, new forms and habits of perception, of culture, etc. which have to be optimised.

The social dimension of the organisation, the formal and informal social environment, the more or less formalised interactions, the processes of perception and attribution of others, the influence, power, hierarchy, social support, group activities, all have their particular correlation in the physical structure of the organisation. Faced with the theories that have defended the existence of an optimum structure (one best way) for all organisations, it is becoming more and more evident that the appropriate structure for an organisation depends on multiple contingency factors that have to be established in each situation.

The second perspective contemplates the work of psychologists, together with other professionals, in the diagnostics, intervention and evaluation of the environment in which the organisation acts. To define the relevant dimensions of this environment and study its various segments (market studies, of the viability of projects, etc.), to develop the analysis of its inter-organisational components, the networks of relationships of the organisation with other individual and social agents of the outside environment, the limits between the organisation itself and its environment as well as the principal processes of transaction, the cultural dimensions of the environment, the insertion of multinational organisations in multiple cultures, the corporate image, etc. are aspects in which the knowledge and the professional action of the environmental is useful and relevant.

So a double profile is being defined: that of an organisational psychologist well trained in environmental aspects or that of an environmental psychologist well trained in organisational aspects. In both cases, as a professional integrated in the staff of the organisation or as an external assessor-consultant.

5.6 *Marketing, Education and Environmental Promotion*

This profile occupies a central part of the figure, although a little more orientated towards the sides of the natural resources and urban surroundings. It affects the nucleus of what is the professional activity of the environmental psychologist in any of the other profiles described. All of them have a communicational dimension, for the creation of opinions, for the formation of new behavioural habits and of new attitudes. Therefore they participate to a certain extent in this profile.

From its beginnings Environmental Psychology has sought to intervene in environmental education, although without a great deal of success. This is probably due to the polysemy of the term Environmental Education, which has always been confusing. The label of Environmental Education has been applied to the psychological formation of professional technicians (especially architects) and used to refer to the encouragement of new habits of behaviour, to the understanding of the city or the natural surroundings.

Environmental education as knowledge of the surroundings brings us back to something that in itself is outside what is strictly psychological, although it is frequently referred to in chapters dedicated to cognition in environmental psychology manuals. At best it can be considered as a peripheral aspect that can be useful for the analysis of cognitive or learning processes. However, if Environmental Education refers us to how the experience of place is caused -or how it is provoked- ; how this experience increases understanding of the environment through a process of significative learning; how an identification comes about with the environment after working on it, manipulating it, transforming it and so behaviours and attitudes emerge that are more respectful to the environment; or how to provoke this process from the proposition of alternative models through campaigns of persuasion, then we are right inside the ambit of what is psychological, but speaking of environmental education, marketing and promotion.

A different aspect is the recognition of spaces, if it is addressed to the elicitation of landscapes, structures or 'good manners' that form a part of the group of significations shared by a collective and that inspire comportments. This moves us away from environmental education to bring us into the ambit of social representations and attitudes, and therefore of marketing and promotion.

So the profile of the psychologist in environmental marketing, education and promotion is concentrated on the creation of a social image with respect to the environmental, in the detection, creation or modification of states of opinion, in the modification of attitudes through marketing campaigns proposing alternative models through persuasive channels, or campaigns that give incentives to changes of environmental behaviour. It is concentrated on specific programmes on energy conservation, selective collection or rubbish, water use, prevention of littering and vandalism, reduction of pollution by automobiles, risk prevention and nature conservation.

To conclude this section, I should like to point out that the profiles described seek to cover the ambit of the applications of environmental psychology on a professional level. They could also constitute an annotation of the orientative field for basic research work, to be carried out from universities or research institutes, and at the same time inspire training programmes in their specifically psychological or complementarily interdisciplinary dimension. In this sense, it is necessary to ensure the presence of environmental factors in the generic formation of every psychologist, as one more subject within the learning of the basic psychological processes. Moreover, it is recommendable that the training programmes should contemplate advanced or specialised subjects, orientated towards organisational aspects and towards the management of the natural and urban environments.

6. Summary and conclusions

Environmental Psychology has a long history, without interruption but with a profound evolution. Three stages can be differentiated, which go from the geopsychology of the beginning of the century to the architectural psychology of the sixties to the emerging green environmental psychology. A non-linear evolution, which in Europe has the appearance of discontinuity, but which finds the nexus in an American transition, which it returned enriched. Its phases of evolution respond to the social conditioners of each moment, which mark its differential nuances.

The great socio-economic and technological changes of the twentieth century have altered the social philosophy of each period, creating authentic different cosmologies, which have modified the parameters of science and therefore also of Environmental Psychology. The way of being in the world, the attitude before the environment, its consideration of daily life, which allows us to trace a clear dividing line between an idyllic past and a crude present, have transformed the focalisation of Environmental Psychology and therefore the object of study, the theoretical and methodological perspective. This can be confirmed in the evolution of the studies presented in congresses and in publications. There has been a transition from structural themes and approaches to more symbolic and existential considerations. Quality of life has become the central issue, although at times, due to the assimilation between the quality of life and a standard of living that all too frequently is made, they could seem to be contradictory to a demand for a greater and better preservation of natural resources.

The attempts at conjugation between the quality of life and the preservation of the environment have lead to the emergence of a growing accumulation of legal rules and regulations for production processes, but which also seek to influence the habits of consumption and behaviour of citizens. In both cases, the human and social origin of the ecological imbalances, and the affectation of people as a direct consequence of these imbalances, delineates a work space for social sciences in general and psychology in particular, explicitly or implicitly supported by these regulations. Altogether it helps us to define professional profiles or areas of specific interest which configure ambits of future development in Environmental Psychology.

As well as the basic studies, we have been able to configure at least six applied fields, following the models of the MA program in Barcelona: planning, management and quality of life; public thoroughfares and transport; environmental impact, environmental audits, ecology of the workplace and environmental marketing, education and promotion. These profiles can serve to organise, clarify and define concrete fields of action which help to define the differential 'product' which Environmental Psychology offers and permit the deepening of its professional development, a real 'failed subject'.

Ingenuity has to be avoided along with the repetition of history. Throughout the book we have seen how modern environmental psychology was applied before it was academic, but nevertheless it has developed more in the university than in the professional field. The first contemporary environmental psychology concerned itself with relevant problems by contributing irrelevant solutions. This was probably due to lack of experience and background, or from starting from the dominant theoretical and epistemological principles in the fifties and sixties, which were not adequate for the resolution of the demands, and so lead to a certain level of failure.

Taking refuge in academia permitted a small advance in the basic knowledge through investigation and reflection, without the weight of the social responsibility of an immediate application. However, it also made it fall into the almost leisurely and autocomplacent entertainment of easy themes, with an end to 'fulfilling' the university dynamic which 'obliges' the attainment of an apparently important productivity (generally reiterative and intranscendent) for the survival of its professionals ('publish or die'). For this reason I consider it important that should be attended to in the definition of formation programmes, not only for what is comfortable and known to us should be attended to, but also the demands of a growing labour market that has to respond to social interests.

So Environmental Psychology, as a social construction which answers opportunities, individual interests and to a logic of science, and adopts a social organisation in scientific communities, presents a network of individual and institutional relationships, with some influence between its actors, which can reach the point of forming real invisible colleges. Time, however, has blurred those I.C.'s a little which we remarked upon in 1986 (the first draft of the text of the book), varying the theoretical and methodological differences, which despite everything are maintained, but coexist. The paradigmatic crisis that the psychology of he nineties seems to be going through is not alien to this coexistence. Neither are the extra-scientific aspects, like the fact that the survival of the authors in a same scientific community during a period of more than twenty years -almost the whole of the professional life of the personages themselves- ends up being a constituent element in the very identity of the subjects (especially facing the exterior of the community) which takes preference over the theoretical and methodological differences. On the other hand, Environmental Psychology in Europe is no longer particular to the scientific community which is formed around the IAPS. Many professional collectives, the majority of the conferences and organisations of the psychosocial variety, of psychology of development, of physiological psychology, of health, of work and organisations, concern themselves with and dedicate a space to aspects or environmental conditioners.

To finalise, it could be said that there is a social demand which is ignorant of the fact that what it really needs could be offered by the environmental psychologist. An environmental psychologist member of an interdisciplinary team where all of its members has the necessary training. A team that can made a transdisciplinary professional offer and that must know how and to whom they should direct their proposal. To syntonise these extremes will be the task of the individuals themselves, but especially of the specialised professional associations and/or of the professional colleges.

So the period between 1984 and 1992 can be seen as a period of subtle but profound change, which affects the thematics, the focusing and the profession. The scientific community is becoming more cohesive, at the same time as it is opening out and diversifying, making it even more difficult to speak of a European Environmental Psychology. But above all, this period has been marked by the entrance into what has been called 'Green Environmental Psychology'.

Epilogue

I began the book approaching science as a social construction, subject to social, institutional, individual and historic give and take, which capriciously -and not always rationally- marks the orientation that it takes.

The epistemological and ontological reflection of psychology over the last years has lead to the approaches of what some call Post-modern Psychology (Best and Douglas, 1991; Serrano, 1992; Natter and Schatzki, 1993), for which the relativity and the complexity of the phenomena is a central nucleus, and from which Constructionism emphasises that reality is not reality as such, but rather a construction made through the discourse that people make of this reality, and what has to be studied is that discourse.

I have presented to you here a discourse on Environmental Psychology, and therefore a construction of Environmental Psychology which could certainly be very different. So it is legitimate to ask oneself whether this 'reality' exists, this Environmental Psychology that I have described to you, and in regards to this discourse, to what point am I not constructing it differently from the way it exists ...

But as Galileo Galilei said, when in 1616 the Inquisition made him apostatise from his theory of the movement of the earth around the sun '...E pur si muove...' (nevertheless, it moves...).

APPENDIX

Statistical appendix

Table 1. *Committees composition, organization department, invited lecturers, sponsors and publishers*

	1969 Dalandhui	1970 Kingston	1973 Lund	1974 Surrey I	1976 Strasbourg	1979 Louvain	1979 Surrey II	1982 Barcelona	1984 Berlin
Committee	Canter (P)UK	Honikman (A) S.A. Lee (P) Bridge (A) Bell (A) Harris (A) (All from Kingston)	Küller (?) Acking (A) Edberg Hesselgren (A) Wallinder (All Sweden's)	Canter (P) T. Lee (P) *Steering Comm.* Barbey (A) Langdon (A) Lipman (P,S,A) Stringer (P) Grifliths (P) (All from UK)	Korosec (P)F. Barbey (A) CH Graumann (P) D Kruse (P) D	Simon (P)B Duffy (A) UK Fatouros (A) G Lipman (PS) UK Noschis (P) CH Von Hoogdalen NL Voyé (S)B	Canter (P)UK T. Lee (P)UK Stringer(P) (UK)	Pol (P) Muntañola (A) Siguan (P) Morales(P) *IAPS Board* 2 UK, 1 S, 1 B, 1 G	Krampen (P) D Posner-Landsh Silbereisen Simonis Canter (P) UK Simon (P) B Prak (A) NL
Call	Dep. Arch.	Dep. Arch	Dep. Arch	Dep. Psych.	Dep. Psych.	Dep. Arch. Dep. Sociol.	Dep. Psych.	Dep. Arch. Dep. Psicol.	Dep. Arch.
Invited Lecturers			Appleyard (P) Berkeley, USA Winkel(P) N.Y. USA Canter (P) Guiford, Surrey		Canter (P) UK P. H. Chombart (An) F Raymond (S) F Sansot(S)F Proshansky (P)N.Y. USA Peled (A)Is Moles (P) F Korosec(P)F	Boudon (A) F Jonge (S) NL Rapoport (A) USA		Muntañola (A) Sp Levy-Leboyer (P)F Canter(P) UK	Galtung (?) D Canter (P) UK Von Uexküll (?) D Niit (P) URSS Sánchez (P) Venezuela. Sommer (P)USA

Professional Code: (A) = Architect (P) = Psychologist (S) = Sociologist (?) = No available

(Cont.)

	1969 Dalandhui	1970 Kingston	1973 Lund	1974 Surrey I	1976 Strasbourg	1979 Louvain	1979 Surrey II	1982 Barcelona	1984 Berlin
Sponsors	University Strathclyde (Arch) Architects Journal(Riba)	Kingston Polytechnic Riba	Swedish Building Research Univ. Lund	University of Surrey	Ministére de l'Equipemement Univ. Strasbourg CCI. Centre George Pompidou	Université Louvain-N. Fonds Nat. Recherche Scientif. Fondation <<Roi Bouduin>> Caisse Générale Epg. Comité Hygiène et Confort	University of Surrey	3 Universities of Catalonia Catalan Goverment Banca Catalana Col-legi Arq. Barcelona Barcelona Design Center (BCD) City Hall	Berlin Senator für WirtschaR und Verkehr. Deutsch FürschungrgemeinschaR Community Research/Minist. Economy and Work.
Publisher	Riba	Riba University	Studenttliteratur (Sweden) Dowden, Hutchinson & Ross, USA	Architeaural Press. UK	Catholic University Louvain-la-Neuve CIACO	Catholic University Louvain-la-Neuve CIACO	Abstracts University	Edicions i Publicacions Universitat Barcelona	Abstracts

Composition per countries organiz. committees

UK	19	(43,2%)
Sw	6	(13,6%)
F	1	(2,3%)
B	4	(9,1%)
CH	2	(4,5%)
D	6	(13,6%)
NL	2	(4,5%)
SP	4	(9,1%)
Total	44	(100%)

Invited Lecturers per countries

F	7	30,4%
EEUU	5	21,7%
GB	4	17,4%
D	2	8,6%
ISR	1	4,3%
NL	1	4,3%
SP	1	4,3%
USSR	1	4,3%
Venez.	1	4,3%
Total	23	100%

Countries codes

B	Belgium
CH F	Switzerland
D	Germany
F	France
G	Greece
ISR	Israel
NL	Netherland
S	Sweden
UK	Great Britain
USA	Unitet.Stat.Am.
USSR	Ex-USSR
Venez.	Venezuela

Table 2. *Synthesis of data and index of active participation and collaboration in the nine conferences*

	(1)	(2)	(3)	(4)	(5)	(6)	(7)	(8)	(9)	(10)
Dalandhui 1969	12	-	13	-	13	-	1	60	-	20
Kingston 1970	26	116,7	27	107,6	23	76,9	1,17	124	106,7	20,9
Lund 1973	50	92,3	52	92,5	39	69,5	1,33	83	-49,4	60,2
Surrey I 1974	34	47,1	36	-44,4	23	-69,5	1,56	289		11,8
Strasbourg 1976	82	141,2	88	144,4	63	173,9	1,39	223		36,7
Louvain 1979	59	-39,0	61	-44,2	48	31,2	1,27	153	-45,7	38,5
Surrey II 1979	168	184,7	179	193,4	124	158,3	1,44	250	63,4	67,2
Barna. Program. 1982	184	9,5	(224)*	(33,3)*	(157)*	(25,1)*	(1,42)*	(257)*	2,8	71,6
Barna. Public. 1982	61	-	70		49	-	1,42	257		
Berlín Program. 1984	311	69,0	(335)*	(38,8)*	(228)*	(45,2)*	(1,46)*	314**		(99)*
Berlín Abstr. 1984	189	-	215	-	155	-	1,38	-		
Total:	514		741		538			1753		

(1) Total of different *authors* who intervene in each conference. The total of this column is not its sum, as many participate assiduously.

(2) Percentage of the increase that it represents over the previous conference.

(3) The Number of *signatures*. It is different from (1) and (5) as each one is counted as independent, although it appears in collaboration, in a single work or that a single author participates in more than one work. The total does not include the data that appear in brackets (*).

(4) Ditto. (2)

(5) Number of works (papers + conferences + workshops + posters) presented in each congress (independently of the number of signatures that they carry). The total sum does not include the data in brackets.

(6) Ditto (2)

(7) Index of collaboration: the relation between the number of signatures and the number of works of the unit of analysis (conferences)

$$IC = \text{Number of signatures / number of works}$$

(8) *Total attendance*: active and passive participation at the congress

(9) Ditto (2)

(10) Index of active participation.
$$IAP = \frac{\text{Number of active participants}}{\text{number of assistants}} \times 100$$

* The data in brackets corresponds to the calculation of the Barcelona and Berlin congresses as a whole. In the general treatments (and whenever it is not indicated to the contrary) data corresponding to the publications and the summaries have been used, which are more limited

** Estimated value, as we do not the data referring to Berlin at our disposal. It has been calculated according to the hypothesis of exponential growth.

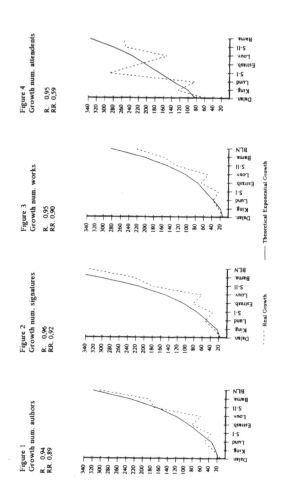

Figure 1
Growth num. authors

R. 0,94
RR. 0,89

Figure 2
Growth num. signatures

R. 0,96
RR. 0,92

Figure 3
Growth num. works

R. 0,95
RR. 0,90

Figure 4
Growth num. attendents

R. 0,95
RR. 0,59

---- Real Growth —— Theoretical Exponential Growth

Table 3. *Frequency of the professions (globally)*

	Dalan O.	Kings 1.	Lund 2.	Surrey I.	Strasb.3.	Lovain 4.	Surrey II.	Barna.* 7	Berlin**8	Total	Total without Berlin**
1- Architect	6(46,1)	4(14,8)	23(44,2)	13(36,1)	20(22,7)	25(41,0)	54(30)	24(34,3)	29(13,5)	198(26,7)	169 32,1 %
2- Psychologist	5(38,5)	11(40,7)	14(26,9)	9(24,9)	31(35,2)	16(26,2)	69(38,5)	35(50,0)	24(11,2)	214(28,9)	190 36,1 %
3- Sociologist	0(-)	0(-)	1(1,9)	1(2,7)	12(13,6)	10(16,4)	4(2,2)	3(4,3)	4(1,9)	35(4,7)	31 5,9 %
4- Engineer	0(-)	0(-)	6(11,5)	4(11,1)	0(-)	0(-)	6(3,3)	0(-)	4(1,9)	20(2,7)	16 3,0 %
5- Geog.& Antrhr.	0(-)	1(5,1)	1(1,9)	3(8,3)	1(1,1)	0(-)	8(4,5)	0(-)	2(0,9)	16(2,1)	14 2,7 %
6- Biol.-Ecol.	0(-)	1(5,1)	0(-)	0(-)	1(1,1)	0(-)	0(-)	0(-)	3(1,4)	5(0,7)	2 0,4 %
7- Archi.& So.Sc.	0(-)	0(-)	3(5,8)	3(8,3)	5(5,7)	2(3,3)	8(4,5)	1(1,4)	3(1,4)	25(3,4)	22 4,2 %
8- Other	1(7,7)	0(-)	3(5,8)	0(-)	1(1,1)	0(-)	4(2,2)	1(1,4)	1(0,5)	11(1,5)	10 1,9 %
0- Not available	1(7,7)	10(57,0)	1(1,9)	3(8,3)	17(19,3)	8(13,1)	26(14,5)	6(8,6)	145(67,4)	217(29,3)	72 13,7 %
Total	13(100)	27(100)	52(100)	36(100)	88(100)	61(100)	179(100)	70(100)	215(100)	741(100)	526

* Data about publications.
** Data about abstracts.
*** We include this column, in the total without Berlin, due to the high number of authors who don't indicate the profession in this conference.

Figure 5. *Evolution of active participation of professions in percentages*

Figure 6. *Evolution of active participation of professions in direct value*

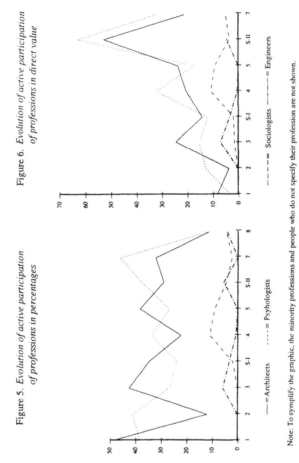

——— = Architects ····· = Psyhologists – – – = Sociologists –·–·– = Engineers

Note: To symplify the graphic, the minority professions and people who do not specify their profession are not shown.

Table 4. *Evolution of the professional composition in Great Britain*

Great Britain	Dalan.	K1ng.	Lund	Sur.I	Strasb.	Louvain	Sur.II	Barna.	Berlin.	Total
1 Architect	5(33,4)	3(21,4)	9(56,2)	7(29,2)	2(25)	6(50)	24(26,7)	3(37,5)	1(9,1)	60(30,9)
2 Psychologist	6(46,1)	6(42,8)	3(18,7)	7(29,2)	3(37,5)	4(33,3)	46(51,1)	3(37,5)	5(45,5)	83(428)
3 Sociologist	-	-	-	1(4,2)	-	-	2(2,2)	(5)	-	3(1,5)
4 Engineer	-	-	1(5,2)	1(4,2)	-	-	1(1,1)	(1)	-	3(1,5)
5 Geog.& Anthr.	-	1(7,1)	1(5,2)	3(12,5)	-	-	2(2,2)	-	-	7(3,6)
6 Biol. Ecol.	-	-	-	-	-	-	-	-	-	-
7 Archit.& Soc.Sc.	-	-	1(5,2)	3(12,5)	3(37,5)	2(16,7)	5(5,5)	-	1(9,1)	15(7,7)
8 Other	1(7,7)	-	2(12,4)	-	-	-	2(2,2)	1(12,5)	-	6(31)
0 Non available	1(7,7)	4(28,6)	-	2(8,4)	-	-	8(8,8)	1(12,5)	4(36,4)	20(10,3)
Total:	13(100)	14	16	24	8	12	90	8	11	194

* The number in brackets indicates the number of signatures on the program. The rest corresponds to the publication. Total does not include the data of the program.

Table 5. *Professional evolution of the participation in USA*

USA	Lund	Surrey I	Strasbourg	Louvain	Surrey II	Barcelona*	Berlin	Total
1 Architect	2	3	-	3	1	(3) 5	7	21
2 Psychologist	4	1	6	-	4	(19) 1	-	16
3 Sociologist	1	-	1	-	-	-	-	2
4 Engineer	-	3	-	-	-	-	-	3
5 Geog.& Anthr.	-	-	-	-	-	-	1	1
6 Biol.-Ecol.	-	-	-	-	-	-	-	-
7 Archit.& So.Sci.	1	-	1	-	-	1	2	5
8 Other	-	-	-	-	-	-	-	-
0 Non available	-	-	3	5	5	(39) 3	32	48
Total:	8	7	11	8	10	(61)10	42	96

* Idem table 4.

Table 6. *Professional evolution in Germany*

Germany	Dalan.	K1ng.	Lund	Sur.I	Strasb.	Louvain	Sur.II	Barna.	Bln.	Total
1 Architect	-	-	-	-	2	1	14	(6)	9	26
2 Psychologist	-	-	-	1	6	-	4	3	6	20
3 Sociologist	-	-	-	-	-	-	-	-	-	-
4 Engineer	-	-	1	-	-	-	1	-	4	6
5 Geog.& Anthr.	-	-	-	-	-	-	5	-	1	6
6 Biol.-Ecol.	-	-	-	-	-	-	-	-	-	-
7 Archi.& So. Sci.	-	-	-	-	-	-	1	-	-	1
8 Other.	-	-	-	-	-	-	2	-	-	2
9 Not available	-	-	-	-	1	-	6	(1)	30	37
Total:	-	-	1	1	9	1	33	3(7)	49	96

* The number in brackets indicates the number of signatures on the program. The rest corresponds to the publication. Total does not include the data of the program.

Table 7. *Professional evolution in the Francophone Area*

Francophone Area	Dalan.	K1ng.	Lund	Sur.I	Strasb.	Louvain	Sur.II	Barna.	Bln.	Total
1- Architect	-	-	1	-	4	3	2	(2) 2	-	10 (15,1)
1- Psychologist	-	-	2	-	11	5	1	(5) 5	2	27 (40,9)
3- Sociologist	-	-	-	-	6	2	-	(5) 1	1	11 (16,7)
4- Engineer	-	-	-	-	-	-	-	(1) -	-	-
5- Geog.& Anthr.	-	-	-	-	1	-	-	-	-	1 (1,5)
6- Biol.-Ecol.	-	-	-	-	1	-	-	-	-	1 (1,5)
7- Archi.& So.Sci.	-	-	-	-	2	-	-	-	-	2 (3)
8- Other	-	-	-	-	1	-	-	-	-	1 (1,5)
9- Not available	-	-	-	-	5	-	-	(4) 1	7	13 (19,7)
Total:	-	-	3	-	31	10	3	(17) 9	10	66 (100)

* The number in brackets indicates the number of signatures on the program. The rest corresponds to the publication. Total does not include the data of the program.

Table 8. *Professional evolution in Sweden*

Sweden	Dalan.	KIng.	Lund	Sur.I	Strasb.	Louvain	Sur.II	Barna.	Bln.	Total
1 Architect	-	1	9	-	-	1	1	(3) 1	2	15
2 Psychologist	-	4	3	-	2	1	2	(2) 2	2	16
3 Sociologist	-	-	-	-	-	-	-	-	-	-
4 Engineer	-	-	1	-	-	-	-	-	-	1
5 Geog.& Anthr.	-	-	-	-	-	-	-	(1) -	-	-
6 Biol.-Ecol.	-	-	-	-	-	-	-	-	-	-
7 Archi.& So.Sci.	-	-	-	-	-	-	-	-	-	-
8 Other	-	-	1	-	-	-	-	-	-	1
9 Non available	-	3	1	-	-	-	-	(2) -	-	4
Total:	-	8	15	-	2	2	3	(8) 3	4	37

* The number in brackets indicates the number of signatures on the program. The rest corresponds to the publication. Total does not include the data of the program.

Table 9. *Evolution of the institutional link*

	Dalan.	King.	Lund	Sur.i	Strasb.	Louvain	Sur.II	Barna.	Berlin.	Total	Tot.without Berlin
1 Archi.Univ.	8(61,5)	18(66,6)	17(32,7)	13(35,1)	30(34,1)	37(60,6)	37(20,7)	32(45,7)	40(18,6)	232(31,3)	192(36,7)
2 Psych.Univ.	2(15,3)	2(7,4)	11(21,1)	10(27,7)	21(23,8)	6(9,8)	59(32,9)	24(34,3)	18(8,4)	153(20,6)	135(25,8)
3 Other Univ.	-	2(7,4)	4(7,7)	4(1 ,1)	14(15,9)	10(16,4)	31(17,3)	6(8,6)	16(7,4)	87(11,7)	71(13,6)
4 Arch.Prof.	-	-	5(9,6)	-	5(5,7)	4(6,5)	16(8,9)	2(2,8)	-	32(4,3)	32(6,1)
5 Psyc. Prof.	-	-	-	-	2(2,3)	-	-	1(1,4)	-	3(0,4)	3(0,6)
6 Of.Inst.Arc.	1(7,7)	2(7,4)	1(1,9)	1(2,7)	-	-	-	-	1(0,5)	6(0,8)	5(0,9)
7 Other Inst.	1(7,7)	-	14(26,9)	8(22,2)	10(11,4)	3(4,9)	29(16,2)	3(4,3)	5(2,3)	73(9,8)	68(13,0)
8 Other	-	1(3,7)	-	-	1(l,1)	-	-	1(1,4)	2(0,9)	5(0,7)	3(0,6)
0 Not avail.	1(7,7)	2(7,4)	-	-	5(5,7)	1(1,6)	7(3,9)	1(1,4)	133(61,9)	150(20,2)	17(3,2)
Total:	13(100)	27(100)	52(100)	36(100)	88	61(100)	179(100)	70(100)	215(100)	741(100)	526(100)

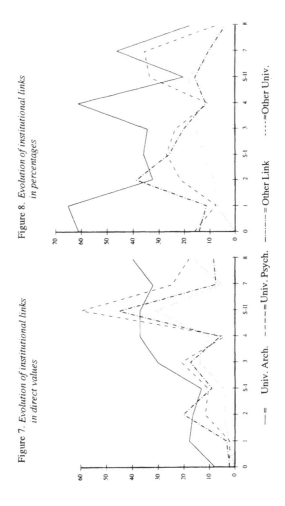

Figure 7. *Evolution of institutional links in direct values*

Figure 8. *Evolution of institutional links in percentages*

Table 10. *The institutional link of professions*

	Archi. Univ.	Psych. Univ.	Other Univ.	Prof. Archi.	Prof. Psych.	Of.Ins. Archi.	Other Of. Inst.	Other Inst.	Not Available	Total
1- Architect	146	2	6	25	1	2	14	-	2	198
2- Psychologist	39	141	12	1	2	1	14	2	2	214
3- Sociologist	3	3	26	-	-	-	2	-	1	35
4- Engineer	2	1	4	-	-	1	10	1	1	20
5- Geog.& Anthr.	1	2	11	-	-	-	1	-	1	16
6- Biol.- Ecol.	-	-	4	-	-	-	1	-	-	5
7- Archit.& So.Sci.	13	-	8	-	-	-	4	-	-	25
8- Other	1	-	5	-	-	-	3	2	-	11
9- Not avaible	27	4	11	6	-	2	24	-	143	217
Total:	232	153	87	32	3	6	73	5	150	741
	(31,3)	(20,6)	(11,7)	(43)	(0,4)	(0,8)	(9,8)	(0,7)	(20,2)	(100)

Table 11. *Author's distribution and signatures according to production*

	Author	*%*	*Signatures*	*%*	*Acum. %*
Great assiduous producers: 2	0,4	28	3,7	3,7	
Medium assiduous producers:	18	3,5	89	12,0	15,7
Small producers:	98	11,0	228	46,5	46,5
Transient producers :	396	77,0	396	53,4	100
Total:	514	100	741	100	

Table 12. *Author's distibution and signature-works (globally)*

(1)	(2)	(3)	(4)	(5)	(6)	(7)	(8)	(9)
18	1	1	0,2	0,2	18	18	2,4	
10	1	2	0,2	0,4	10	28	1,3	3,7
6	1	3	0,2	0,6	6	34	0,8	4,5
7	3	6	0,6	1,2	21	55	2,8	7,3
6	2	8	0,4	1,5	12	67	1,6	8,9
5	2	10	0,4	2,0	10	77	1,3	10,2
4	12	22	2,3	4,3	48	125	6,5	16,7
3	28	50	5,4	9,7	84	209	11,3	28
2	68	118	13,2	22,9	136	345	18,3	46,3
1	396	514	77,0	100	396	741	53,4	100
Total:	514		100		741		100	

Total works: 538

(1) Number of works presented by the same author.
(2) Number of authors with the number of works described in (1).
(3) Accumulated authors.
(4) Percentage on (2).
(5) Accumulated signatures (ref. to (6)).
(6) Total of signatures, product of (1)x(2).
(7) Accumulated signatures (ref. to (6)).
(8) Percentage on (6).
(9) Accumulated percentage.

Table 13. *Works by countries.*

		Tot. %	Ac. %	0	1	2	S.1	3	4	S.II	7*	8
Great producer	U.K.	132(24,5)	24,5	13	14	13	15	5	10	47	(14)6	8
	U.S.A.	86(16,0)	40,5			4	5	7	5	30	(36)5	30
	Germany	60(11,1)	51,6			1	1	7	1	6	(7)2	43
	France	57(10,6)	62,2			2		24	8	6	(13)7	10
	Sweden	34(6,3)	68,5		6	15		2	2	2	(5)3	4
Medium producer	Spain	18(3,3)	71,8					1		1	(34)11	5
	Belgium	16(3,0)	74,8					4	6		(8)3	3
	Canada	15(2,8)	77,6				1	1	3	5	(1)2	3
	Australia	13(2,4)	80,0				1		1	7	(3)1	3
	Japan	13(2,4)	82,4							3	(5)1	9
	Switzerland	13(2,4)	84,8					5	2	1	(3)1	4
	Holland	12(2,2)	87,0					1	1	1	(4)1	8
	Italy	11(2,0)	89,0						1	2	(2)	9
	Turkey	11(2,0)	91,0			1		2	2	4	(6)1	
	Israel	10(1,9)	92,9					1	2	4	(3)	2
	Greece	9(1,7)	94,6		1	1		2	1		(4)2	3
Small producer	South Africa	1(0,7)	95,3									
	U.R.S.S.	4(0,7)	96,0							1		3
	Denmark	3(0,6)	96,6		1	2						
	India	3(0,6)	97,2							1		2
	Poland	3(0,6)	97,8							1	(1)1	1
	Mexico	2(0,4)	98,2						1		(4)1	
	New Zealand	2(0,4)	98,6								(1)1	1
Occasional producer	Argentina	1(0,2)	98,8					1				
	Luxembourg	1(0,2)	99,0						1			
	Nigeria	1(0,2)	99,2							1		4
	Venezuela	1(0,2)	99,4								(1)	1
	China	1(0,2)	99,6							1		
	(Ireland)	-	-									
	N.A.	2(0,4)	100		1					1		
	(New Guinea)	-	-								(1)	
	Total:	538(100)		13	23	39	23	63	48	124	49 (157)	156
	Total Countries:			1	4	8	5	14	17	19	(22) 17	21(27)

* The data in brackets indicate real participation The others indicate published works in the proceedings. The columns of total and percentage only registrate the published works.

Table 14. *Résumé of the most dealt-with themes per conference*

	Code	Daland.	Klng.	Lund	Surrey.I	Strasb.	Louvain	SureyII	Barna.*	Berlin.	Total	%
1.	097	3	1	9	2	3	1	32	6(14)	24	81	(15,1)
2.	007	5	7	4	2	3	6	13	10(10)	8	58	(10,8)
3.	027	1	5	5	1	6	3	11	5(5)	10	47	(8,7)
4.	067	1	0	2	0	23	1	3	0(1)	1	31	(5,8)
5.	092	0	0	0	3	1	3	11	2(5)	9	29	(5,4)
6.	077	0	0	0	0	2	8	2	1(1)	8	21	(4,1)
7.	017	1	4	3	1	0	1	1	1(1)	6	18	(3,3)
8.	095	0	0	1	2	1	0	5	4(5)	5	18	(3,3)
9.	177	0	0	0	0	0	0	6	0(3)	5	11	(2,0)
10.	197	0	0	0	0	0	5	5	0(7)	0	10	(1,9)
11.	025	0	0	2	0	0	0	1	1(13)	5	9	(1,7)
12.	072	0	0	0	1	3	3	1	0(1)	1	9	(1,7)
13.	087	0	1	1	0	0	0	1	2(3)	4	9	(1,7)
14.	024	0	0	0	0	0	1	3	1(5)	3	8	(1,5)
15.	062	0	0	0	0	5	0	0	0(2)	3	8	(1,5)
16.	093	1	1	0	0	0	1	2	0(4)	3	8	(1,5)
Other	Other	1	4	12	11	16	15	28	16(75)	60	163	(30)
	Total:	13	23	39	23	63	48	125	49(157)	155	538	(100)

1. Evaluation of the environment in general. Methodological proposals.
2. General approaches about the environment. Relation. Arch. & Soc. Sc.
3. Cognitive aspects, image and significance in general, evolutive aspects.
4. Proxemics, territorial behaviour, appropiation of environment in general.
5. Valuation and experience of housing.
6. Use of environment in general (emphasis in the use).
7. Perception of the environment in general.
   Perceived medium, colour, ilumination.

8. Valuation and experience of school environment and children's settings.
9. Theorical frials on architecture-environment in general
10. Participation in environmental design
11. Cognition, developmental aspects, image, signif.in children.
12. Use of housing (emphasis in its use).
13. Adaptation, stress and misbehaviors regarding envir.
14. Cognitive aspects, maps, image, significance city.
15. Proxemics, territorial behavior and appropiation of housing.
16. Evaluation of neighborhood or suburb.

* Other. See general code and global data in table 18.

** *On brackets*: total of works according to the program. Not included in total sums. *Column without brackets*: total according publication.

206

Table 15a. *Evolution of the perspectives of analysis*

Code*	Dalan.	KIng.	Lurd	Sur.I	Strasb.	Louvain	Sur.II	Barna.	Bln.	Total
0	5	7	4	14	3	7	3	12	11	66 (12,3)
1	1	4	5	1	0	2	2	1	8	24 (4,5)
2	1	6	8	16	6	6	2	8	21	74 (13,8)
3	0	0	2	1	2	1	2	2	7	17 (3,2)
4	0	1	0	3	0	0	1	1	3	9 (1,7)
5	0	0	0	1	0	0	0	0	1	2 (0,4)
6	1	1	2	3	35	2	1	1	6	52 (9,7)
7	0	0	0	3	7	16	2	1	15	44 (8,2)
8	1	2	1	1	0	0	0	3	4	12 (2,2)
9	4	2	11	64	9	6	8	13	43	160 (29,7)
10	0	0	0	2	0	0	0	0	0	2 (0,4)
11	0	0	0	0	0	0	0	1	7	8 (1,5)
12	0	0	3	0	0	0	0	0	5	8 (1,5)
13	0	0	0	0	0	0	0	0	4	4 (0,7)
14	0	0	0	0	0	0	0	0	1	1 (0,2)
15	0	0	0	1	0	0	0	0	0	1 (0,2)
16	0	0	0	0	0	0	0	0	3	3 (0,6)
17	0	0	0	6	0	1	0	0	5	12 (2,2)
18	0	0	1	0	1	1	0	3	8	14 (2,6)
19	0	0	0	6	0	6	0	0	2	14 (2,6)
20	0	0	0	0	0	0	1	3	0	4 (0,7)
21	0	0	1	1	0	0	0	0	0	2 (0,4)
22	0	0	1	2	0	0	1	0	1	5 (0,9)
Total	13	23	39	125	63	48	23	49	155	538 (100)

* See equivalence of codes in table 18.

Table 15b. *Evolution of the types of environment analysed*

Code*	Dalan.	KIng.	Lund	Sur.I	Strasb.	Louvain	Sur.II	Barna.	Bln.	Total
0 Non Environment	0	1	2	2	0	2	2	0	10	19 (3,5)
1 Work environ.	0	1	0	0	4	1	3	0	1	10 (1,9)
2 Housing	0	0	0	5	11	9	14	4	21	64 (11,9)
3 Neighborhood	1	2	0	0	0	4	2	0	7	16 (3,0)
4 Urban environ.	0	0	2	2	2	4	4	5	16	35 (6,5)
5 School environ.	1	1	3	4	1	1	9	10	14	44 (8,2)
6 Clinical envir.	0	0	2	0	0	0	6	0	2	10 (1,9)
7 General envir.	11	18	30	10	38	26	79	30	80	322 (59,9)
8 Street-Squares	0	0	0	5	1	3	0	1	10 (1,9)	
9 Green nat.envi.	0	0	0	0	2	0	3	0	3	8 (1,5)
Total	13	23	39	23	63	48	125	49	155	538 (100)

Table 16 (Simplified). *Thematic interest in the most productive countries (approaches)*

	00	01	02	03	04	05	06	07	08	09	10	11	12	13	14	15	16	17	18	19	20	21	22	Total
02 Germany	4	1	4	3	1	0	7	4	1	20	0	5	1	1	1	0	3	1	2	2	0	0	0	61
05 Belgium	3	0	3	0	0	0	4	2	0	3	0	0	0	0	0	0	0	0	1	0	0	0	0	16
06 Canada	1	0	0	1	0	0	1	0	0	5	0	0	0	0	0	0	0	0	3	2	0	0	2	15
08 Spain	4	0	7	0	1	0	0	0	1	3	0	0	0	0	0	0	0	1	0	0	1	0	0	18
09 French	5	2	7	1	0	0	17	4	2	14	0	0	0	3	0	0	0	1	0	1	0	0	0	57
10 U.K.	24	6	17	2	3	0	7	7	4	47	1	1	1	0	0	0	0	2	3	3	1	1	2	132
12 Holland	2	1	2	0	0	0	0	1	0	5	0	0	0	0	0	0	0	0	1	0	0	0	0	12
13 Israel	2	0	1	0	0	0	1	1	0	3	0	0	1	0	0	0	0	0	0	1	0	0	0	10
14 Italy	3	1	5	1	0	0	0	1	0	0	0	0	0	0	0	0	0	0	0	0	0	0	0	11
15 Japan	0	1	1	1	0	0	0	2	1	3	0	0	1	0	0	0	0	0	0	0	0	0	0	10
22 Sweden	6	7	8	1	1	0	0	5	0	4	0	0	1	0	0	0	0	0	0	0	0	1	0	34
23 Switzerland	0	0	2	2	0	0	2	6	0	1	0	0	0	0	0	0	0	0	0	0	0	0	0	13
24 Turkey	0	0	0	1	0	0	3	1	0	4	0	0	0	0	0	0	0	1	0	0	1	0	0	11
25 U.S.A.	6	3	7	4	1	1	6	5	3	34	1	1	2	0	0	0	0	5	4	3	0	0	0	86
Other	6	2	9	0	2	1	4	5	0	13	0	1	1	0	0	1	0	1	0	2	1	0	1	50
Total	66	24	73	17	9	2	52	44	12	159	2	8	8	4	1	1	3	12	14	14	4	2	5	536 NA:2

Table 17. *Analysed environment in the most productive countries*

Country	0	1	2	3	4	5	6	7	8	9	Total
Germany	4	1	9	1	6	1	1	36	0	2	61
Belgium	1	0	2	1	3	0	0	8	1	0	16
Canada	0	1	2	1	0	4	0	7	0	0	15
Spain	0	0	0	0	0	7	0	11	0	0	18
France	1	2	9	0	3	4	0	34	3	1	57
U.K.	4	5	15	2	5	10	3	86	2	0	132
Holand	0	0	0	1	2	1	1	7	0	0	12
Israel	0	0	1	0	1	1	1	6	0	0	10
Italy	1	0	1	0	1	1	0	7	0	0	11
Japan	0	0	2	0	2	0	0	6	0	0	10
Sweden	2	0	0	3	1	2	0	26	0	0	34
Switzerland	0	0	5	0	2	0	0	5	0	1	13
Turquey	0	0	2	1	0	1	1	6	0	0	11
U.S.A.	5	0	8	4	3	8	2	50	2	4	86
Other	1	1	8	2	6	4	1	26	1	0	50
Total:	19	10	64	16	35	44	10	321	9	8	536 N.A: 2

Table 18. *Frequency of subjects in each conference*

Code	Dalan.	Kings.	Lund	Strasb.	Louvain	Sur.I	Sur.II	Barna.	Berlin	Total
000	0	0	0	0	0	1	0	0	1	2
010	0	0	0	0	0	0	0	0	0	0
020	0	0	0	0	0	0	0	0	1	1
030	0	0	0	0	1	0	0	0	1	2
040	0	0	0	0	0	0	0	1	0	1
050	0	0	0	0	0	0	0	1	0	1
060	0	0	0	0	0	0	1	0	0	1
070	5	7	4	3	6	2	13	10	6	58
080	0	0	0	0	0	0	0	0	0	0
090	0	0	0	0	0	0	0	0	0	0
100	0	0	1	0	1	0	0	0	0	2
110	0	0	0	0	0	0	0	0	0	0
120	0	0	0	0	0	0	0	0	0	0
130	0	0	0	0	0	0	0	0	0	0
140	0	0	1	0	0	1	0	0	2	4
150	0	0	0	0	0	0	0	0	0	0
160	0	0	0	0	0	0	0	0	0	0
170	1	4	3	0	1	1	1	1	6	18
180	0	0	0	0	0	0	0	0	0	0
190	0	0	0	0	0	0	0	0	0	0
200	0	1	1	0	0	0	0	0	0	2
210	0	0	0	0	0	0	0	0	0	0
220	0	0	0	0	2	1	1	1	0	5
230	0	0	0	0	0	0	0	0	0	0
240	0	0	0	0	1	0	3	1	3	8
250	0	0	2	0	0	0	1	1	5	9
260	0	0	0	0	0	0	0	0	0	0
270	1	5	5	6	3	1	11	5	10	47
280	0	0	0	0	0	0	0	0	1	1
290	0	0	0	0	0	0	0	0	2	2
300	0	0	0	0	0	0	0	0	1	1
310	0	0	0	0	0	0	0	0	0	0
320	0	0	0	2	0	0	0	1	2	5
330	0	0	0	0	0	0	0	0	0	0
340	0	0	0	0	1	0	0	0	1	2
350	0	0	0	0	0	0	0	0	1	1
360	0	0	2	0	0	0	0	0	0	2
370	0	0	0	0	0	2	1	1	2	6
380	0	0	0	0	0	0	0	0	0	0
390	0	0	0	0	0	0	0	0	0	0
400	0	0	0	0	0	0	0	0	0	0
410	0	0	0	0	0	0	0	0	0	0
420	0	0	0	0	0	0	0	0	0	0
430	0	1	0	0	0	0	0	0	0	1
440	0	0	0	0	0	0	0	0	0	0
450	0	0	0	0	0	0	2	1	1	4
460	0	0	0	0	0	0	0	0	0	0
470	0	0	0	0	0	1	1	0	2	4
480	0	0	0	0	0	0	0	0	0	0
490	0	0	0	0	0	0	0	0	0	0
500	0	0	0	0	0	0	0	0	0	0
510	0	0	0	0	0	0	0	0	0	0
520	0	0	0	0	0	0	0	0	0	0
530	0	0	0	0	0	0	0	0	0	0
540	0	0	0	0	0	0	0	0	0	0

Code	Dalan.	Klngs.	Lund	Strasb.	Louvain	Sur.I	Sur.II	Barna.	Berlin.	Total
550	0	0	0	0	0	0	0	0	0	0
560	0	0	0	0	0	0	0	0	0	0
570	0	0	0	0	0	0	0	0	1	1
580	0	0	0	0	0	0	1	0	0	1
590	0	0	0	0	0	0	0	0	0	0
600	0	0	0	0	0	0	0	0	0	0
610	0	0	0	2	0	0	0	0	0	2
620	0	0	0	5	0	0	0	0	3	8
630	0	0	0	0	1	0	0	0	0	1
640	0	0	0	2	0	0	0	0	2	4
650	0	1	0	0	0	1	0	1	0	3
660	0	0	0	0	0	0	0	0	0	0
670	1	0	2	23	1	0	3	0	1	31
680	0	0	0	3	0	0	0	0	0	3
690	0	0	0	0	0	0	0	0	0	0
700	0	0	0	0	0	0	0	0	0	0
710	0	0	0	0	0	0	0	0	0	1
720	0	0	0	3	3	0	1	0	1	9
730	0	0	0	0	1	0	0	0	2	3
740	0	0	0	0	2	1	0	0	2	4
750	0	0	0	0	1	1	0	0	1	3
760	0	0	0	0	0	0	0	0	1	1
770	0	0	0	3	8	0	2	1	8	22
780	0	0	0	0	0	0	0	0	0	0
790	0	0	0	1	0	0	0	0	0	1
800	0	0	0	0	0	0	0	0	0	0
810	0	1	0	0	0	0	0	0	0	0
820	0	0	0	0	0	0	0	0	0	0
830	0	0	0	0	0	0	0	0	0	0
840	0	0	0	0	0	0	0	0	0	0
850	1	0	0	0	0	0	0	1	0	2
860	0	0	0	0	0	0	0	0	0	0
870	0	1	1	0	0	0	1	2	4	9
880	0	0	0	0	0	0	0	0	0	0
890	0	0	0	0	0	0	0	0	0	0
900	0	0	0	0	0	0	0	0	0	0
910	0	0	0	1	0	0	3	0	0	4
920	0	0	0	1	3	3	11	2	9	29
930	1	1	0	0	1	0	2	0	3	8
940	0	0	1	0	0	1	1	1	1	5
950	0	0	1	1	0	2	5	4	5	18
960	0	0	0	0	0	0	5	0	1	6
970	3	1	0	3	1	2	32	6	24	81
980	0	0	0	2	1	0	2	0	0	5
990	0	0	0	1	0	0	3	0	0	4
100	0	0	0	0	0	0	2	0	0	2
101	0	0	0	0	0	0	0	0	0	0
102	0	0	0	0	0	0	0	0	0	0
103	0	0	0	0	0	0	0	0	0	0
104	0	0	0	0	0	0	0	0	0	0
105	0	0	0	0	0	0	0	0	0	0
106	0	0	0	0	0	0	0	0	0	0
107	0	0	0	0	0	0	0	0	0	0
108	0	0	0	0	0	0	0	0	0	0
109	0	0	0	0	0	0	0	0	0	0
115	0	0	0	0	0	0	0	0	2	2
111	0	0	0	0	0	0	0	0	0	0

Code	Dalan.	Klng.	Lund	Strasb.	Louvain	Sur.I	Sur.II	Barna.	Berlin.	Total
112	0	0	0	0	0	0	0	0	1	1
113	0	0	0	0	0	0	0	0	0	0
114	0	0	0	0	0	0	0	1	0	1
115	0	0	0	0	0	0	0	0	0	0
116	0	0	0	0	0	0	0	0	0	0
117	0	0	0	0	0	0	0	0	3	3
118	0	0	0	0	0	0	0	0	0	0
119	0	0	0	0	0	0	0	0	1	1
120	0	0	0	0	0	0	0	0	3	3
121	0	0	0	0	0	0	0	0	0	0
122	0	0	0	0	0	0	0	0	0	0
123	0	0	0	0	0	0	0	0	0	0
124	0	0	0	0	0	0	0	0	1	1
125	0	0	0	0	0	0	0	0	0	0
126	0	0	0	0	0	0	0	0	0	0
127	0	0	3	0	0	0	0	0	1	4
128	0	0	0	0	0	0	0	0	0	0
129	0	0	0	0	0	0	0	0	0	0
130	0	0	0	0	0	0	0	0	0	3
131	0	0	0	0	0	0	0	0	0	0
132	0	0	0	0	0	0	0	0	0	0
133	0	0	0	0	0	0	0	0	0	0
134	0	0	0	0	0	0	0	0	1	5
135	0	0	0	0	0	0	0	0	0	0
136	0	0	0	0	0	0	0	0	0	0
137	0	0	0	0	0	0	0	0	1	1
138	0	0	0	0	0	0	0	0	0	0
139	0	0	0	0	0	0	0	0	0	0
140	0	0	0	0	0	0	0	0	0	0
141	0	0	0	0	0	0	0	0	0	0
142	0	0	0	0	0	0	0	0	0	0
143	0	0	0	0	0	0	0	0	0	0
144	0	0	0	0	0	0	0	0	1	1
145	0	0	0	0	0	0	0	0	0	0
146	0	0	0	0	0	0	0	0	0	0
147	0	0	0	0	0	0	0	0	0	0
148	0	0	0	0	0	0	0	0	0	0
149	0	0	0	0	0	0	0	0	0	0
150	0	0	0	0	0	0	0	0	0	0
151	0	0	0	0	0	0	0	0	0	0
152	0	0	0	0	0	0	0	0	0	0
153	0	0	0	0	0	0	0	0	0	0
154	0	0	0	0	0	0	0	0	0	0
155	0	0	0	0	0	0	0	0	0	0
156	0	0	0	0	0	0	0	0	0	0
157	0	0	0	0	0	0	1	0	0	1
158	0	0	0	0	0	0	0	0	0	0
159	0	0	0	0	0	0	0	0	0	0
160	0	0	0	0	0	0	0	0	1	1
161	0	0	0	0	0	0	0	0	0	0
162	0	0	0	0	0	0	0	0	0	0
163	0	0	0	0	0	0	0	0	0	0
164	0	0	0	0	0	0	0	0	1	1
165	0	0	0	0	0	0	0	0	0	0
166	0	0	0	0	0	0	0	0	0	0
167	0	0	0	0	0	0	0	0	1	1

Code	Dalan.	KIngs.	Lund	Strasb.	Louvain	Sur.I	Sur.II	Barna.	Berlin.	Total
168	0	0	0	0	0	0	0	0	0	0
169	0	0	0	0	0	0	0	0	0	0
170	0	0	0	0	1	0	0	0	0	1
171	0	0	0	0	0	0	0	0	0	0
172	0	0	0	0	0	0	0	0	0	0
173	0	0	0	0	0	0	0	0	0	0
174	0	0	0	0	0	0	0	0	0	0
175	0	0	0	0	0	0	0	0	0	0
176	0	0	0	0	0	0	0	0	0	0
177	0	0	0	0	0	0	0	0	5	11
178	0	0	0	0	0	0	0	0	0	0
179	0	0	0	0	0	0	0	0	0	0
180	0	0	0	0	0	0	0	0	1	1
181	0	0	0	1	0	0	0	0	1	2
182	0	0	0	0	0	0	0	0	1	1
183	0	0	0	0	0	0	0	0	1	1
184	0	0	0	0	0	0	0	1	1	2
185	0	0	0	0	0	0	0	0	1	1
186	0	0	0	0	0	0	0	0	0	0
187	0	0	1	0	1	0	0	2	2	6
188	0	0	0	0	0	0	0	0	0	0
189	0	0	0	0	0	0	0	0	0	0
190	0	0	0	0	0	0	0	0	1	1
191	0	0	0	0	0	0	0	0	0	0
192	0	0	0	0	1	0	1	0	1	3
193	0	0	0	0	0	0	0	0	0	0
194	0	0	0	0	0	0	0	0	0	0
195	0	0	0	0	0	0	0	0	0	0
196	0	0	0	0	0	0	0	0	0	0
197	0	0	0	0	5	0	5	0	0	10
198	0	0	0	0	0	0	0	0	0	0
199	0	0	0	0	0	0	0	0	0	0
200	0	0	0	0	0	1	0	0	0	1
201	0	0	0	0	0	0	0	0	0	0
202	0	0	0	0	0	0	0	0	0	0
203	0	0	0	0	0	0	0	0	0	0
204	0	0	0	0	0	0	0	0	0	0
205	0	0	0	0	0	0	0	1	0	0
206	0	0	0	0	0	0	0	0	0	0
207	0	0	0	0	0	0	0	2	0	2
208	0	0	0	0	0	0	0	0	0	0
209	0	0	0	0	0	0	0	0	0	0
210	0	0	0	0	0	0	0	0	0	0
211	0	0	0	0	0	0	0	0	0	0
212	0	0	0	0	0	0	0	0	0	0
213	0	0	0	0	0	0	0	0	0	0
214	0	0	0	0	0	0	0	0	0	0
215	0	0	0	0	0	0	0	0	0	0
216	0	0	0	0	0	0	0	0	0	0
217	0	0	1	0	0	0	1	0	0	2
218	0	0	0	0	0	0	0	0	0	0
219	0	0	0	0	0	0	0	0	0	0
220	0	0	0	0	0	0	0	0	0	0
221	0	0	0	0	0	0	0	0	0	0
222	0	0	0	0	0	0	0	0	0	0
223	0	0	0	0	0	0	0	0	0	0
224	0	0	0	0	0	0	0	0	0	0

Code	Dalan.	Kings.	Lund	Strasb.	Louvain	Sur.I	Sur.II	Barna.	Berlin.	Total
225	0	0	0	0	0	0	1	0	0	1
226	0	0	0	0	0	0	0	0	0	0
227	0	0	1	0	0	1	1	0	1	4
Total:	13	23	39	63	48	23	125	49	155	538

Thematic Code

The used thematic classification system is composed of a double entry code. The first two digits mean the analysis perspective and type of the theme from a disciplinary consideration. The third digit relates to the type of analyzed environment. The combination of digits provide the used categories.

First two digits: Approaches

00 General approach. Theoretical essays concerning architecture and social sciences relationship.
01 Environmental perception. Perceived environment, with no emphasis on cognitive aspects. Colour and lighting, auditive perception.
02 Cognition. Evolutionary aspects of thought and intelligence development. The image of environment. Maps. Significance.
03 Personality and environment.
04 Ecological psychology. Behaviour Settings. Behaviour in specific environments.
05 Ecological behaviour. Environmental degradation, pollution, etc.
06 Proxemics. Territorial behaviour. Crowding. Space appropiation. Privacy.
07 Use of space. Behaviour focused on environmental use.
08 Environment and stress. Functional adaptation to the environment. Deviant behaviour due to environment pressure. Deliquency. Environmental impact over the person.
09 Assessment and experience of environment. Systems, scales and methods of evaluation specifically regarding environments.
10 Other themes of psychology and psychiatry.
11 Other themes of urban sociology, general sociology, economy and history.
12 Other themes of culture, philosophy, semiology, aesthetics and art.
13 Anthropology.
14 Themes of ethology.
15 Themes of biology, medicine, physiology.
16 Themes of ecology.
17 Practical and theoretical essays of architecture (not included in 00).
18 Environmental design and planning.
19 Design participation.
20 Methodological proposal not releated directely to the environment.
21 Technology, solar architecture, etc.
22 Education.
23 Geography.
99 Other themes.

Third digit: Analyzed environment.

1 Place of work. Organization.
2 House, housing, home.
3 District, neighbourhood.
4 Town and population in rural environments.
5 School and children space.
6 Hospitals. Welfare institutions.
7 General. Environments out of specific categories.
8 Public space. Streets, squares.
9 Green spaces. Natural environment.
0 Not related to the environment.

Figure 9. *Bradford Areas applied to the quotations (without self-quotations)*

N. quotations			Quoted authors		Received quot. expr. on signatures**		
Auth.	N.Quot.	Tot.Quot.	Total	%	Total	%	
46	1	46 46	101	4,88	1.044	28,75	1.ª área
a	16	15-27 322					
5	84	5 a 14 676					
4 a 2			454	22,15	1.093	30,1	2.ª área
1			1.494	72,9	1.494	41	3.ª área
Total		2.049 100	3.631	100			

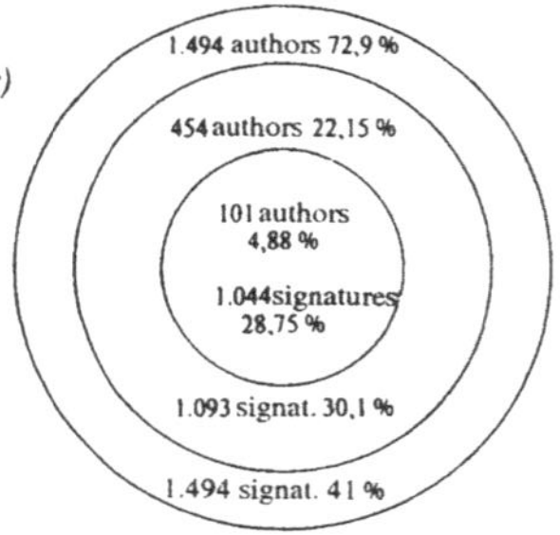

1.494 authors 72,9 %

454 authors 22,15 %

101 authors
4,88 %

1.044 signatures
28,75 %

1.093 signat. 30,1 %

1.494 signat. 41 %

* The data in brackets corresponds to the calculation of the Barcelona and Berlin congresses as a whole. In the general treatments (and whenever it is not indicated to the contrary) data corresponding to the publications and the summaries have been used, which are more limited
** Estimated value, as we do not the data referring to Berlin at our disposal. It has been calculated according to the hypothesis of exponential growth.

Table 19. *Emited references from each conference to the other*

Conference	Emited references
0	- -
1	- 2: 0, 0
2	- 7: 0, 0, 5 to 1
SI.	- 4: 0, 0, 1, 2
3	- 8: 1, 7 to 2
4	- 26: 0, 1, 1, 2, 2, 6 to S.I, 7 to 3, 4 to 4, 4 to SII.
7	- 5: 0, 1, 2, 3, 4
8	- 6: 7, 5 to 8
Total:	58

Note: We don't dispose of data regarding the 6th. conference, Surrey II due to the only edition of small abstracts, without reference.

Table 20. *Received quotations for each conference from the other*

Conference	Received quotations from other conferences	Received quotations withaut self-quot.
0	- 8: 1, 1, 2, 2, SI., SI., 4, 7	- 7
1	- 10: 5 from 2, SI., 3, 4, 4, 7	- 10
2	- 11: SI., 7 to 3, 4, 4	- 6
SI.	- 6: 6 from 4	- 5
3	- 8: 7 from 4, 7	- 6
4	- 5: 4 to 4,7	- 2
S.II	- 4: 4 to 4	- 0
7	- 1: 8	- 1
8	- 5: 5 to 8	- 2
Total:	58	

Table 21. *The most visible authors in the SSCI*

	IAPS Works	Other	Self Quot.	Total	Visibility Index
T. Lee	3 (0.5)	527	15	545	2.73
D. Canter	27 (8.3)	250	25	302	2.48
A. Lipman	3 (1.4)	199	7	209	2.32
R. Küller	15 (36.6)	26	-	41	1.61
B. Honikman	2 (12.5)	14	-	16	1.20
M. Krampen	-	13	-	13	1.11
P. Korosec	-	3	-	3	0.47
Mikellides	-	1	-	1	0
	5O (4.2)	1.033	47	1.130	

Note: The percentage is calculated on the total of each author.

Table 22. *Summary of the most visible authors in each cultural area, with MVI*

British A.		French A.		Sweden		Germanic A.	
1 Sommer	1,22	1 Alexander	0,96	1 Küller	0,81	1 Proshansk y	0,63
2 Lee	1,14	2 Raymond	0,96	2 Osgood	0,79	2 Mckechnie	0,59
3 Hall	0,96	3 Bourdieu	0,93	3 Lynch	0,69	3 Seiwert	0,43
4 Canter	0,94	4 Haumon	0,88	4 Hesselgren	0,67	4 Krampen	0,40
5 Wools	0,94	5 Castells	0,81	5 Berlyne	0,59	5 Harmut	0,35
6 Bannister	0,93	6 Lugassy	0,76	6 Hering	0,56	6 Hampel	0,31
7 Proshansky	0,91	7 Palmade	0,74	7 Canter	0,55	7 Beckman	0,15
8 Lynch	0,90	8 Godard	0,74	8 Hartman	0,47	8 Richter	0,15
9 Rapoport	0,90	9 Rapoport	0,72	9 Guilford	0,46	9 Muller	0,15
10 Kelly	0,88	10 Bachelard	0,71	10 Craik	0,41		
				11 Tannenbaum	0,41		

USA Participation		Spain		Greece		Authors shared by Europeans	
1 Ittelson	0,85	1 Canter	0,79	1 Castells	0,47	1 Canter	2,28
2 Rivlin	0,76	2 Kaplan		2 Rapoport	0,47	2 Rapoport	2,09
3 Sommer	0,74	3 Barker	0,57			3 Lynch	1,69
4 Hall	0,71	4 Rapoport				4 Proshansky	1,54
5 Proshansky	0,70	5 Muntañola				5 Bourdieu	1,28
6 Sanoft	0,70					6 Castells	1,28
7 Piaget	0,65					7 Barker	1,17
8 Appleyard	0,63					8 Küller	1,12
9 Newman	0,61					9 Piaget	0,50
10 Barker	0,55						

MVI (*Modified* Index of visibility).- The VI proposed by Platz considers the number of quotations independently from the number of authors or studies that emit them. To avoid the bias that this implies we have introduced a modification that seeks to correct it. The MVI consists in the logarithm of the total number of quotations received with the auto-quotations eliminated, multiplied by the total number of authors who emit the quotations (once those who make auto-quotations and their co-signatories have been eliminated) and by the number of studies that emit them (also eliminating the auto-quotations), taking the cubed root of this result. The calculation is realised according to the formula:

$$IMV = \log_{10} \sqrt[3]{\text{quotation} \times \text{authors} \times \text{works}}$$

Table 23. *Summary of authors who receive three quotations or more, in two or more cultural areas*

	U.S,A.	U.K.	French A.	Sweden	German A.	Spain	Greece.	Total Europe + USA	Total Europe
Appleyard	0,63	0,59						1,22	0,59
Barker	0,55	0,60				0,57		1,72	1,17
Bourdieu			0,93	0,35				1,28	1,28
Canter	0,35	0,94		0,55		0,79		2,63	2,28
Castells			0,81				0,47	1,28	1,28
Craik	0,20			0,41				0,61	0,41
Goffman	0,35	0,85						1,2	0,85
Hall	0,71	0,96						1,67	0,96
Korosec	0,15		0,15					0,30	0,15
Küller		0,31		0,81				1,12	1,12
Lynch	0,40	0,90	0,10	0,69				2,09	1,69
Newman	0,61	0,59						1,2	0,59
Piaget	0,65	0,41	0,10					1,16	0,51
Osgood	0,15				0,79			0,94	0,79
Proshansky	0,70	0,91				0,63		2,24	1,54
Rapoport	0,0	0,90	0,72				0,47	2,09	2,09
Sanoff	0,70	0,63						1,33	0,63
Sommer	0,74	1,22						1,96	1,22

Table 24. *Received quotations from the 13 journals with more impact per conference*

Title	Code	Dalan	King.	Lund	Sur.I	Strb.	Lov.	Sur.II.	Barna	Bln.T	Total
Environment and Behavior	9	-	1	13	9	1	-	-	11	15	58
Architects Journal	39	1	7	4	4	-	5	-	2	-	23
Psychological Review	35	3	-	3	2	-	1	-	2	3	14
Human Factor	15	1	-	1	2	-	-	-	9	-	13
Journal of the Amerc. Inst. Planners	41	1	3	3	-	-	2	-	1	3	13
American Psychologist	2	0	1	1	1	1	1	-	5	2	12
Journal of Social Issues	3	1	-	1	3	1	4	-	2	-	12
Psychometrika	57	1	-	8	1	-	1	-	1	-	12
J. Experimental Psychology	78	3	3	2	1	-	-	-	-	2	11
Riba Journal	114	-	6	1	1	-	3	-	-	-	11
Ergonomics	16	2	2	1	-	-	1	-	3	-	9
Annual Rev. Of Psychology	18	-	-	2	-	-	1	-	5	1	9
Human Relations	76	1	-	1	1	1	3	-	-	2	9
Total:		14	23	41	25	4	30	-	41	28	

Table 25. *Year of publication of artiches in the most quoted journals*

	NA	57	58	59	60	61	62	63	64	65	66	67	68	69	70	71	72	73	74	75	76	77	78	79	80	81	82	83	84	Total
Env. and Beh.	2													7	11	1	14	2	6		3	6		2	3				1	58
Arch. Journ.								1	1	1		1	4	2	2	6	1		1		1	1			1					23
Psi. Rev.	2	5		1		1								1			1	1	1				1							14
Hum. Fact.			1									1	1		1	2	4	1								2				13
J.Am.Inst.Plan.	1		1			1	1					1	2	3	1	1	1													13
Am. Psych.							2			1	1				1						1	1	1	3		1				12
J.S. Iss.	3							1			4				1						1	2								12
Psychometrika	1					1	1	5				2			2															12
J. Exp. Psych.	3	1	1	1				1		1						1												3		11
Riba Journ.											1	2	2		2		3	1												11
Ergon.			1									1	1	1			3					1		1						9
Ann.R. Psych.									1									2					2			1	3			9
Human Relations	1													6					1							1				9
(J. Env. Psy)																										1	3	2		6
	13	6	3	3	0	1	5	4	6	4	6	8	17	13	21	11	27	7	1	8	2	8	10	4	3	9	7	4	1	212

Other NA=6

⌐————— 38 - 17,92% —————⌐   ⌐———104 - 49,06%———⌐   ⌐————— 56 - 26,4% —————⌐

Bibliographical appendix

Historical classics

Allesch, G.J. (1925) 'Die ästhetische Erscheinungweise der Farben'. *Psychologie Forschung*. 6.

Allport, G. (1937) *Personality: a psychological interpretation*. New York. Holt.

Angyal, A. (1931) 'Uber die Raumlage Vorgestellter Orter. Archiv für Gesamte Psychologie 73'. In: Howard & Templeton (Eds.) (1966) *Human Spatial Orientation*. New York. Wiley.

Anonyme. (1936) *Réglementation des marchés et foires de la Ville de Strasbourg Bibliothèque Nation. de Strasbourg*.

Bartlett, F.C. (1932) *Remembering*. Cambridge University Press.

Bedford, T. (1936) 'The warmth factor in comfort at work'. In: *Med. Res. Council, (Industrial Health Board), Report 76*. London. HMSO.

Beebe-Centre, J.G. (1932) *Pleasantness and unpleasantness*. New York. Van Nostrand.

Bender, W.R.G. (1933) 'The effect of pain and emotional stimuli and alcohol upon pupillary reflex activity'. *Psychological monographs*. 44, 1-32.

Birkhoff, G.D. (1933) *Aesthetic measure*. Cambridge, Mass. Harvard University Press.

Bridges, K.M.B. (1929) 'The Occupational Interests and Attention of Four-Year-Old Children'. *Journal of Genetic Psychology*. 36.

Burgess, E.W. (1923) 'The Growth of the City: An Introduction to Research Projekt'. In: Park, Burgess & McKenzie (1925) *The City*. Chicago.

Burgess, E.W. (1929) 'Urban areas'. In: Smith & White (1929) *Chicago, an Experiment in social Science Research*. Chicago.

Cambria, F. (1927) 'Discuss Theatre Design and Decoration in an interview by Thomas C. Kennedy'. *Motion Picture News June*. 2185-2189.

Campbell, N.R. (1928) *Measurement and Calculation*. Longmans, London.

Chapin, P.S. (1935) *Contemporary American Institutions*. New York. Harper.

Claparede, E. (1943) 'L'orientation Lontaine'. In: J.Wampner (1943) *Nouveau Traite de Psychologie 8 (Whole n.3)*. Paris. Presses Universitaires de France.

Cohen, M.R. & Nagel, E. (1934) *Introduction to logic and scientific method*. London. Routledge.

Collingwood, R.G. & Myres, J.N.L. (1936) *Roman Britain and the English Settlements*. Oxford. Clarendon Press.

Coppier, L. (1932) *De la Maurienne à la Tarentaise*. Chambéry, dardel.

Dewey, J. (1916) *Democracy and Education*. New York. Macmillan Company.

Dorim, A. (1927) 'Showmanship and Interior Decoration'. In: *Motion Picture News May*.

Dumezil, G. (1944) *Naissance de Rome*. Paris. Gallimard.

Duncker, K. (1945) 'On problem solving'. *Psychological Monographs*. 62 (5), 270.

Eberson, J. (1927) 'New Theatres for Old: Originator of the Atmospheric Style Discusses the Formula in which Art...'. *Motion Picture News December*. 29.

Eliot, T.S. (1932) *Selected Essays*. Faber & Faber.

Eysenck, H.J. (1940) 'The general factor in aesthetic judgements'. *The British Journal of Psychology*. Cambridge (1941) 31.

Farwell, L. (1930) 'Reactions of Kindergarten, First and Second Grade Childrens to Contructive Play Materials'. *Genetic Psychology Monographs*. 8.

Firey, W. (1945) 'Sentiment and symbolism as ecological variables'. *American sociological review*. 10, 140-148.

Freud, S. (1922) *Group Psychology and the Analysis of the Ego*. Hogarth Press.

Freud, S. (1928) 'Le problème économique du masochisme'. *Revue Française de Psychologie II*. 2.

Garnier, T. (1918) *La cité industrielle*. Paris.

Gershun, A. (1935) 'The Light Field'. *Journal of Maths. and Physics*. Svetovoe Pole, Moscow. 18, 51.

Gisberttz, W. (1936) *Das deutche Kleinsiedlungsrecht-Hand-buch der Kleinsiedelung. 2 Bde.*. Berlin.

Gordon, K. (1923) 'A Study of aesthetic judgements'. *Journal of Experimental Psychology*. VI, 1.

Henon, P.J. (1928) 'The Architect's Service to the Industry: an Interview with'. *Motion Picture News December*. 29.

Hering, E. (1878) *Zur Lehre vom Lichtsinne*. Wien.

Hering, E. (1920) *Grundezuge der Lehre vom Lichtsinne*. Leipzig.

Hougten, F.C. & Young, C.P. (1923) 'Determining lines of equal comfort'. *Trans. Am. Soc. Heat Vent. Engineers*. 29, 163-76.

Howard, E. (1902) *Gardencities of Tomorrow*. London.

Hull, CL. (1943) *Principles of behavior*. New York. Appleton Century Crafts.

James, W. (1909) *Précis de psychologie*. Paris. R.Rivière.

Kampffmeyer, H. (1026) *Siedlung und Kleingarten*. Wien.

Lamb, T. An Interview with;(1928) 'Good Old Days to these Better New Days'. *Motion Picture News June*. 30.

Lee, S.C. (1929) 'Stretching the Building Fun and the Plot Area'. *Motion Picture News December*. 28, 31.

M.P.N. (1929) 'Conservative Design Has Lasting Appeal'. *Motion Picture News June*. 22, 43-52.

Markelius, S. (1930) *The Swedish periodical Byggmästaren*.

Maslow, A.N. (1943) 'A theory of human activation'. *Psychological Review*. 50, 370-396.

Murray, H. (1938) *Explorations in personality*. New York. Oxford University Press.

McDowell, M.S. (1937) 'Frecuency of Choise of Play Materials by Pre-school Children'. *Child Development*. 8.

Odgen, C.K. & Richards, I.A. (1923)(1945) *The Meaning of Meaning*. Routledge & Kegan Paul.

Patern, M. & Newhall, S.M. (1943) 'In Child Behaviour and Development'. In: Barker et alt. (Eds.). McGraw Hill.

Pevsner, N. (1936) *Pionniers of Modern Design, from William Morris to Walter Gropius*. London.

Philipe, J. (1904) *L'Image Mentale*. Paris. Alcan.

Piaget, J. (1929) *The Child's Conception of the World*. Routledge.

Piaget, J. (1936) *La naissance de l'intelligence chez l'enfant*. Neuchâtel. Delachaux et Niestlé.

Plant, J.S. (1937) *Personality and the Cultural Pattern*. Cambridge. Harvard University Press.

Rapp, G. & C.W. (1923) 'Better Theatres Section'. *Exhibitors Herald*. 16, , May 26.

Robert, J. (1939) *La maison rurale dans les Alpes françaises*. Arrault Ed.

Robertson, H. (1924) *Principles of Architectural composition*. London. London: Architectural Press.

Ruskin, J. (1907) *The seven lamps of architecture*. London. Dent.

Sargent, S.S. (1940) 'Thinking processes various levels of difficulty'. *Archives of Psychology.* New York. 249, whole n.

Simmel, G. (1903) 'Die Grobstädte und das Geistesleben'. In: Petermann, Th. (Hrsq). *Die Grobstadt.* Dresden. 187-206.

Smuts, J.C. (1926) 'Concepts of Space and Time/ and General Concept of Holism and Concept of Fields'. *Holism and Evolution.* London. McMillan. 22-34/ 84-117.

Sorokin, P.A. (1937) *Social and cultural dynamics.* London. Allen & Unwin.

Trowbridge, C.C. (1913) 'On Fundamental Methods of Orientation and Imaginary Maps'. *Science, 38.* 888-897.

Updergraaf, R. & Herbst, E.K. (1933) 'An Experimental Study of the Social Behavior Stimulated in Young Child.by Certain Play Materials'. *Journal of Genetic Psychology.* 42.

Usher, R.D. & Hunnybun, N.K. (1933) 'Overcrowding as a Factorin Personality Maladjustement'. *Mother and Child.* 4.

Vailland, R. (1955) *Trois cent vingt-cinq mille francs.* Paris.

Van Alstyne, D. (1932) *Play Behaviour and Choise of Play Materials of Pre-School Children.* University of Chicago Press.

Wallon, H. (1931) 'Comment se dévelope chez l'enfant la notion de corps prope'. *Journal de Psychologie.* Paris. P.U.F. 705-48.

Wallon, H. (1941) *L'evolution psychologique de l'enfant.* Paris. A.Colin.

Wallon, H. (1949) *Les origines du caractère chez l'enfant.* Paris. P.U.F.

Wirth, L. (1938) 'Urbanism as a way of life'. *The American Journal of Sociology.* 44.

Wittenborn, J.R. (1943) 'Factorial equations for tests of attention". *Psychometrika.* 8, 19.

Wulzinger, K. & Watzinger, C. (1924) *Damaskus, die islamischeStadt.* Leipzig. Berlin u.

Functional classics

Allport, F. H. (1955) *Theories of Perception and the Concept of Structure*. Wiley.

American Public Health Association (1948) *Planning the Neighborhood*. Chicago Public Administration Service.

Anderson, K. Ladd, G. and Smith H. (1954) 'A study of 2.500 Kansas High School Braduates'. *Kansas Studies in Education*. Univ. of Kansas. Vol. 4.

Arnheim, R. (1954) *Art and Visual Perception*. University of California Press. /London: Faber 1956.

Ashby, W. R. (1956) *An introduction to Cybernetics*. London: Chapman & Hall.

Ashworth, W. (1954) *The Genesis of Modern British Town Planning. A Study*. London: Routlogge & Kegan Paul.

Attneave, F. (1950) 'Dimensions of similarity'. *American Journal of Psychology*. vol. 63, pp. 515-556.

Attneave, F. (1957) 'Transfer of Experience with a Class Schems to Ident.'. *Journal of Experimental Psychology*. vol 54.

Bachelard, G. (1950) *La formation de l'esprit scientifique*. Paris: Vrin.

Bachelard, G. (1957) *La poétique de l'espace*. Paris: P.U.F.

Baker, A., Davies, R. L. and Sivadon P. (1959) 'Psychiatric Services and Architecture'. *Public Health Papers*. Geneva: World Health Organization. No. 1.

Barker, R. (1953) 'On the Nature of the Environment'. *Journal of Social Issues*.

Bartlett, F. C. (1958) *Thinking*. London: Unwin.

Bedford, T. (1958) 'Research on heating and ventilation in relation to human comfort'. *Heat Pip. Air Condit.*. pp. 127-34.

Bennet, G. K., Seashore, H. G., and Wesman, A. G. (1959) *Differential Aptitude Tests*. New York: Psychological Corporation.

Bernstein, B. (1958) 'Some sociological determinants of perception'. *British Journal of Sociology*. vol. IX, pp. 159-174.

Blackshaw, M. Et al. (1959) *Utilisation de l'espace dans les logements*. Genève. Nations Unies.

Blake, R. et al. (1956) 'Housing Architecture and Social Interaction'. *Sociometry*. 19 No. vol.II, pp. 133-139.

Bloch, R. (1958) *Les origines de Rome*. Paris: P.U.F.

Brenan,T. (1948) 'Preface to midland city by Brennan' In: in Smith, J.G. & Sargent Florence, P. (1948), *Wolverhampton social and industrial survey*. London. Dobson.

Broadbent, D. E. (1958) *Perception and communication*. London: Pergamon.

Bruner, J. (1951) 'Personality Dynamics and the Process of Percelving' In: in R. R. Blake & G. V. Ramsey. *Perception. An Approach to Personality*. New York: Ronald Press.

Bruner, J. S. (1957) 'On Perceptual Readines'. *Psychological Review*. vol. 64, pp. 123-152.

Bruner, J. S. (1957) 'On Perceptual Readines'. *Psychological Review*. vol. 64.

Bruner, J. S. (1957) 'On Perceptual readiness'. *Psychological Review*. vol. 64, pp. 123-152.

Bruner, J. S., Goodnow, J. J., and Austin, G. A. (1956) *A Study of Thinking*. New York: Wiley.

Brunswik, E. (1947) *Systematic and Representative Design of Psychological Experiments*. Berkeley: University of California Press.

Bursill, A. E. (1958) 'Restriction of peripheral vision during exposure to hot and humid conditions'. *Quart. J. Exp. Psychol..* vol. 10, no.3, pp. 113-29.

Butler, G. (1959) *Introduction to Community Recreation*. New York: National Recreation Association.

Campbell, D. T. & Fiske, D. W. (1959) 'Convergent and Discriminant Validation by the Multitrait-Multimethod Matrix'. *Psychological Bulletin*. vol. 56.

Cassirer, E. (1955) 'The Philosophy of Symbolic Forms'. Yale University Press.vol. 3.

Cassiror, E. (1957) 'The Philosophy of Symbolic Forms'. *The Phenomenology of Knowledge*. New Haven: Yale University Press.vol. 3.

Cattell, R. B. (1950) *Personality: A Systematic, Theoretical and Factorial Study*. New York: NcGraw-Hill.

Chaloner, W. H. (1954) 'Robert Owen, Peter Drinkwater and the Early Factory System in Manchester, 1788-1800'. *Bulletin of the John Rylands Library*. vol. 37, pp. 78-102.

Chance, M. R. A. (1956) 'Social structure in a colony of Macaca mulatta'. *Brit. J. Anim. Behav..* vol. 4, pp. 1-13.

Chapman, D. (1955) *The Home and Social Status*. London: Routledge & Kegan Paul.

Chombart de Lauwe, P. H. (1956) *La Vie quotidienne des families ouvrières*. Paris: CNRS.

Chombart de Lauwe, Y. M. J. (1959) *Psychopathologie sociale de l'enfant inadapté*. Paris: Centre National de la Recherche Scientifique.

Chomsky, N. (1957) *Syntactic Structure*. Mouton Publisher.

Chrenko, F. A. (1953) *Probit analysis of subjective reaction to thermal stimuli..* Brit. J. Psychol.

Chrenko, F. A. (1953) 'Heated Ceilings and Comfort Journal of the Institution of Heating and Ventilation Engineers' In: (1964) *Threshold Intensities of Thermal Radiation Evoking Sensation of Warmth Journal of Physiology*.

Dean, J.P. (1951) 'The Ghosts of Home Ownership'. *Journal of Social Issues*.

Duffy, E. (1957) 'The psychological significance of the concept of 'arousal' or 'activation''. *Psychological Review*. vol.64, n.5, pp.266-75.

Dumezil. (1949) *Les mythes romains*. Paris. Gallimard.

Easterbrook, J.A. 'The effect of emotion on cue-utilisation and the organisation of behaviour'. *Psychological Review*. vol.66, n.131, 1959. pp.183-201.

Edwards, A.L. (1957) 'Techniques of attitude scale construction'. New York. Appleton-Century-Crofts.

Ekman, G. (1955) 'Dimensions of emotion'. *Acta Psychologica*. 11, 279-288.

Eldem, S.H. (1954) 'Türk evi plan tipleri'. Istanbull.T.Ü. Mim. Fak.

Erginbas, D. (1954) 'Diyarbakir evleri'. IstanbulMin. Fak.

Erikson, E. H. (1951) 'Sex differences in the play configuration of preadolescents'. *American Journal of Orthopsychiatry*. 21.

Eser, L. (1955) *Kütahya Evleri*. Istanbull.T.Ü. Mim. Fak.

Euler, C. & Soderberg, U. (1956) 'The relation between gamma motor activity and the electroencephalogram'. *Experimentia*. vol.12, pp. 278-9.

Eves, H. & Newson, C.V. (1958) *An introduction to the Foundation and Fundamental Concepts of Mathematics*. New York. Holt, Rinehart & Winston.

Eysenck, H.J. (1959) *The maudsley personality inventory*. London. University of London Press.

Fava, S.F. (1958) 'Contrasts in neigbourhood: New York city and a suburban county'. In: W.M.Dobriner (ed.)(1958) *The suburban community*. New York. Putnam's. 122-131.

Festinger, L. (1957) *A theory of cognitive dissonance*. Standford. Standford University Press/ London: Tavistock 1959.

Festinger, L., Schachter, S. & Back, K. (1950) *Social pressures in informal groups*. New York.

Harper/ London: Tavistock 1959.

Foote, N. (1951) 'Identification as the basis for a theory of motivation'. *American Sociological Review*. 16: 14-21.

Form, W.H. (1951) 'Stratification in low and middle income housing areas'. *Journal of Social Issues*. 7.

Frend, S. (1951) 'Le mois el le ça. Psychologie collective et analyse du mois'. *Essais de Psychologie*. Paris. Payot.

Gibson, J. (1950) *The perception of the visual world*. Boston-New York. Houghton Mifflin Company/ Cambridge, Mass.: The Riv.

Goffman, E. (1959) *The presentation of self in everyday life*. New York. Doubleday.

Gorb, P. (1951) 'Robert Owen as a businesman'. *Bulletin of Business Historical Society*. 25, 127-148.

Gropius, W. (1955) *Scope of total architecture*. New York.

Guilford, J.P. (1949) 'A factorial approach to the analysis of variances inaestheticjudgements'. *The American Psychological Association. Journal of Experimental Psychology*. vol.39.

Guilford, J.P. (1952) 'When not to factoranalyze'. *The American psychological association. Psychological Bulletin*. vol. 49.

Guilford, J.P. (1954) *Psychometric methods*. New York, Toronto, London. McGraw-Hill.

Guiroud, P.L.R. (1955) *La sémantique*. Paris. Que-sais-je? P.U.F.

Gullahorn, J.T. (1952) 'Distance and friendship as factors in the gross interaction matrix'. *Sociometry 15th. Feb.-May*. n.1-2 (I-II) 123-134.

Gurvitch, G. (1958) *La multiplicité des temps sociaux*. Paris. Centre de documentation universitaire.

Hall, E. (1959) *The hidden dimension*. New York. Garden City Doubleday, 1959, 1966.

Hall, E.T. (1959) *The silent language*. New York. Doubleday & Co.

Hall, E.T. (1959) *The silent language*. New York. Doubleday Company.

Handel, L.A. (1950) *Hollywood looks at it audience; A report of film audience research* . Urbana: University of Illinois reprinted by Arno Press 1976.

Hatt, P. (1946) 'The concept of natural area'. *American Sociological Review*. 11/1946, s.423-427.

Hess, R.D. & Handel, G. (1959) *Family worlds: Apsychological approach to family life*. Chicago. University of Chicago Press.

Hesselgren, S. (1952) *Hesselgrens color atlas*. Stockholm.

Hesselgren, S. (1954) *Subjective color standarization*. Stockholm.

Hesselgren, S. (1954) *Arkitecturens uttrycksmedel*. Stockholm. Almquist & Wiskell/Gebers.

Heyl, B.C. (1957) *New bearings in aesthetic and art criticism*. New York. Yale University Press.

Hinde, R.A. (1959) 'Some recent trends in ethology'. In: S, Koch (ed.)(1959) *Psychology: A study of Science 2*. New York. McGraw-Hill.

Hochberg, J.E. (1957) 'Effects of the gestalt revolution: the Cornell symposium on perception'. *Psychological Review*. 64, 2 73-84.

Hoffmeyer-Zlotnik, J. (1976) *Gastarbeiter in Sanierungsgebiet*. Hamburg.

Hollingshead, A.B. & Redlich, F.C. (1958) *Social class and mentalillness*. New York. Wiley.

Homans, G. (1950) *The Human Group*. New York. Harcourt, Braco & World.

Hopkinson, R.G. (1957) *Evaluation of glare*. Ill. Eng.

Hungerland, H. (1953) 'An anlysis of some determinants in the perception of works of art'. *The Journal of Aesthetics and Art criticism*. 54.

Johansson, T. (1952) 'Farglara'. In: Maleri (1952) *Natur och Kultur*. Hantverkets bok. 235-324.

Jones, E. (1950) 'Early development of female sexuality'. In: Papers on Psychoanalysis (1950). London. Baillères.

Kafesçioglu, R. (1955) *Kuzey-Bati Anadolu Ahsap Ev ve Yapilari*. Istanbul. I.T.Ü. Mim. Fak.

Kelly, G. (1955) *The psychology of personal constructs*. New York. Norton.

Kelly, G.A. (1955) *The psychology of personal constructs*. New York. W.W. Norton & Co.

Keuthe, J.L. () 'Social Schemas'. *Journal of Abnormal and Social Psychology*. 64: 31-8-59.

Klein, M. (1959) *La psychanalyse des enfants*. Paris. PUF.

Klose, H. (1957) *Fünfzig. Jahre Staatilcher Naturschutz*. Giessen. Ein Rückblick auf den Weg der deutschen Naturschu.

Kuper, L. (1953) *Living in towns*. Cresset Press.

Kuper, L. (1953) 'Blueprint for living together'. In: L.Kuper (ed.)(1953) *Living in towns*. London. Cresset Press.

Lacan, J. (1949) *Le stade du miroir comme formateur de la fonction du Je. Revue Française de Psychanalyse*.

Lacey, J.I. (1959) 'Psychological approaches to the evaluation of psychotherapeutic process and outcome'. *Research in psychotherapy*. Washington, D.C.. American Psychological Associationvol. 1.

Lagache, D. (1955) *Le modèle psychanalytique de la personnalité*. Paris. PUF.

Lagache, D. (1957) 'Fascination de la conscience par le moi, la psychanalyse et la structure de la personnalité'. *La Psychanalyse*. III, 1957 et IV, 1958.

Laing, R.D. (1959) *The divided self*. London. Tavistock 1959/ Penguin Books 1965.

Langer, S.K. (1942) *Philosophy in a new key, a study in the symbolism of reason, rite, and art*. New York. The New American Library of World Literature.

Langer, S.K. (1953) 'Philosophy in a new key'. *Feeling and Form, a Theory of Art*. London. Routledge & Kegan Paul Ltd.

Lantz, H.R. (1953) 'Population density and psychiatric diagnosis'. *Sociology and Social Research*. 37, 322-327.

Larson, C. (1949) 'School size as a factor in the adjusment of high school seniors'. State College of Washington. Bull. n. 511 Youth Series n. 6.

Lee, T.R. (1957) 'On the relation between the school journey and social and emotional adjustment in rural infant children'. *British Journal of Educational Psychology*. vol.27, pp. 101-14.

Levy-Strauss, C. (1958) *Antropologie structurale*. Paris. Librairie Plon.

Lewin, K. (1951) *Field theory in social science*. New York. Harper & Row.

Lewis, H.E. & Masterton, J.P. (1957) *Sleep and walkefulness in the Arctic*. Lancet.

Lindblom, C.M. () 'The science of "mudding through"'. *Public Administration Review*. 19, 79-99.

Lockwood, D. (1958) *The black coated worker*. George Allen and Unwin Ltd.

Long, Norton, E. (1958) 'The local community as an ecology of games'. *American Journal of Sociology*. vol.64.

Loring, W. (1956) 'Housing characteristics and social disorganisation'. *Social Problems 3*.

Lynch, K. (1959) 'A walk around the block'. *Landscape*. 8, pp. 24-34.

Lynch, K. & Rodwin, L. (1958) 'A theory of urban form'. *Journal of the American Inst. of Planners*. vol.24, n. 4

Mackworth, N.H. (1950) *Researches on the measurement of human performance*. London. Meds. Res. Council, Spec. Report.

Magde, C. (1950) 'Private and public spaces'. *Human Relations 3*. 187-199.

Malraux, A. (1957) *La Métamorphose des dieux*. Lausanne. La guilde du livre.

Mancipoz, F. (1948) 'La lutte pour les marais de Bourgoin'. *Evocations*. Nov. Dec.

Maslow, A, H. (1954) *Motivation and personality*. New York. Harper & Row.

Maslow, A.H. & Mintz, N.L. (1956) 'Effects of esthetic surroundings'. *Journal of Psychology*. 41, 2 247-254.

Mercer, B.E. (1956) *The American Community*. New York. Random House.

Merton, R. (1948) 'The social psychology of housing'. In: W.Dennis (ed.)(1948) *Current Trends in Social Psychology*. Pittsburgh. University of Pittsburgh Press.

Merton, R.K. (1957) 'The sociology of knowledge'. In: Merton(1957) *Social Theory and Social Structure*. Glencoe.

Mintz, N.L. (1956) 'Effects of aesthetic surroundings 2'. *Journal of Psychology*. 41, 459-66.

Mitrani, N. (1958) 'Attitudes et symboles technobureaucratiques'. *Cahiers Internationaux de Sociologie*. vol.XXIV.

Moon, P. & Spencer, D.E. (1951) 'Modelling with light'. *Journal of the Franklin Institute*. 251, 453.

Moreno, J.L. (1951) *Sociometry, experimental method and the science of sociology*. New York. Beacon House.

Morris, C. (1946) *Signs, lamguage and behavior*. Prentice Hall.

Morris, T. (1957) *The criminal area*. London. Routledge.

Neutra, R. (1954) *Survival through design*. Oxford University Press.

Ogden, C.K. & Riichards, I.A. (1949) *The meaning of meaning, a study of the influence of language upon though and the science of symbolism*. London. Routledge & Kegan Paul Ltd.

Oran, S. (1954) *Orta Anadolu Köylerinde Bir Aile Tarin Isletmesi Binalari*. Istanbul. I.T.Ü. Mim. Fak.

Orwell, G. (1949) *Nineteen eighy-four*. London. Secker

Osgood, C.E. (1958) *Method and theory in experimental psychology*. New York. Oxford University Press.

Osgood, C.E., Suci & Tannenbaum. (1957) *The measurement of meaning*. Urbana. University of Illinois Press.

Osmond, E. (1974) *Understanding understanding*. New York. Harper & Row.

Osmond, H. (1957) 'Functions as the basis of psychiatric ward design'. *Mental Hospitals*. 8: 23-30.

Parsons, T. (1951) *The Social System*. London. Routledge & Kegan Paul/ New York: Free Press.

Paulsson, G. & Paulsson, N. (1956) *Tingens bruk och prägel*. Stockholm. Kooperativa förbundets bokförlag.

Pepler, R.D. (1958) 'Warmth and performance; an investigation in the tropics'. *Ergonomics*. vol.2 pp.63-88.

Piaget, J. & Inhelder, B. (1948) *La représentation de l'espace chez l'enfant*. Paris. PUF.

Poincaré, H. (1952) *Science and hypothesis*. London. Dover.

Popper, K. (1959) *Logic and scientific discovery*. London. Hutchinson.

Radcliffe, A.R. & Fordre, D. (1950) *African system of kinship and marriage*.

Rauda. (1957) *Lebendige Städtebauliche Raumbildung*.

Riesman, D. (1950) The lonely crowd. Yale University Press.

Rossi, P. (1955) *Why families move*. Glencoe, IL. Free Press.

Roy, K. (1950) 'Parents attitudes towards their children'. *Journal of Home Economy*. 42.

Sandstrom, C.I. (1951) Orintation in the present space. Stockholm. Almqvist & Wiksell.

Sassure, F. (1916) 'Cour de linguistique générale'.In: *Cours de linguistique générale*. Paris. Editions de Payot (1916, 1949).

Schnore, M.M. (1959) 'Individuals patterns of psychological activity as a function of task differences and degree of arousal'. *Journal of Experimental Psychology*. vol.58 n.2 pp. 117-28.

Schutz, W, C. (1958) *FIRO: A three dimensional theory of interpersonal behaviour*. New York. Holt, Rinehart and Winston.

Sechehaye, M. (1951) *Autobiography of a schizophrenic girl*. New York.

Seyle, H. (1956) *The stress of life*. New York. McGraw-Hill.

Shannon, C. & Weaver, W. (1949) *The mathematical theory of communication*. Urbana. University of Illinois Press.

Shils, E. (1966) 'Privacy: its constitution and vicissitudes'. *Law and Contemporany Problems*. 31: 281-306.

Siegel, S. (1956) *Non-parametric statistics for the behavioral sciences*. New York. McGraw-Hill.

Simmel, G. (1950) The sociology of Georg Simmel. New York. Free Press.

Simmel, G. (1955) *Conflict and the web of group afiliations*. New York. Free Press.

Since Gray, P. (1947) *Some factors affecting the design of small dwellings*. London. Ministry of

Works/The Social Survey.

Sommer, R. (1959) 'Studies in personal space'. *Sociometry 22.* n.3(III) 247-260.

Sommer, R. & Ross, H. (1958) 'Social interaction on a geriatric ward'. *International Journal of Social Psychiatry.* 4, 128-133.

Spinley, B.M. (1953) *The deprived and the privileged: Personality development in English society.* London. Routledge & P. Kegan.

Spitz, R. (1953) *La première année de la vie de l'enfant, génèse des premières relations objectales.* Paris. PUF.

Spitz, R. (1958) 'On the genesis of superego components'. *Psychoanalytic study of the child.* XIII.

Sprott, W.J.H. (1958) *Human Groups.* Pelican Books.

Strauss, A. (1956) *The social psychology of G.H.Mead.* Chicago. University of Chicago Press.

Teichner, W.H. & Wehrkamp, P.F. (1954) 'Visual-motor performance as a function of short duration ambient temperature'. *Journal of Experimental Psychology.* vol.47, pp.447-50.

Tomsu, L. (1954) *Bursa Evleri.* Istanbul. I.T.Ü. Mim. Fak.

Torgerson, W.S. (1958) *Theory and methods of scaling.* New York. Wiley.

Townsend, K.C. (1953) *Imtroduction to experimental method.* New York. McGraw-Hill.

Tyrwhitt, J. (1955) 'The moving eye'. *Explorations Forum.* University of Toronto. pp.115-119.

Unita Residentiale Al Dm. (1957) *Casabelle Continua 215.*

Vinacke, W.E. (1952) *The Psychology of thinking.* New York. McGraw-Hill.

Waldram, J.M. (1954) 'Studies in interior lighting'. *Transactions of the illuminating engineering society.* London. 19, 95.

Walker, H. & Lev, J. (1953) *Satatistical interference.* New York. Henry Holt & Co.

Wallon, H. (1949) *Les origenes des caractère chez l'enfant.* Paris. PUF.

Webb, C.G. (1959) 'An analysis of some observations of thermal comfort in an equatorial climate'. *Brit. J. Ind. Med.* vol.16, n.4, pp.297-310.

Werner, H. (1948) Comparative psychology of mental development. N.Y. International Universities Press.

Whorf, B.L. (1956) *Language thought and reality.* M.I.T. Press.

Whyte, W.H. (1956) *The organization man.* New York. Doubleday.

Winch, P. (1958) *The idea of a social science.* London. Routledge and Kegan Paul.

Wittgenstein, L. (1953) *Philosophical investigations.* Oxford. Blackwell.

Young, J.Z. (1951) *Doubt and certaining in science.* Oxford. O.U.P.

Young, M. & Willmott, P. (1957) *Family and kinship in East London.* Harmondsworth. Penguin Books.

Zevi, B. (1957) *Architecture as space.* New York. Horizon Press.

BIBLIOGRAPHICAL REFERENCES

Abdelhadi, A. (1984) 'Paterns of life style a housing project: Cairo 1983' In: Krampen (1984).

Abderhalden. (ed.) (1924) *Handbuch der biologischen arbeitsmethoden*. Berlin: Urban & Schartzenberg

Abel, C. (1979) 'Constructing the Architectural Others' In: Canter (Ed) (1979).

Acking, Carl-Axel. (1973a) 'Closing of the Conference' In: Küller (1973).

Acking, Carl-Axel. (1973b) 'Opening of the Conference' In: Küller (1973).

Acking, C.A. (1974) 'Opening of the conference' In: Küller. (1974).

Acking, C.A. (1976) *Hur miljön upplevs vid nedsatt synförmaga. (Rapport R31.)*. Stockholm. The Swedish Council for Building Research.

Acking, C.A., & Küller, R. (1970) 'Perception of the human environment' In: Honikman(1970).

Acking, C.A. & Küller,R. (1972) 'The perception of an interior as a function of its color'. *Ergonomics*. 15, 645-54.

Acking,C.A. & Sorte,G.J.(1972) 'Metoder för presentation av planerad miljö'. Unpublished, Lund Institute of Technology.

Acking, C.A. & Sorte, G.J. (1973) 'How do we verbalize what we see?'. *Landscape architecture*. 64, 470-5.

Adorno, T. W. et al (1950) *The Authoritarian Personality*. Harper & Row.

Albrechts, L. & Lombaerde, P. (1982) 'The use of differentiated habitat as an alternative for the zoning principle: applied to an acono' In: Pol, Muntañola, Morales (Ed. 1984).

Alcaiz,M. (1983) 'Sondeo de opinión sobre el conocimiento de las Juntas de Distrito' Valencia. Ayuntamiento de Valencia.

Alcober, C., Casas, E., Cruz, J.L., Lamadrid, C. & Vidal, B. (1992). 'Basura Limpia-Basura Sucia: Recogida de Residuos Sólidos Urbanos'. In: Corraliza (Ed. 1993).

Alexander, R. S. (1979) 'Environmental Modifiers: Temporary and Manipulable Environments for Experiment' In: Surrey(1979).

Allott, K. J., Ladd, F. C. & Hall, G. (1976) 'Premier mariage et premier logement' In: Korosec (Ed) (1978).

Allport, F. H. (1955) *Theories of Perception and the Concept of Structure*. Wiley.

Althabe,G. (1977) 'Legnotidien en procès'. *Dialectiques*.

Altman,I. (1981) 'Refletions on Environmental Psychology: 1981'. *Human Environments*. Irnvine, Cal.2, (1): 5-11.

Alzina,J. & Sureda, J. (1980) *La nostra història. Elements didàctics per l'estudi de la Història a partir del medi*. I.C.E. Universitat de Palma.

Ambrose, I., & Canter, D. (1979) 'The Design of total Institutions: Organisational Objectives as Evaluation Criteria' In: Canter (Ed) (1979).

Ambrose, I. & Ostergaard, M. & Huiid, R. (1989) 'Evaluation of Housing in Use 'Blangstedgard''. *Danish Building Research Institute. Horsholm*

Ambrose, I. (1990) 'Users' evaluations danish experimental housing'. The *Danish Building Research Institute*. Denmark. Paper presented to the conference Les Entretiens de l'Habitat, 3,4, & 5 octobre. Lille, France.

Ambrose, I. & Christiansen, U. Ed. (1991) 'Management and Implementation of Ecological Measures in Human Settlements'. *Danish Building Research Institute*.

Amenos, J.M. & Tharrats, Ll. (1985) 'Dinámica eto-espacial en función de la variable ambiental' In: *Cuadernos de Psicología*. Universidad Autónoma de Barcelona.

American Public Health Association (1948) *Planning the Neighborhood*. Chicago. Public Administration Service.

Amerigo, M. & Aragonés J.I. (1989). 'Satisfacción Residencial en un Barrio remodelado: Predictores físicos y sociales'. *Revista de Psicología Social.*, 3, pp. 61-70.

Amerigo, M. (1990). *Satisfacción Residencial. Una Aproximación Psicosocial a los Estudios de Calidad de Vida*. Madrid. Universidad Complutense.

Amerigo, M. (1992). 'A Model of Residential Satisfaction'.In XII IAPS.

Andel, J.V. (1984) 'Effects of the redevolopment of an elementary school-yard' In: Krampen (1984).

Anderson, J. R. & Weidemann, S. (1979) 'An Instrument to Measure Residents' Perception of Residential Quality' In: Surrey(1979).

Anderson, K. Ladd, G. and Smith H. (1954) 'A study of 2.500 Kansas High School Braduates'. *Kansas Studies Education*. Univ. of Kansas. Vol. 4.

Ando, Y. (1984) 'Efects of intense noise during human fetal life' In: Krampen (1984).

Angell,J.R. (1929) 'Discurs en el IXTH Internacional Congres of Psychology'.In: Montoro. (1982).

Anguera, M.T. (1979) 'Observaciones de la conducta espacial'. In proceeding IV Congr. Nacional de Psicol. de Pamplona.

Anguera, M.T. (1980a) *Desarrollo de Estrategias de Observación y Mapeo Conductual como técnicas básicas*.

Anguera, M.T. (1980b) 'The episode of behavior as molar observational unit behavior group'. Easter, Englandpaper Meeting of Experim. Anal. of Behav. Group.

Anguera,M.T. (1980c) 'Towards an approximation between Ecological Psycho. & Experimental Psycho.Incidence of the method'. Leipzig. Paper XXII International Congres of Psychology.

Anguera, M. T. (1982a) 'Evaluación del modelo secuencial de apiñamiento mediante la técnica de simulación' In: Pol, Muntañola, Morales (Ed. 1984).

Anguera, M.T. (1982b) 'Fiabilidad de la Codificación en estudios naturales'. Santiago de Compostela. Comunicación en VII Congreso Internac. de Psicol.

Anguera, M.T. (1983) 'Adaptacióm del análisis de panel a la evaluación de contextos a través de su represent. espacial' In: Fernandez Ballesteros. *Evaluación de Contextos*. Publicaciones Universidad de Murcia.

Anguera, M.T. & Blanco,A. (1982) 'Del Mapeo Conductual al Mapeo Cognitivo'. Barcelona. In VII IAPS conference.

Anguera, M.T., Blanco, A. & Gabucio, F. *Mapes Cognitius: Sistematització Conceptual i possibilitats tècniques per una Viabilitat Metodològica*. Universitat de Barcelona.

Anthony, Ph. D. & Kathryn H. (1984a) International house: 'Home away from home?' In: Krampen (1984).

Anthony, Ph. D. & Kathryn, H. (1984b) 'The Role the of Home Environment Family Conflict: Therapists' viewpoints' In: Krampen (1984).

Appleyard, D. (1973) 'Professional Priorities for Environmental Psychology' In: Küller(1973).

Appleyard, D. (1979) 'The environment as a social symbol'. (P.5) In: Canter (1979).

Aragonés, J.I. (1984) 'Mapa Cognitivo de Madrid'. In III Encuentro Nacional Psicología Social. 335-339.

Aragonés, J.I. (1985) *Mapas cognitivos de ambientes urbanos: Un estudio empírico sobre Madrid*. Madrid. Tesis doctoral(1983). Publicaciones de la Universidad Complutense.

Aragonés, J.I. & Arredondo, J.M. (1986) Structure of Urban Cognitive Maps. *Journal of Environmental Psychology*.

Aragonés, J.I. et alt. (1983) 'Mapas Cognitivos y Representación del Espacio'. *Estudios de Psicología*. 14/15, 35-38.

Aragonés, J.I. & Corraliza, J.A. (1986). *Conducta y Ambiente*. Proceedings of I Jornadas de Psicología Ambiental. Madrid. Comunidad de Madrid

Argulloll, R. & Trias, E. (1992). *El cansancio de Occidente*. Barcelona, Destino.

Aranguren, J.L. (1984) 'Filosofía, Ecología, Ecologismo'. *Información Ambiental*. n.1.

Arnheim, R. (1954a) *Art and Visual Perception*. University of California Press.

Arnheim, R. (1954b) *Art and Visual Perception*. London: Faber & Faber.

Arnheim, R. (1956) *Art and Visual Perception*. London:Faber.

Ashby, W. R. (1956) *An introduction to Cybernetics*. London: Chapman & Hall.

Ashworth, W. (1954) *The Genesis of Modern British Town Planning. A Study*. London: Routlogge & Kegan Paul.

Atasoy, A. & Erkman, U. (1976) 'L'appropriation de l'espace comme une donnée du design architectural' In: Korosec(1976).

Attneave, F. (1950) 'Dimensions of similarity'. *American Journal of Psychology*. vol. 63, pp. 515-556.

Attneave, F. (1957) 'Transfer of Experience with a Class Schems to Ident.'. *Journal of Experimental Psychology*. vol 54.

Auburn, T. C., Jones, D. M. & Chapman, A. J. (1979) 'Social Aspects of Working Noisy Environments' In: Surrey(1979).

Augoyard, J.F. (1979) *Pas à pas, essai sur le cheminement quotidien en milieu urbain*. Paris. Sevil.

Axia, V., Baroni, M. R. and Mainardi Peron, E. (1984) 'How Children Remember External and Internal Familiar Places' In: Krampen (1984).

Axia, V., Baroni, M. R. and Mainardi Peron, E (1985) 'Ricostruzioni Verbali e Memoria Infantile di Ambienti'. In: W. Fornasa e M. Montanini Manfrendi, *Memoria e sviluppo mentale*, Franco Angeli, Milano. pp.. 150-157.

Axia. G.,Baroni. M. R., e Mainardi Peron. E. (1988) 'Representation of familiar places in children and adults: Verbal reports as a method for studying environmental knowledge'. *Journal of Environmental Psychology*. 8, pp. 123-139.

Axia, G., Baroni. M. R. e Mainardi Peron. E., (1990) 'Cognitive Assessment of Classrooms in Childhood, Early and Late Adulthood, *Children's Environments Quarterly*. 7, pp. 17-25.

Axia, G., Mainardi Peron, E. e Baroni M. R.,(1991) *Environmental Assessment Across the life Span*. In: T. Gärling e G. W. Evans (Eds.), *Environment, Cognition and Action. An Integrated Approach*. New York, Oxford University Press. pp. 221.

Azgaldov, G.G. (1978) *Quantitative measures and problems of architectural beauty*. Moscow. Stroiizdat.

Bachelard, G. (1950) *La formation de l'esprit scientifique*. Paris:Vrin.

Bachelard, G. (1957) *La poétique de l'espace*. Paris: P.U.F.

Baggs, S.A. (1979) 'Attitudes Towards the Use of Underground Environments with Specific Reference to Housing' In: Surrey(1979).

Bagley, Ch. (1974) 'The Built Environment as an Influence on Personality and Social Behaviour: A Spatial Study' In: Canter & Lee (Eds.). (1974).

Bagnara, S. (Ed.) (1976) 'La psicologia ambientale, problemi e prospettive'. *Giornale italiano di psicologia*. 1, 49-73.

Bagnara, S. & Misiti, R. (Eds.) (1976) *Psicologia ambientale*. Bologna. Il Mulino.

Baker, A., Davies, R. L. and Sivadon P. (1959) 'Psychiatric Services and Architecture'. *Public Health Papers*. Geneva:World Health Organization.No. 1.

Baker, Alan M. (1979) 'A Simulation of Urban Environmental Learning' In: Surrey(1979).

Bakshtein, J.M. (1981) 'Some social-psychological problems organizing cultural services the city'. In: T.Niit, et al. (Eds.) (1981) *People and environments: Psychological problems.* Tallinn. EOOP233-5.

Balboni, P. et al. (1978) *La percezione dell'ambiente: L'esperimento di Venezia.* Venezia. Ciedart.

Baniassad, E. (1970) 'The need to incorporate the social sciences into the teaching of architectural design is strongly' In: Honikman (1970).

Barbey, G. (1976) 'L'appropriation des espaces du logement: Tentative de cadrage théorique'. In: Korosec. (1976).

Barbey, G. (1984a) 'Phénomene de tertiarisation accélérée du centre historique des villes.' In: Krampen 1984.

Barbey, G. (1984b) 'Le logement de masse: implantation dans l'espace et contrôle social'. In: Krampen 1984.

Barbey, G. (1984c) 'Effets de sublimation dans l'architecture domestique'. In: Krampen 1984.

Barbey, G. & Gelber, C. (1973) *The relationship between the built envir.and human behaviour: A surrey and anal. of the exist.lit.* Lansanne. Federal Institute of Technology.

Barilleau, E. E. et Lombardo, J. D. (1976) 'Appropiation de l'espace dans des édifices multifamiliaux au cours de la période des vacances sco' In: Korosec(1976).

Barker & Wrigth. (1955) *Midwest and its children. The psychological ecology of american town.* New York. Harper and Row.

Barker, R. (1953) 'On the Nature of the Environment". *Journal of Social Issues.*

Barker, R. (1968) *Ecological psychology: Concepts and methods for studying environment of human behaviour.* Standford. Standford University Press.

Baroni, M. R., Cornoldi, C., De Beni, R., D'Urso, V. Mainardi Peron, E., Palomba, D. e Stegano, L.(1989) *Emozioni in celluloide.* Raffaello Cortina Editore, Milano, pp. 199.

Baroni, M. R., Remo, J. e Mainardi, Peron, E. (1980) 'Memory for natural settings: Role of diffusse and focused attention'. *Perceptual and Motor Skills.* 51, pp. 883-889.

Baroni, M. R.,e Mainardi Peron. E. (1991) 'Conveying environmental knowledge through language: Methodological issues. *Attidel International workshop on 'Home environment: Physical space and psychological processes'.* Cortona. (in corso di publicazione).

Bartlett, F.C. (1932) *Remembering.* Cambridge. University Press.

Bartlett, F. C. (1958) *Thinking.* London:Unwin.

Bateman, M.D., Burtenshaw & Duffett, A. (1974) 'Environmental Perception and Migration: A Study of Perception of Residential Areas South Hamps' In: Canter & Lee (eds.)(1974).

Bayazit, N., Yonder, A., Ozsoy, A. (1976) 'Trois niveaux de comportement reliés au besoin d'intimité dans l'appropriation des espaces d'habit.' In: Korosec(1976).

Beaux, D. (1976) 'Essai de description globale de l'environnement vécu -milieu physique, activités et. ....' In: Korosec. (1976).

Beaux, D. (1979a) 'Perceptual Coherence Along Public Bahavioural Circuits' In: Surrey(1979).

Beaux, D. (1979b) 'Cohérence perceptuelle au cours du circuit comportamental' In: Simon (ed.)(1979).

Bechini, A. (1986) 'Percepción del barrio de la Sagrera (Barcelona) y su valoración en función de las variables socioculturales. In Proceeding I Jornadas de Psicología Ambiental.

Becker, H. & Collier, J-M. (1979) 'Intimité et espaces de transition'. In: Simon. (1979).

Bedford, T. (1958) 'Research on heating and ventilation relation to human comfort'. *Heat Pip. Air Condit.* pp. 127-34.

Beer, Ingeborg, West-Krampen 1984 'Grosse Ideale für grosse siedlunger. Arkitektur und menschliches handeln am beispiel des siedlung' In: Berlin(1984).

Belyayeva. (1977) *Architectural-spatial environment of the city as the object of visual perception.* Moscow. Stroiizdat.

Bennet, G. K., Seashore, H. G., and Wesman, A. G. (1959) *Differential Aptitude Tests.* New York:Psychological Corporation.

Berglund, B. (1974) 'Quantitative and qualitative analysis of industrial odors with human observers'.*Annals of the New York Academy of Sciences.* 237, 35-51.

Berglund, B. (1976) 'Scaling loudness, noisiness, and annoyance of community noises'. *Journal of Acoustic Society of America*. 60, 1119-25.

Berglund, B. (1977) 'Quantitative approaches environmental studies'. *International Journal of Psychology*. 12, 111-23.

Berglund, B. et. al. (1971) 'On the principle of odor interaction'. *Acta psychologica*. 35, 255-68.

Berglund, B. et. al. (1973a) 'Multidimensional analysis of twenty-one odors'. *Scandinavian Journal of Psychology*. 14, 131-7.

Berglund, B. et. al. (1973b) 'A quantitative principle of perceived intensity summation odor mixtures'. *Journal of Experimental Psychology*. 100, 29-38.

Berglund, B. et. al. (1975) 'Scaling loudness, noisiness, and annoyance of aircraft noise'. *Journal of Acoustical Society of America*. 57, 930-4.

Berglund, B. et. al. (1976) 'Psychological processing of odor mixtures'. *Psychological Review*. 83, 432-41.

Berglund, B. et. al. (1981) 'Loudness (or annoyance) summation of combined noises'. *Journal of Acoustical Society of America*. 70, 1628-34.

Berglund, B. et. al. (1982a) 'A longitudinal study of air contaminants a newly built preschool'. *Environment International*.

Berglund, B. et. al. (1982b) 'Olfactory and chemical charact. of indoor air: Towards a psychophysical model for air quality'. *Environment International*.

Berglund, B. et.al. (1971) 'Individual psychophysical functions for 28 odorants'.*Perception and Psychophysics*. 9, 379-84.

Berglund, U. (1981) 'Loudness and annoyance from community noises'. In: A. Schick (ed.)(1981) *Akustik zwischen physik und psychologie*.Stuttgart. Klettcotta. 27-32.

Krampen 1984 *Abstracts of 8 IAPS conference. Environment and human action*. Berlin.

Berlyne, D.E. (1971) *Aesthetic and psychobiology*. N.Y. Appleton Century Crofts.

Bernal (1964) *La ciencia de la ciencia*. (1968) Méjico. Grijalbo. v.o. Souvenir Press.

Bernáldez, F. G., Ruiz, J. P. & Ruiz, M. (1984) 'Landscape Perception and Appraisal: Ethics, Aesthetics and utility' In: Berlin(1984).

Bernard, Y. (1979) 'Aesthetic Evaluation of Landscape' In: Canter (Ed) (1979).

Bernard, Y. (1984) 'Structuration, organisation, animation de l'espace domestique'. In: Krampen 1984.

Bernard, Y. (1991a) 'Evolution of lifestyle and dwelling practices in France'. *The Journal of Architectural and Planning Research*. Vol. 8, n°3.

Bernard, Y. (1991b) 'Rapport D'activite'. Unité de Recherche Associée au C.N.R.S. París.

Bernard, Y. (1991c) 'Environmental Psychology in France'. Journal of Environmental Psychology. 11, 277-285.

Bernard, Y. (1992a) 'North American and European Research on Fear of Crime'.*Applied Psychology: and International Review. 41(1), 65-75.*

Bernard, Y (1992b) 'Les sciences sociales et le sentiment d'insécurite'. In Y.Bernard et M.Segaud *La ville inquiète* Editions de l'Espace Européen, Paris

Bernard, Y., & Guerpillon, P. (1982) 'Perception du paysage. Dimensions Subjectives de la ...' In: Pol, Muntañola, Morales (Ed. 1984).

Bernard, Y. & Lebeau, M. & Giuliani, M. & Bonnes, M. (1987) 'Practiques de l'habitat et monde sociaux'. *Psychologie Française*. n° 32, 1-2.

Bernard, M. & Ledoux, Y. & Morlie, J.L. & Noel, F. (1979) "Réhabilitation résidentielle et recherche appliquée". In: Simon. (1979).

Bernard, Y, Levy-Leboyer, C. (1987) 'La Psychologie de l'Envioronnement en France'. In *Psychologie en France* Julliet 1987, Tome 32 - 1-2 pp. 7-16

Bernard, Y. & Segaud, M. (1992) 'La ville inquiète'. Editions de l'espace Européen. París.

Berndt & Berndt. (1965) *Alboriginal man Australia*. Sidney. Angus & Robertson.

Bernstein, B. (1958) 'Some sociological determinants of perception'. *British Journal of Sociology*. vol. IX, pp. 159-174.

Best (1969) 'Directions-finding large buildings' In: Canter (Ed.)(1969).

Best, S., Douglas & Kelluer, D. (1991). *Postmodern Theory*. NY. Guilford Press.

Bianchi, E. (1980a) 'La percezione dell'ambiente: Una rassegna geografica'. In: R. Geipel & M.C. Bianchi (Eds.) (1980), pp. 35-50.

Bianchi, E. (1980b) 'Da Lowenthal a Downs a Frémont: aspetti della geografia della percezione'. *Rivista Geografica Italiana*. 87, 75-87.

Bianchi, E. (1980c) 'Spazi soggettivi geografia: esempi tratti dalla cartografia e dai resoconti di viaggio'. In: F.Perussia (ed.) (1980). 143-87.

Bianchi, E. (1981) 'Teratologia e geografia. L'homo monstruosus autori dell'antichita classica'. *Acme*. 34, 227-49.

Bianchi, E. (1982) 'La rappresentatione cognitiva dell'ambiente come problema geografico'. *Ricerche di Paicologia*. 22-23, 269-89.

Bianchi, E. & Perusia, F. (1978) *Il centro di Milano: percezione e realità*. Milano. Unicopli.

Bianchi, E. & Perusia, F. (1981) 'Risultanze sulla costruzione della città come immagine: casi italiani'. *Sociologia urbana e rurale*. 3, 101-7.

Bianchi, E. & Perussia, F. 'Immagini dell'inquinamento lacustre differentimento lacustre differenti contesti ambient.'. *Memorie della Societa Geografica Italiana*.

Bishop Reid, A. (1979) 'Temporal Aspects of Place' In: Surrey(1979).

Bishop, J. & Foulsham, J. (1973) 'Children's images of Harwich' In: Küller(1973).

Blackshaw, M. Et al. (1959) *Utilisation de l'espace dans les logements*. Genève. Nations Unies.

Blake, R. et al. (1956) 'Housing Architecture and Social Interaction'. *Sociometry*. 19 No. vol.II, pp. 133-139.

Blanch,J.M. (1983) *Psicologías sociales. Aproximación histórica*. Barcelona Ed. Hora.

Blanco, A. (1983) *Análisis Cuantitativo de la Conducta en sus contextos naturales*. Tesis Doctoral. Universidad de Barcelona.

Blauw, P.W. (1984) 'Social contacts an urban and suburban environment'. In: Krampen 1984.

Bloch, R. (1958) *les origines de rome*. Paris:P.U.F.

Bloch, V. (1978) 'Discurs en el XXI Internacional Congres of Psychology'.In: Montoro. (1982).Paris.

Boix & Ferrando, P. (1981) 'Mortalitat urbana: un estudi epidemiològic descriptiu de la mortalitat a la Ciutat de València'. València. Universitat de València. Doctoral Disertation.

Bonaiuto, M. & Bonaiuto, P. (1984) 'Functional choice of Dress and the Influence of Clothing on Self-Image' In: Krampen (1984).

Bonaiuto, P., Romano, M. & Bonaiuto, F. (1984) 'Phenomena of reduction or increase perceptual irregularity of architectural structures and environment' In: Berlin(1984).

Bonnes-Dobrowolny, M. (1980) 'Profilo critico dell'emergente psicologia ambientale'. *Richerche di Psicologia*. 12, 107-35.

Bonnes-Dobrowolny, M. (1991) 'Urban ecology applied to the city of Rome'. *Progress Report*. n°4.Bonnes-Dobrowolny, M., Secchiaroli, G. (1979a) 'The city as a 'Socio-Spatial Schema'' In: Surrey(1979).

Bonnes, M. (1992) Recorded personal interview, by Pol. Thessaloniki, July 1992.

Bonnes-Dobrowolny, M. & Secchiaroli, G. (1979b) 'Il centro di Milano: spazio e significato nella rapresentazione cognitiva del centro' AP. 2, 233-55.

Bonnes-Dobrowlny, M.& Secchiaroli G.(1984) 'The Cognitive Construction of the Urban Environment: a Multidimensional and Categorial Approach' In: Berlin(1984).

Bonnes-Dobrowolny, M. & Secchiaroli, G. (1982) 'Aspetti "sociospaziali" della rapresentazione cognitiva del centro cittadino'. *Ricerche di psicologia*. 22-23, 155-9.

Bonnes-Dobrowolny, M. e Secchiaroli, G. (1992) 'Psicologia ambientale'. *La Nuova Italia Scientifica*.

Book, A. (1981a) 'Maintenance of environmental orientation during locomotion'. In: *Umea Psychological Reports Supplement Series No.8*. Umea. University of Umea. (Doctoral dissertation).

Book, A. (1981b) 'Maintenance of orientation during locomotion unfamiliar environments'. *Journal of Experimental Psychology: Human Perception & Performance*. 7, 995-1006.

Borsi, F. (1983) *El poder i l'espai. L'escena del princep*. Diputacions de Barcelona, València, Madrid & Sevila.

Boselie, F. (1984) 'Designing for visual clarity and beauty'. In: Krampen 1984.

Boudon, Ph. (1979) 'Espace, lieux' In: Simon(1979).

Bourdieu. (1970) 'The berber house or the world reversed'. *Social Science Information*. 9, 2.

Bradford. (1948) *documentation*. London. Crosby Lockwood.

Brauer, R.L. (1974) 'The importance of individual differences buildings'. In: Canter and Lee. (1984).

Brennan, T. (1948) *Midland City: Wolverhampton Social and Industrial Survey*. London: Dobson Books.

Broadbent, G. & Ward, A. (eds.)(1969) *Design methods architecture*. London. Lund Humphries, Architectural Ass. paper 4.

Broadbent, O.E. (1955) 'Noise: Its effect on bahaviour'. *The royal society of health journal*. 75 (8) August.

Broadbent, O.E. (1958) 'Effect of noise on an "intellectual "task'. *Journal of the acoustical society of America*. 30 (9): 824-827, Sept.

Broadbent, O.E. (1964) *Perception and comunication*. Oxford. Pergamon Press.

Broadbent, O.E. & Gregory, M. (1967) 'The relationship between need achievement, neuroticism and introversion' (Abstract) *Bulletin of British Psychological Society*. 20 (67): 16A.

Broadbent, O.E. & Heron, A. (1962) 'The effects of a subsidiary task on performance involving immediate memory by young & old men'. *British Journal of Psychology*. 53 (2): 189-198.

Brodin, C. (1973) 'A study of preferencies for simulated outdoors environments with different intensities of feeling' In: Kuller (1973).

Bronfenbrenner, U. (1976) *Ökologische Sozialisationsforschung*. Stuttgar: Klett.

Bronfenbrenner, U. (1977) 'Toward an experimental ecology of human development'. In: *American Psychologist*, 32, pp. 513-531.

Brower, S.N. (1982) 'Informal social control of space residential areas'. In: Pol, Morales, Muntañola. (1984).

Brown, J. (1979) 'A Multidimensional Scaling Analysis of House Buying Behaviour' In: Canter (Ed) (1979).

Bruce, A. (1970) 'Housewife attitudes towards shops and shopping centres' In: Honikman (1970).

Brullet, M. (1982) 'Comunnicació a les jornades en motiu del 50 aniversari del Patronat Escolar de Barcelona'. Ajuntament de Barcelona.

Brundrett, G. W. (1979) 'A Behavioural Approach to Environmental Criteria' In: Canter (Ed) (1979).

Bruner, J. (1951) 'Personality Dynamics and the Process of Percelving' In: R. R. Blake & G. V. Ramsey. *Perception. An Approach to Personality*. New York: Ronald & Press.

Bruner, J. S. (1957a) 'On Perceptual Readines'. *Psychological Review*. vol. 64, pp. 123-152.

Bruner, J. S. (1957b) 'On Perceptual Readines'. *Psychological Review*. vol. 64. pp. 123-152.

Bruner, J.S. (1990). *Actos de Significado. Más allá de la Revolución Cognitiva*. Alianza Ed. Madrid

Bruner, J. S., Goodnow, J. J., and Austin, G. A. (1956) *A Study of Thinking*. NY. Wiley.

Brunswik, E. (1947) *Systematic and Representative Design of Psychological Experiments*. Berkeley:University of California Press.

Brunswik, E. (1956) *Perception and the pepresentative design of psychological experiments*. University California Press.

Brusa, C. (1978) *Geografia e percezione dell'ambiente. Varese vista dagli operatori dell'Ente Publico locale*. Torino. Giappichelli.

Brusa, C. (1979) 'L'immagine di una città secondo la cartografia turistica: l'esempio di Parma'. *Bolletino della Associazione Italiana di Cartografia*. 46, 23-9.

Brusa, C. (1980) 'La geografia della percezione quale strumento di educazione ambientale'. *Rivista Geografica Italiana*. 87, 46-60.

Bryan. (1968) 'The explosion published information myth or reality'. *Australian Library Journal*. 17, 389-401.

Bucsescu, D. & Stringer, P. (1979) 'Place and Two Case Studies Milton Keynes' In:

243

Surrey(1979).

Bunting, T. E. & Semple, T. McL. (1979) 'Measuring Dimensions of Environmental Personality Children' In: Surrey(1979).

Burisch, M. (1979) 'Can Housing Quality be Measured Objectively?' In: Canter (Ed) (1979). P. 17.

Bursill, A. E. (1958) 'Restriction of peripheral vision during exposure to hot and humid conditions'. *Quart. J. Exp. Psychol.* vol. 10, no.3, pp. 113-29.

Burton I Kebler (1960) 'The half-life of some scientific and technical literatures'. *Amer. Doc.* 11, 18-22.

Butler, G. (1959) *Introduction to Community Recreation.* New York: National Recreation Association.

Cadwallader, M. (1979) 'The Process of Neighborhood Choice' In: Canter (Ed) (1979).

Caid, M. & Cazalis, L. (1973) 'The therapeutic effects of an architectural activity a children home' In: Kuller (1973).

Cakin, S. (1979a) 'Qualitative appraisal of jointly used spaces' In: Simon (Ed.) (1979).

Cakin, S. (1979b) 'Strategies of Judgement Design Evaluation' In: Canter (Ed) (1979).

Cakir, A. (1973) 'A study of discomfort glare by floodlighting for color-TV' In: Kuller (Ed.) (1973).

Campbell, D. T. & Fiske, D. W. (1959) 'Convergent and Discriminant Validation by the Multitrait-Multimethod Matrix'. *Psychological Bulletin.* vol. 56.

Cano, G. (1982) 'La llum i els colors de l'escola: dades i continguts de les recerques tocant a llurs efectes' In: Pol, Morales, Muntañola(Ed)(1984).

Canter, D. (1969a) 'Should we treat building users as subjects or objects?' In: Canter (Ed.) (1969).

Canter, D. (1969b) 'Architectural Psychology. An Introduction' In: Canter (Ed.) (1969).

Canter, D. (1969c) *The psycological implications of office size.* Doctoral Thesi. University of Liverpool.

Canter, D. (Ed.) (1969) *Architectural Psychology.* Conference Dalandhui. London. Riba.

Canter, D. (1970) 'The place of Architectural Psychology' In: Honikman (Ed.) (1970).

Canter, D. (1972) 'Edificios y personas: Breve panorama de investigaciones realizadas' In: Llorens, T. (Ed.) (1972) *Hacia una psicologia de la arquitectura.* Barcelona. La Gaya-Ciencia.

Canter, D. (1973b) 'Evaluating Buildings: Emerging scales and the salience of building elements over constructs' In: Küller (1973).

Canter, D. (1975) 'Contribución a la toma de decisiones ambientales'. In: D. Canter y Stringer (Eds.) (1978), *Interacción Ambiental.* Madrid, IEAL.

Canter, D. (1976) 'Une procedure pour l'exploration de l'appropriation de l'espace' In: Korosec (Ed) (1978).

Canter, D. (1979a) 'The Psychology of Houses and Housing: Pitfalls and Potentials' In: Canter (Ed) (1979).

Canter, D. (1979b) 'A Mapping Sentence for People and Places' In: Canter (Ed) (1979).

Canter, D. (1979c) 'Welcome and Introduction' In: Canter (Ed) (1979).

Canter, D. (1979d) 'Y a-t-il des lois d'interaction environnementale?' In: Simon (Ed.) (1979).

Canter, D. (1980) *Fires and Human Behavior.* London, Wiley.

Canter, D. (Ed) (1979) *Abastracts International Conference on Environmental Psychology.* Guilford, University of Surrey

Canter, D. (1982) 'Social Action design: Nine questions to bridge the map' In: Pol, Muntañola, Morales (Eds.) (1984).

Canter, D. (1984) *Recorded personal Interview by Pol.* Guilford, Febrer 1984..

Canter, D. (1992a) 'Psicologia ambiental: Futbol, assassins, foc, facetes i professió. Entrevista a David Canter' (Interview done by E.Pol) In *Psicologia.Text i context.* n.5 Juny 1992. pp 26-29.

Canter, D. (1992b) Recorded personal interview, by Pol. Thessaloniki.Canter, D., Breaux, J. & Sime, J. (1979) 'Human Behavior Fires' In: Canter (Ed) (1979).

Canter,D., Comber, M. & Uzzell, D. (1989) *Football in its place*. London, Methuen.

Canter, D., Gilchrist, Miller & Roberts. (1974) 'An Empirical Study of the Focal Point the living Room' In: Canter & Lee (Eds.) (1974).

Canter, D. & Kyung Hoi Lee. (1974) 'A Non-Reactive Study of Room Usage Modern Japanese Apartments' In: Canter & Lee (Eds.) (1974).

Canter,D. & Lee,T. (eds.)(1974) *Psychology and the built environment (Proceedings of Surrey 1974)*. London. Architectural Press (1975).

Canter, D., Sánchez-Robles, J. C. & Watts, N. (1974) 'A Scale for the Cross-Cultural Evaluation of Houses' In: Canter & Lee (Ed.) (1979).

Canter, D., Stringer,P. et al. (Eds.)(1975) *Interacción ambiental*. Madrid. IEAL (1978).

Caparrós, A. (1980a) *Los paradigmas en psicologia*. Barcelona. Horsori.

Caparrós, A. (1980b) 'Problemas histográficos de la Historia de la psicologia'. *Revista de historia de la psicologia*. València. 3-4, 393-414.

Caparrós, A. (1982) 'Psicologia diferencial, ¿ciencia o tecnologia?'. *Estudios de psicologia*. 9, 16-23.

Carpintero. (1980) 'La psicologia, pasado, presente y futuro'. *Revista de historia de la psicologia*. 1, 33-55.

Carpintero. (1981a) 'La ciencia de la ciencia y la investigación psicológica en el mundo contemporaneo' In: Carpintero.&.Peiró (eds.)(1981).

Carpintero,H. (1981b) 'Aplicaciones de la metodologia bibliométrica: Una introducción' In: Carpintero. & Peiró (eds.)(1981).

Carpintero & Pascual & Peiro. (1977) 'La literatura científica en la psicología actual' *Análisis y modificación de conducta*.

Carpintero & Peiró. (1979a) *Estudio bibliométrico de la literatura periódica sobre psicología en lengua inglesa: Am.J. (1part)*. València. Mimeo.

Carpintero & Peiro. (1979b) *Estudio bibliométrico de la literatura periódica sobre psicologia en lengua inglesa:(2 part.)*. València. Mimeo.

Carpintero. & Peiro. (eds.) (1981) *Psicologia contemporánea*. València. Alfaplus.

Carreiras, M. (1986) 'Mapas cognitivos: Revisión crítica'. *Estudios de Psicología. 26*.

Carreiras, M. (1992) 'Mapas cognitivos y orientación espacial'.In: J. Mayor y J. L. Pinillos. Alhambra (ed.1992).

Carreiras, M. & Codina, B. (1990) 'Mapas Cognitivos en deficientes visuales y evidentes'. *Psicológica*. 11, 153-165

Carreiras, M. & Codina, B, & Escribano, J. L. & Rodríguez. P, & Sanchez, V. (1991) 'Sistema computerizado de medidas de la representación interna del espacio'.*Psicológica*. 12, 357-372.

Carreiras, M. & Codina, B. (1992) 'Spatial cognition of the blind and sighted: visual and amodal hypotheses'. *European Bulletin of Cognitive Psychology*. Vol. 12, n°1, pp. 51-78

Carreiras, M. & Gärling, T. (1990) 'Descrimination of cardinal compass directions'. *Acta Psicológica*. 73, pp. 3-11.

Carreiras, M. & de Vega, M.(1984) 'Mapas cognitivos: Influencia de la tipicidad semántica y de la densidad métrica en la asimetría de las distancias'. *Revista de Investigación Psicológica*. 2, 1.

Carstairs, M. B. & Collins, G. M. (1979) 'People and Their Homes' In: Surrey(1979).

Cason & Lubotsky. (1936) 'The influence and dependence of psychological journal on each other' In: Quoted by Peiro & Carpintero. (1981) *Psychological Bulletin*. 33, 95-103.

Cassirer, E. (1955) 'The Philosophy of Symbolic Forms'. Yale University Press. vol.3.

Cassirer, E. (1957) 'The Philosophy of Symbolic Forms'. *The Phenomenology of Knowledge*. New Haven:Yale University Press.vol. 3.

Castellet, J.M. & Molas, J. (1967) *Poesia Catalana del s. XX*. Barcelona. Edicions 62.

Castells, M. (1973) *Les luttes urbaines et le pouvoir politique*. Paris. Maspero.

Catling, S. (1979) 'Children's Maps of Place' In: Surrey(1979).

Cattell, R. B. (1950) *Personality: A Systematic, Theoretical and Factorial Study*. New York:NcGraw-Hill.

Cazalis, L (1979) 'Conflicts existants mis en évidence et les conflicts appraissants au cours de la contruction d'institutions' In: Simon (Ed.)(1979).

Chaloner, W. H. (1954) 'Robert Owen, Peter Drinkwater and the Early Factory System Manchester'. *Bulletin of the John Rylands Library*. vol. 37, pp. 78-102.

Chance, M. R. A. (1956) 'Social structure a colony of Macaca mulatta'. *Brit. J. Anim. Behav.* vol. 4, pp. 1-13.

Chapman, D. (1955) *The Home and Social Status*. London: Routledge & Kegan Paul.

Choay,F. (1965) *L'urbanisme, utopies et réalites*. Paris. Sevil.

Choay,F. (1972) *Semiologie et urbanisme: les seus de la ville*. Paris. Sevil.

Chombart de Lauwe, M. J. (1959) *Psychopathologie sociale de l'enfant inadapté*. Paris: Centre National de la Recherche Scientifique.

Chombart de Lauwe, M. J. (1976) 'L'appropriation de l'espace par les enfants: processus de socialisation' In: Korosec(1976).

Chombart de Lauwe, M.J. (1979) 'Espaces d'enfants'. *Futuribles*. 25.

Chombart de Lauwe, M.J. (1984) Recorded personal interview, by Pol. Paris. December 1984.

Chombart de Lauwe, M.J. et. al. (1976) *Enfant et jeu: les practiques des enfants durant leur temps libre en function des types d'envir....* Paris. Editions de CNRS.

Chombart de Lauwe, P. H. (1956) *La Vie quotidienne des families ouvrières*. Paris: CNRS.

Chombart De Lauwe, P.H. (1976) 'Appropriation de l'espace et changement social'. In: Korosec. (1976).

Chombart de Lauwe, P.H. et al. (1976) *Transformations du l'environnement, des aspirations et des valeurs*. Paris. Editions du CNRS

Chombart de Lauwe, P.H. (1982) *La fin des villes: mythes ou réalités*. Paris. Calmann-Lévy

Chombart de Lauwe, P.H. (1983) 'Oppression subversion and self-expression daily life'. *International Social Journal*.

Chombart de Lauwe, P.H. (1984) Recorded personal interview, by Pol. In: Paris. December 1984.

Chombart de Lauwe, P.H. et al. (1959) *Famille et habitation*. Paris. Publications du CNRS.

Chomsky, N. (1957) *Syntactic Structure*. Mouton Publisher.

Chrenko, F. A. (1953) *Probit analysis of subjective reaction to thermal stimuli*. Brit. J. Psychol.

Chrenko, F. A. (1964) 'Heated Ceilings and Comfort Journal of the Institution of Heating and Ventilation Engineers'. *Threshold Intensities of Thermal Radiation Evoking Sensation of Warmth Journal of Physiology*.

Churchman, A. (1979) 'Public participation as a means of dealing with conglicting views:acase study Israel' In: Simon(1979).

Churchman, A., Ginsberg, Y. (1979) 'High Rise Housing - Advantages and disadvantages for Residents' In: Surrey(1979).

Churchman, A. (Ed) (1986) *Environments in transition* Abstracts of 9th IAPS Conference. Haifa, Fac. of Arch. and Planning. Technion-Israel Institut of Technology

Clamp, P.E. (1984) 'Five accounts of perception - a guide to understanding the literature of landscape quality'. In: Pol, Morales, Muntañola. (1984).

Claparéde, E. (1929) 'Esquisse hiostorique des congres internationaux de psychologie' In: Quoted by Montoro (1982) *IX International Congres of Psychology*.

Clark, B.L. (1964) 'Multiple authorship trends scientific papers'. *Sciencie*. 143, 822-824.

Clarke, L. (1973) 'Explorations into the nature of environment codes: The relevance of Bernstein's theory of codes'' In: Kuller (Ed.) (1973).

Clouten, N. (1979) 'Monitoring the Use of Space by Stereoscopic Time-Lapse Photography' In: Canter (Ed) (1979).

Cole, J.R. & Cole, R. (1973) *Social stratification science*. Chicago. University of Chicago Press.

Colectivo (1983) '1er. Encuentro sobre descentralización y participación'. Valencia. Noviembre 1983.

Colectivo (1984) 'La vivienda en Valencia'. Valencia. Ayuntamiento de Valencia.

Collings, J. (1979) 'An Ethogenical Approach to Environmental Analysis' In: Canter (Ed) (1979).

Colom, A.J. (1980) 'Educació i territori. Perspectives didàctiques a traves de models territori-culturals' In: Alzina, J. & Sureda, J. (1980) *La nostra Història. Elements didàctics per l'estudi de la Història a partir del medi.* I.C.E. Universitat de Palma.

Colom, A.J. (1983) 'Concepto de Educación Ambiental'. *Teoría de la Educación I. El problema de la Educación.* Murcia. Ed. Límites. 27-42.

Colom, A.J. & Sureda, J. (1980) *Hacia una teoría del medio educativo.* I.C.E. Universitat de Palma.

Colom, A.J. & Sureda, J. (1981) 'Un caso de isomorfismo entre las Ciencias ambientales y las Ciencias de la Educación...'. *Educació i Cultura.* Palma de Mallorca.

Conan, M. H. (1984) 'Les representations de la ville et de la nature: un essai d'utilisation de l'analyse psychosociologique' In: Krampen (1984).

Conference. The East-West colloquium in Environmental Psychology. See Niit, T., Raudsepp, M., Liik, K. (Ed) 1991

Conference. *I Encuentro Iberoamericano de Psicología Social* (1980). Dep. Social Psychology. University of Barcelona.

Conference. *II Encuentro Iberoamericano de Psicología Social* (1981). Dep. Soc. Psych. Univ. Complutense, Madrid.

Conference. *III Encuentro Nacional de Psicología Social (1983).* Dep. Soc. Psych. Univ. La Laguna.

Conference. IV Encuentro P.S. - *I Congreso Nacional de Psicología Social* (1985). Dep. Soc. Psych. Unv. Granada.

Conference IAPS. Conference of Dalandhui, U.K. (1969). See Canter (Ed) 1969d.

Conference IAPS. Conference of Kingston ,U.K. (1970). 1th IAPC. See Honikman (Ed) 1970.

Conference IAPS. Conference of Lund, Sweden, (1973). 2th IAPC. See Küller (Ed) 1973.

Conference IAPS. Conference of Surrey-I, U.K. (1974). ICEP. See Canter & Lee (Ed) 1974.

Conference IAPS. Conference of Strasbourg, France (1976). 3th IAPC. See Korosec (Ed) 1978.

Conference IAPS. Conference of Louvain-la-Neuve, Belgium (1979). 4th IAPC. See Simon (Ed) 1979.

Conference IAPS. Conference of Surrey-II, U.K. (1979). ICEP. Unpublished. See Canter 1979 (Ed)

Conference IAPS. Conference of Barcelona, Spain (1982). 7th IAPS. See Pol, Muntañola & Morales (Ed) 1984

Conference IAPS. Conference of Berlin, Germany (1984). 8th IAPS. See Krampen (Ed) 1984.

Conference IAPS. Conference of Haifa, Israel (1986). 9th IAPS. See Churchman (Ed) 1986.

Conference IAPS. Conference of Delft, Netherlands (1988). 10th IAPS. See Van Hoogdalem et al. (Ed) 1988

Conference IAPS. Conference of Ankara, Turkey (1990). 11th IAPS. Pamir, Imamoglu, Teymur (Eds) (1990)

Conference IAPS. Conference of Thessaloniki, Greece (1992). XII-IAPS. See Mazis et al. (Ed) 1992

Conference School & Env. *Jornadas sobre Ambientes Escolares* (1973). ICE: Universidad de Barcelona. INCIE, UNESCO.

Conference Schol & Env. Jornadas (1979) *Ecología y Educación.* ICE: Universidad de Barcelona.

Conference School & Env. I Jornades *Entorn Escolar. Problemàtica Psicològica, Educativa i de Disseny.* (1979). See Pol & Morales (Ed) (1980).

Conference School & Env. II Jornades sobre l'Entorn Escolar 1980. See Pol & Morales (Eds) (1981) *Imatge de l'escola, Interacció Ambiental, vers una nova normativa.* ICE Universitat de Barcelona-BCD.

Conference School & Env. *Jornades sobre Edificació Escolar (1981).* Generalitat de Catalunya.

Conference School & Env. III Jornades sobre l'Entorn Escolar 1982. Pol, Morales, Muntañola (Eds) (1984)

Conference School & Env. Jornadas sobre *Calidad Ambiental en la Escuela (1983).* Madrid. Ministerio de Educación.

Conference School & Env. *Jornadas sobre Organización, Entorno y Educación (1983)*. Dept. de Pedagogía, Universitat de Barcelona.

Conference School & Env. *Jornadas sobre Educación Ambiental (1983)*. Diputació de Barcelona y M.O.P.U., Sitges.

Conference School & Env. *Jornades d' Educació Ambiental* (1983). ICE Universitat de Barcelona.

Conference School & Env. IV Jornades sobre l'Entorn Escolar (1984) *De l'Arquitecte al Mestre, del Dissenyador a l'usuari*. ICE Univ. Barcelona.BCD. Facul. de Psicologia.

Conference of I Jornadas *de Psicología Ambiental (1986)*. See Aragonés & Corraliza (1986).

Conference of II Jornadas de Psicologia Ambiental (1989) Palma de Mallorca. See Pol & Pich (Ed) (1989)

Conference of III Jornadas de Psicologia Ambiental (1991) Sevilla. See De Castro (Ed) (1991).

Conference of Jornadas de Psicologia Ambiental. (1992) Orellana. See Corraliza (Ed) (1993).

Cooper, I. (1979) 'Design and Use of Primary School Buildings' In: Canter (Ed) (1979).

Cooper, I. (1979) *Chaucer infant and nursery school-environmental appraisal: users' evaluations of the*. Cardiff.Depart. of Educ. & Science. Unpub.internal report.

Cooper, I. (1982) 'Design and use of British primary school buildings: an examination of goverment-endorsed advice' In: versio prim. a Surrey-II(1979) *Design studies*. vol3 n.1, January.

Cooper, I. (1984) Recorded personal Interview by Pol. Cambrige.

Cooper. (1984) *Invest.how lay people assess environ.in buildings:some method.issues design-orientated research*. Cambrige, U.K. Martin Centre Arch. & Urb. Studies Unpubl.work.paper.

Corraliza, J.A. (1978a) 'El estudio de las dimensiones afectivas del ambiente'. In: Fernandez-Balletros (Coord.), *El ambiente. Análisis Psicológico*. Madrid, Pirámide. pp. 103-124.

Corraliza, J.A. (1978b) *La experiencia del ambiente. Percepción y significado del medio construido*. Madrid, Tecnos.

Corraliza, J. A. (Ed.) (1993). *El comportamiento en el Medio Construido y Natural*. Proceedings of meeting on environmental psychology research in Spain. Orellana, Junta de Extremadura.

Courgean, D. (1972) 'Les réseave de relation entre personnes. Etude d'un millieu rural'. *La population*. 4-5.

Couwenbergh, J. P. (1982) 'Propos sur l'évaluation de l'environnement construit' In: Pol, Muntañola, Morales (Eds.)(1984).

Couwenbergh, J. P. (1984) 'Pour une conception labyrintique de l'espace. Ou comment concilier privatisation et communaute' In: Krampen (1984).

Covarrubias, J. C. (1979) 'L'intelligibilité et la banalité de espaces urbains en tant que conditionneurs des exp. conflict.' In: Simon (Ed.)(1979).

Craik, K.H. (1968) 'La compresión del ambiente físico cotidiano' In: Proshansky. et al. (eds.) 1970 *Journal of the Am. Inst. of planners*. 34.

Craik, K.H. (1973) 'Environmental Psychology'. *Annual Review of Psychology*. 24, 403-421.

Craik, K.H. (1977) 'Multiple scientific paradigms environmental Psychology'. *International Journal of Psychology*. 12, 2 147-157.

Craik, K.H. (1985) Recorded personal Interview by Pol. California. V. Berkeley.

Craik, K.H. (1987) 'Aspects internationaux de la psychologie de l'environnement'. *Psychologie Française*. Tome 32, 1-2.

Craik, K.H. & Canter, D. (1981) 'Environmental Psychology'. *Journal of Environmental Psychology*. n.1, vol.1, p.p.1-11.

Crane. (1969) 'Social structure a group of scientists: A test of the "Invisible College" Hypothesis'. *American Sociological Review*. 34, 335-352.

Crane. (1972) *Invisible Colleges*. Chicago. University Chicago Press.

Crane. (1977) *Invisible Colleges*. Chicago. Chicago University Press.

Croome, D. (1979) 'Human Confort Buildings' In: Canter (Ed) (1979).

Crunelle, M. (1982) 'Compte remdu d'une recherche personnelle' In: Pol, Muntañola, Morales (eds.) (1984).

Cuttle, Ch. (1973) 'The sharpness and the flow of light' In: Kuller (Ed.) (1973).

Cutle, Ch. (1973) 'The sharpness and the flow of light' In: Kuller (Ed.) (1973).

Daish, J. R. and Joiner, D. A. (1984) 'An environnment-behaviour model for design teaching and practice' In: Krampen (1984).

Daish, J., Gray, J., Kernohan, D. (1982) 'Fitting the missing link' In: Pol, Muntañola, Morales (Eds.) (1984).

Daley. (1968) 'Psychological research in architecture: the myth of quantifiability' *Architects Journal*. 148, (34): 339-341.

Daley. (1969) 'A philosophical critique of behaviorism in architectural design' In: Broadbent. & Wars. (eds.) (1969).

Daley, J. (1970) 'Technology models in architecture - acritique' In: Honikman (1970).

Darder, P. & Pérez, J.A. (1983) *El Quafe, Qüestinari d'anàlisi funcional de l'escola*. Barcelona. Rosa Sensat.

Darder, P. & Pérez, J.A. (1985a) *Anàlisi del Funcionament dels Centres privats d'EGB de Catalunya*. Barcelona. Departament d'Ensenyament. Generalitat de Catalunya.

Darder, P. & Pérez, J. A. (1985b) *Organització i Avaluació de l'Escola*. Barcelona. Edicions 62.

Darder, P. & Pérez, J. A. (1985c) *La Evaluación de la Escuela*. CEAC.

Darder, P. & Pérez, J. A. (1985d) *Els Grups de Classe*. Barcelona. Edicions 62.

Darke, R. (1979) 'Public Participation and Planning: Councillors' and Officers' Views' In: Canter (Ed) (1979).

Darke, J. & Darke, R. (1981a) 'Open letter to Howard Harris and Alan Lipman' *Arch Psych. News*. 11 (2): 38-39.

Darke, R. & Darke, J. (1981b) 'Towards a sociology of the built environment?' *Architectural Psychology Newsletter*. XII, 1: 8-16.

Darke, J., Lawson, B., Spencer, CH. (1979) 'Surveying the Users: Some People are Never Satisfied' In: Canter (Ed) (1979).

Dasgupta, S. K., Mukherjee, P. M., Sarkar, J. (1979) 'Rural Housing and Environmental Development: A Study on the Demonstrative Impact on Indian Villagers' In: Canter (Ed) (1979).

Davis, G. & Altman, I. (1976) 'Territoires, intimité et contrôle dans les environnements de travail' In: Korosec (Ed) (1978).

Davis, G. & Szigeti, F. (1979) 'Functional and technical programming. When the owner/sponsor is a large or complex organization' In. Simon (Ed.) (1979).

Dawson, J. L. M. (1979) 'Environmental Psychology: A Bio-social Approach' In: Canter (Ed) (1979).

De Carlo. et al. (1981) *Energia nucleare: indagine su atteggiamenti e opinioni*. Padova. Liviana.

De Latour Dejean, C. H. (1976) 'Les systèmes relationnels et leur impact sur l'espace'. In: Korosec. (1976).

Dearinger, J. A. (1979) 'Measuring Preferences for Natural Landscapes' In: Canter (Ed) (1979).

De Castro, R. (Ed.) (1991). *Psicología Ambiental: Intervención y Evaluación del Entorno*. Proceedings of III Jornadas de Psicología Ambiental. Sevilla. Arquetipo.

Delahousse, A. (1979) 'Polyvalent Effects of Spatial Setting According to Social Norms' In: Canter (Ed) (1979).

Denis M. M., Groshens, M. C. (1982) 'L'Architecture Rurale Française: l'Alsace' In: Barcelona (1982).

Dennis. (1954) 'Productivity among american psychologists" In: Quoted by Montoro (1982) *American Psychologist*. 9, 191-194.

Departamento de Pedagogia de la Universidad de Palma (1978) 'El medio como experiencia educativa' *Cuadernos de Pedagogía*. Barcelona. 41.

Desaulty, M., Gardent, J. Guinard, J. & Robin, Ch. (1976) 'Quelques Modalites d'appropriation des Espaces Exterieurs et leurs bases sensori-motrices'. In Korosec (Ed) (1978).

Dewey. (1979) *Bibligrafical Classification*.

Díaz-Veiga, P. (1987) 'Evaluación del Apoyo Social'. In: R. Fernández-Ballesteros (Coord.) (1987), pp. 125-149.

Djorklid, P. (1984) 'Children's outdoor environment from the perspective of environmental and developmental psychology' In: Krampen (1984).

D'Ors, E. (1911) *La Benplantada*. Barcelona, Edicions 62 (1980).

D'Ors, E. (1915) *Gualba, en les mil veus*. Barcelona, Edicions 62 (1980).

D'Ors, E. (1923) *Tres horas en el museo del Prado*. Madrid, Aguilar (1971).

D'Ors, E. (1961) *El valle de Josafat*. Madrid, Espasa Calpe.

D'Ors, E. (1966) *Las ideas y las formas*. Madrid, Aguilar.

Dourmanov, Vl. (1991). 'Spatial Image as a Genetic Basis in the Development of Phisical Environment. (The Example of Housing)'. In Niit, Raudsepp & Liik (Ed) 1991. pp. 117-122.

Down. & Stea. (eds.) (1973) *Image and Environment*. Chicago. Eduard Arnold.

Dozio, M. J. & Feddersen, P. (1976) 'Quelle est l'utilité de la notion de l'appropriation de l'espace pour la conception architecturale et urbanistique?' In: Korosec (Ed) (1978).

Driessen, F. M. H. M. (1984) 'Changes in preferences with respect to the residential environment'. In: Krampen 1984.

Dridze, T. (1991). 'Social Diagnostics and Social Communication in a Strategy of Decision-Making in Town Planning'. In Niit, T., Raudsepp, M., Liik, K. (Ed) 1991. pp.17-32.

Duffy, F. & Worthington, J. (1976) 'La création des espaces des organisations' In: Korosec (Ed) (1978).

Dumesnil, C. (1982) 'Interior environmental research: A review of the Literature' In: Pol, Muntañola, Morales (Eds.) (1984).

Durlak, Jerome T. & Lehman, Joan. (1974) 'User Awareness and Sensitivity to Open Space: A Study of Traditional and Open Plan Schools' In: Canter & Lee (1974).

Eco, H. () 'L'estructura absent' In: Lumen.

Edberg, Gösta. (1973) 'Applied Research in the field of architectural psychology' In: Kuller (Ed.) (1973).

Edge, A. & Donald, I. (1984) Recorded personal Interview by Pol. Birmingham.

Edwards, M. (1973) 'Comparison of Some Expectations of a Sample of Housing Architects with known data' In: Canter & Lee (1973).

Edwards, M. (1979a) 'The development of a projective technique for interviewing architects' In: Canter (Ed) (1979).

Edwards, M. (1979b) 'The development of a projective technique for interviewing architects about their work' In: Simon (Ed.) (1979).

Edwards, M. (1979c) 'Does it matter if architects do not really know what people do at home and consequently design for them as for themselves' In: Simon. (1979).

Ekhart, Hahn. (1984) 'Ökologischer stadtumbau - eine zentrale aufcabe präventiver umweltpolitik' In: Krampen (1984).

El - Badwan. (1984) 'Veränderung des damazener Stadthauses' In: Krampen (1984).

Ellis, P. (1979a) 'Conceptions of Outside Spaces in Housing Areas' In: Canter (Ed) (1979).

Ellis, P. (1979b) 'Architects' and residents' differing conceptions of outside spaces in housing areas' In: Simon (Ed.) (1979)P. (1981) 'Towards a social psychology of the environment- '*Architec Psych. News*. 11 (2) 81: 44-46.

Ellis, P. & Duffy, F. (1982) 'Design, Environment and Social Sciences' In: Pol, Muntañola, Morales (Eds.) (1984).

Endler. et al. (1975) 'Productivity and scholarly impact (citation) of British, Canadian and U. S. departments of psychology' *American Psychologists*. 12, 1978, 1064-1082.

Engfer, A. (1979) 'Sozialökologische Badingungen alterlichen Erzienhungsuerhaltens' In: K.A. Schneewind & Th. Herrmann (Eds.) *Das Trierer Erziehungsstilsymposion*. Göttingen: Hogrefe.

Ensfer (1979) Quoted by Kruse y Graumann (1984 preprint).

Espe, H.(1979) 'Differences of Impressions Between Classicistic and National Socialistic Architecture' In: Canter (Ed) (1979).

Espe, H. (1984a) 'Some findings with a german version of the environmental response inventory' In: Krampen (1984).

Espe, H. (1984b) 'Relationships between ideal room concepts and personality' In: Krampen (1984).

Espe, H. (1984c) 'Room evaluation, moods and personality'. In: Pol, Morales, Muntañola. (1984).

Eubank-Ahrens, M. F. A. (1984) 'Mirkungskontrolle einer Wohnumfeldverbesserung: Verhalten und Erleben im öffentlichen Freiraum' In: Krampen (1984).

Fahle, B. (1979) 'User Participation as Environmental Social Research, Planning and Education' In: Canter (Ed) (1979).

Fahle, B. & Schmidt, A. (1976) 'Environnement, perception et comportement dans une zone piétonnière' In: Korosec (Ed) (1978).

Fatourus, D. A. (1973) 'Perceptual ecology and the organization of physical environment' In: Kuller (1973).

Feltz, C. (1979) 'La forme de la maison, enjeu social. Pratiques des utilisateurs et pratiques des architects' In: Simon (1979).

Feodor. & Koroiev. (1954) 'Spatial composition in a project and in nature' In: (1961). Moscow.

Ferguson, R. (1979) 'The Hospital as a Therapeutic Environment' In: Canter (Ed) (1979).

Fernández Ballesteros, R. 'Evaluación y medio ambiente: el caso ambiental' In: *Revista de Estudios Territoriales*.

Fernández Ballesteros, R. (1980) *Psicodiagnóstico. Concepto y Metodología*. Madrid. Ed. Cincel Kareluz.

Fernández Ballesteros, R. (1981) 'Evaluación en Psicología Ambiental' In: en F. Jimenez Burillo (Ed.) *Psicología y Medio Ambiente*. Madrid. CEOTMA.

Fernández Ballesteros, R. (1982) 'El contexto en evaluación psicológica' In: en R. Fernandez Ballesteros (Ed.) *Evaluación de Contextos*. Servicio de Publicaciones de la Universidad de Murcia.

Fernández Ballesteros, R. (Ed.) (1982) *Evaluación de Contextos*. Servicio de Publicaciones de la Universidad de Murcia.

Fernández Ballesteros, R. (Ed.) (1983) *Psicodiagnóstico*. Madrid UNED, Unidades Didácticas.

Fernández Ballesteros, R. (1983) 'Evaluación del Ambiente' In: Fernández Ballesteros, R. (Ed.) *Psicodiagnóstico*. Madrid. UNED.

Fernández-Ballesteros, R. (Coord.) (1987) *El Ambiente. análisis Psicológico*. Madrid, E. Pirámide.

Fernández Ballesteros, R. & Huici, C. (1975) 'Evaluación del clima de grupo' no publicado. Protocolo, hoja de respuestas.

Fernández Ballesteros, R & Sierra, B. 'Estudio factorial sobre la percepción del ambiente escolar' In: en R. Fernández Ballesteros (Dir.) *Evaluación de Contextos*. Servicio de Publicaciones de la Universidad de Murcia.

Fernández Ballesteros, R. & Sierra, B. (1982) 'Escala de ambiente escolar' Madrid. TEA. Protocolo, perfil, hoja de respues.

Fernández Ballesteros, R. & Suici, C. (1974) 'Evaluación del clima Familiar' In: N   o publicado.

Fernández Ballesteros, R. et alt. (1982) 'Influencia de mobiliario en la conducta interpersonal de ancianos institucionalizados' In: Fernández Ballesteros (Dir.) *Evaluación de Contextos*. Servicio de Publicaciones de la Universidad de Murcia.

Fernández-Dols. & Rodriguez-Sanabra, F. (1982) 'Guía documental de la Psicología ambiental' *Estudios de Psicología*. 9, 142-157.

Ferrand, A. (1976) 'La pratique spatiale des groupes de jeunes 'Ségrégation et appropiation symbolique' In: Korosec. (1976).

Festinger. et al. (1950) *Social Pressures in Informal Groups*. N. Y. Harper and Row.

Finlayson, K. A. (1984) 'The use of community involvement procedures in the upgrading of an informal housing area' In: Krampen (1984).

Fischer, G. N. (1976) 'Travail industriel et appropriation de l'espace' In: Korosec (Ed.) (1976).

Fischer, G. (1967) *Psychosociologie de l'espace*. Paris. Puf.

Fischer, G. (1980) *Espace Industriel et liberté: L'autogestion clandestine*. Paris. Puf.

Fischer, M. (1978) 'Ökolofische Bedingungen für Verhalteensanffälligkeiten in des Schule' In: Lohmann & Minsel (Eds.) *Störungen in Bereixh der Schule*. München: Urban & Schwarzenberg, 157-181.

Fischer. (1979) Phänomenologische Analysen der Person-Umwelt-Beziehung. In: S.H. Filip (Ed.), *Selbstkomzept-Forschung. Probleme, Befunde, Perspektiven*, pp. 47-73. Stuttgart, F.R.G.: Klett-Cotta.

Flade, Dr. Antje & Darmstadt, BRD. (1984) 'Evaluierung von Wohnungen im Hinblick auf Familiengerechtigkeit' In: Krampen (1984).

Foote, K. E. (1984) 'Color in public spaces: toward a communication based theory of the urban built environment' In: Krampen (1984).

Fourcade, Lesieur. (1976) 'L'induction des postures: supports spatiaux et emprise psychosociale' *Travail Humain*. 39.

Fowles, D. L. & Ames, I. (1984) 'A tradeoff technique to determine user preference: comparing university-owned family housing physical features' In: Krampen (1984).

Francis, M. (1984) 'A comparison of different meanings attached by users to a public park and community garden' In: Krampen (1984).

Fridgen, J. D. (1979) 'Implicit Theories of Recreation Environments' In: Canter (Ed) (1979).

Fritz, H-J. (1984) 'Zur langfristigen entwcklung von büroarbeitsräumen'. In: Krampen 1984.

Gaetti. & Venini. (1982) 'L'incidenza della zona di residenza urbana sull'elaborazione cognitiva dell'immagine di città' *Ricerche di Psicologia*. 22-23, 187-201.

Gaillar, C. et al. (1971) *Sciences humaines et environnement. Orientations bibliographiques*. Paris Institut de l'environnement.

Galindo, M.P. & Gilmartín, M.A. (1993). 'Los Estudios de Preferencia Ambiental: principales problemas teóricos y conceptuales'. In Corraliza (Ed.1993.)

Galing, T. (1980) 'A comparison of multidimensional scaling with the semantic differential technique as methods for structural analysis of environmental perception and cognition' *Umea Psychological Reports*. n. 155.

Garbrecht, D. (1976) 'L'utilisation des espaces verts urbains' In: Korosec (Ed) (1978).

García, J. (en curso) *Mediadores Sociocognitivos en el entorno Construido*. Univ. Autónoma de Barcelona. Departamento de Psicología.

Garfield, E. (1978a) 'The 100 articles most cited by social scientits, 1969-1977' *Current Contents*. 32, 5-14.

Garfield, E. (1978b) 'The most cited SSCI authors' *Current Contents*. 45, 5-15.

Garfield, E. (1979) *Citation Indexing. Its theory and application in science, technology and humanities*. N. Y. Wiley.

Garfield. et al. (1978) 'Citation data as science indicators' In: in Elkana et al. *Toward a Metric of Science*. N. Y. Wiley.

Gärling, T. (1969) 'Studies in visual perception of architectural spaces and rooms. II' *Scandinavian Journal of Psychology*. 10, 257-68.

Gärling, T. (1970a) 'Studies in visual perception of architectural spaces and rooms. III' *Scandinavian Journal of Psychology*. 11, 124-31.

Gärling, T. (1970b) 'Studies in visual perception of architectural spaces and rooms. IV' *Scandinavian Journal of Psychology*. 11, 133-45.

Gärling, T. (1972a) 'Studies in visual perception of architectural spaces and rooms' *Reports from the Psychological Laboratories, the University of Stockholm*. Doctoral Dissertation, Depart. Psych., Univ. Stockholm. Supplement Series n. 15.

Gärling, T. (1972b) 'Studies in visual perception of architectural spaces and rooms. V' *Scandinavian Journal of Psychology*. 13, 222-7.

Gärling, T. (1973) 'Some applications of multidimensional scaling methods to the structural analysis of environmental perception and cognition' In: Kuller (1973).

Gärling, T. (1976a) 'A multidimensional scaling and semantic differential technique study of the perception of environment settings' *Scandinavian Journal of Psychology*. 17, 323-32.

Gärling, T. (1976b) 'The structural analysis of environmental perception and cognition: A multidimensional scaling approach' *Environment and Behavior*. 8, 385-415.

Gärling, T. (1980a) *Environmental orient. during locomotion:Exper. studies of human processing of inform. about spatial layout of the environment.* Stockholm The Swedish Council for Building Research (Document D24).

Gärling, T. (1980b) *Perception of the spatial layout of environments during locomotion. (Rapport R159).* Stockholm The Swedish Council for Building Research.

Gärling, T. (1982) 'Swedish environmental psychology' *Journal Environmental Psychology*. 2 (3): 233-252.

Gärling, T. (1985) 'Childre's environments, accidents, and accident prevention: A conceptual analysis'. *Children's Environments Quarterly*, 2 4-8.

Gärling, T., Book, A. & Lindberg, E. (1979) 'Maintenance of Orientation in the Horizontal Plane During Locomotion' In: Canter (Ed) (1979).

Gärling, A.& Gärling, T (1990) Parent's residential satisfaction and perceptions of children's accident risk. *Journal of Environmental Psychology*, 10, 27-36

Gärling, T, Gärling, A, Mauritzson-Sandberg, E. & Björnstig, U. (1989). Child safety in the home: Mothers' perceptions of danger to young children. *Architecture and Behaviour*, 5, 293-303.

Gärling, T.& Gärling, E. (1988). 'Distance minimization in downtown pedestrian shopping behavior'. *Environment and Planning*2 A, 20, 547-554.

Gärling, T, Garvill, J., Lindberg, E. & Montgomery, H. (1992) 'Residential choices and beliefs about future life satisfaction: Test of a modified multiattribute utility (MAU) framework'. Stokholm: Swedish Building Research Council.

Gärling, T., Küller, R., Sivik, L. & Sorte, G. J. (1974) 'Man-environment research. General programme for environmental psychology'. Lund unpublished paper, Lund Institute of Technology.

Gärling, T., Küller, R., Sivik, L. & Sgorte, G. J. (1976) *Man-environment research. General programme for environmental psychology. (Summary S19).* Stockholm The Swedish Council for Building Research.

Gärling, T., Lindberg, A. & Book, A. (1979) 'The acquisition and use of an internal representation of the spatial layout of the environment during locomotion' *Man-Environment Systems*. 9, 200-8.

Gärling, T., Lindberg, E & Book, A. (1982) 'Cognitive mapping of large-scale environments, action plans, orientation, and their interrelationships'. Irvinepaper presented at the II Annual Irvine Symposium on Envir. Psych., University of California.

Gärling, T., Lindberg, E., Book, A. & Saisa, J. (1982) 'The spatio-temporal sequencing of everyday activities: How people manage to find the shortest path to travel in their home town'. Barcelona paper presented 7th. IAPS.

Gärling, T, Lindberg, E., Carreiras, M. & Book, A. (1986) 'The spatiotemporal sequencing of maps'. *Journal of Environmental Psychology*, 6, 1-18.

Gärling, T. & Toomingas, A. (1975) *Psychological investigation of the auditive perception of space. Architectural Psychology. (Rapport R24).* Stockholm The Swedish Council for Building Research.

Gärling, T. & Valsiner, J. (1973) 'Parental Concern about children's traffic safety in residential neighborhoods' In: Kuller (1973).

Garvey. (1979) *Communication: the essence of science.* Pergamon.

Gasparini, A. (1980) 'La percezione ambientale in sociologia' *Rivista Geografica Italiana*. 87, 114-20.

Gaudin, J. P. (1979) *L'amenagement de la société: politiques, savoirs, representat. sociales, la product. de l'espace aux 19ème et 20éme siècles.* Paris. Anthropos.

Gavinelli, C. (1979) 'Modern Use of Space in Chinese Environment' In: Canter (Ed) (1979).

Gavinelli, C. (1984) 'Les pieces postmodernes et la reformation de la vie actuelle' In: Krampen (1984).

Gehl, J. (1970) 'An investigation of satisfaction in living areas'. In: Honikman. (1970).

Gehl, J. (1970) 'A social psychological dimension of architecture'. In: Honikman. (1970).

Gehl, J. (1971) *Livet mellen husene*. Copenhagen. Arkitektens forlag.

Gentileschi, M. L. (1980) 'Variazione dell'immagine del territorio e processi decisionali degli immigrati di ritorno'. Nota preliminare In: R. Geipel & M. Cesa-Bianchi (eds.) (1980). pp. 149-59.

Germain. (1948) 'El XII Congreso Internacional de Psicologia' In: Quoted by Montoro (1982) *Revista de Psicologia General y Aplicada*. 3, 611.

Gerngross-Haas, G. (1979) 'Some Problems Related to the Evaluation of "Innovative" Buildings' In: Canter (Ed) (1979).

Gibson, J. J. (1950) *The perception of the visual world*. Boston. Houghton-Mifflin.

Gibson, J. J. (1950) *La percepción del mundo visual*. Buenos Aires. Ediciones Infinito.

Gibson, J. J. (1966) *The sense consideral as perceptual system*. Houghton-Mifflin.

Gil Corell, M. (1982) *¿Qué es el medio ambiente?*. Valencia. Diputación Provincial.

Gimenez, E. & LLorens, T. (1970) 'La imagen de la Ciudad de Valencia' *Hogar y Arquitectura*. N. 86, PP. 13-144.

Ginsburg. (1934) 'The residential building'. Quoted by Niit, Heidmets & Kruuwall. 1981.

Giuliani, M. V. (1984) 'Home Interior Decoration: Psychosocial models and aesthetic theories' In: Krampen (1984).

Giuliani, M. V. (1991) 'Towards an analysis of mental representations of attachment to the home'. *The Journal of Architectural and Planing Research*. 8, 2, 133.

Giuliani, M. V., Rullo. G. (1986) 'The context of the environmental evaluation: The evaluator's role'. *IAPS 9th International Conference*. In: Israel (1986).

Giuliani, M. V., Rullo, G. (1988) 'Un approccio psicologico allo studio delle relazioni tra individuo e ambiente'. Roma (1988).

Giuliani, M. V., Rullo, G., Bacaro, C. (1987) 'Structures familiales et modèles territoriaux'. *Collection Habitat et Sociétés*. Editions L'Harmattan. París (1987).

Giuliani, M. V., Rullo, G., e Bonnes, M. (1987) 'L'ambiente domestico: Tipologie cross-culturali e indicatori soggesttivi di somiglianza'. *Studi e Ricerche sul Territorio*. Edizioni Unicopli. Milano (1987)

Giuliani, M. V., Rullo, G., Bove, G. (1990) 'Socializing and privacy spaces inside homes: an empirical study'. *Culture/Space/Histoy*. IAPS 11.Pamir et al. (1990).

Glaeser, B. (1984) 'Humanökologisches handeln' In: Krampen (1984).

Glassel, J. (1984) 'Naturhaus Berlin Ökologisches wohnhaus für die Stadt' In: Krampen (1984).

Glazytxev. (1981) 'Art design modelling of environmental behavior' In: T. Niit, et al. (eds.) *People and Environment: Psychological Problems*. Tallinn. EOOP. pp. 169-72.

Gleichmann, P. R. & Fritz, H. J. (1976) 'Bureaux paysagers - Un rapport intermédiaire' In: Korosec (Ed.) (1976).

Godelier, M. (1982) *Les sciences de l'homme et de la société en France*. Paris. La Documentation Française.

Gómez de Benito et al. (1976) 'Psicologia y medio urbano' *Psicologia servicio público*. Madrid Pablo del Rio.

González, P. (1973) *Proyecto Semiótico para la señalización de los transportes de Barcelona*. Sección Diseño Gráfico. Escuela Massana Barcelona.

González, P. (1974) 'Psicología y Vivienda' *Atlas de Decoración*. Cia. Internacional Editora.

González, P. (1975) *Ejercicios con el espacio y la vivencia emocional*. Sección Diseño Gráfico. Escuela Massana Barcelona.

González, P. (1979) 'El Espacio privilegiado en la Dinámica de Grupos'.

González-Bernáldez, F. (1983) 'La educación ambiental: evaluación crítica y perspectivas'. Sitges Jornadas Educación Ambiental. MOPU-Diputació de Barcelona.

González-Bernáldez, F. (1984) 'Applying Landscape perception by the public to Urban Planning: the Spanish MAB experience'. Suzdal, USSR paper in Intern. Experts Meeting on Urban Planing. UNESCO.

González-Bernáldez, F. & Parra, F. (1979) 'Dimensions of Landscape Preferences from Pairwise Comparisons'. Nevada paper in Conf. Applied Techn. for Analy. &Manag. Visual Resource.

Goodey, B. (1970) 'Perception gaming and dephi: experiential approaches to environmental education' In: Honikman (Ed.) (1970).

Gotman, A. (1984) 'Logement, parente, heritage'. In: Krampen (1984).

Gottesdiener, H. (1979) 'Space Organisation of Exhibits and Information Gathered by Visitors' In: Canter (Ed) (1979).

Gottesdiener, H. (1984) 'L'ecrit, l'organisation de l'espace et les tableaux regardes dans un musee' In: Krampen (1984).

Graff, C. (1976) 'Aliénation ou identification: le rôle de l'espace dans la maison' In: Korosec (Ed.) (1976).

Graflich, C. (1984) 'Sinnhafte architextur empfindung und wahrnehmung von gebauter wohnumwelt' In: Krampen (1984).

Gratacós, R. (1984) 'Space perception by blind Children. Proposal for the learning of the Ancien City of Barcelona'. Berlin paper to 8 IAPS Conference. In Krampen (1984)

Gratacós, R. (1986) 'Percepción del espacio de la ciudad en los niños ciegos'. comunicación presentada a las I Jornadas de Psicología Ambiental. Madrid, Nov. 1986.

Graumann, C.F. (1974) 'Psychology and the world of things' *Journal of Phenomenological Psychology*, 4.

Graumann, C. F. (1976) 'Le concept d'appropriation (Aneignung) et les modes d'appropriation de l'espace' In: Korosec (Ed.) (1976).

Graumann, C. F. (1976) 'Rapport sur le séminaire consacré aux "Modes d'appropiation de l'espace"'. In: Korosec. (1976).

Graumann, C.F. (Hg.) (1978) *Ökoloische Perspective in der Psychologie*. Bern: Huber.

Graumann, C.F., Kruse & Lantermann (Hrsgr) (1985) U*mwelt Psychologie (Ökopsychologie) - Ein Hnadbuch in Schlüsselbegriffen*. München: Urban & Schwarzenberg.

Green, C. H. (1979) 'Acceptable Risk as a Problem Definition' In: Canter (Ed) (1979).

Griffiths, I. D. (1969) 'Thermal Comfost; a behavioural approach' In: Canter (Ed.) (1969).

Griffiths, I. D., Delauzun, F. & Swan, M. (1979) 'Study of Professional Attitudes to Architectural Quality in Housing' In: Canter (Ed) (1979).

Grivel, F. (1984) 'Sensibilite de l'home a des niveaux de temperature ambiante compris entre 17 et 33o. C maintenus pendant de courtes durees' In: Krampen (1984).

Groat, L. (1979) 'Post-Modernism and the Multiple Sorting Task' In: Canter (Ed) (1979).

Groat, L. N. (1984) 'Strategies for contextual design: a partial order scalogram analysis' In: Krampen (1984).

Groning, G. & Wolschke, J. (1984) 'Jugendbewegung und naturverständnis' In: Krampen (1984).

Gruais, M. (1976) 'La structuration de l'espace à trois dimensions et la perception de la forme par de jeunes enfants de 4 à 5 ans' In: Korosec (Ed.) (1976).

Gruais, M. (1979) 'Relations des comportements de l'habitant et du concepteur devant l'évolution de l'habitat en site défini' In: Simon (Ed.) (1979).

Gruska, A & Mazerat, B. (1975) *Inventaire des travaux et centres de recherche portant sur les representations, attitudes, aspirations et practiques...* Paris.

Guardia, J. (1992). 'Algunas Propuestas Metodológicas a la Investigación Ambiental'. In Corraliza (Ed. 1993).

Guardia, J. & Pol, E. (1992). 'Elaboration and Analysis of a questionnaire for the apprecaition of Quality of Life in the City.' In Proceedings XII IAPS.

Guerrand, R. H. (1976) 'Appropriation des lieux d'habitat et couches défavorisées'. In: Korosec. (1976).

Guifford, R. *Environmental Psychology. Principes and Practice.*Boston, Allyn and Bacon

Gustavsson, M. & Mansson, K. (1980) 'The accessibility of buildings for the blind. Evaluation of designs with regard to people having impaired ability to orient themselves' (Rapport R93). Stockholm. The Swedish Council for Building Research.

Haeckel. (1866) *General Morphology of Organisms.* (Quoted by Kruse & Graumann. 1984. Preprint).

Hall. (1973) *La Dimensión Oculta.* Madrid. IEAL.

Hamel & Andel. (1982) *Environmental Psychology Documentation System.* Technical University of Eindhoven, Holanda.

Hamilton-Eddy, D. (1970) 'The semantics of security and the dream of community'. In: Honikman. (1970).

Härd, A. (1973) 'Content and contrast in colour' In: Küller (19Härd, A. (1975) 'NCS, a descriptive colour order and scaling system with appication for environmental design' *Man-Environment Systems.* 5, 161-7.

Härd, A. & Sivik, L. (1979) 'Colour-man-environment. A Swedish building research project' *Man-Environment Systems.* 9, 213-6.

Härd, A. & Sivik, L. (1981) 'Natural color system: a Swedish standard for color notation' *Color Research and Application.* 6, 129-38.

Hardie, G. J. (1984a) 'Simulation - a means of engendering user participation in the housing process' In: Krampen (1984).

Hardie, G. J. (1984b) 'The historical and attitudinal context of selp-help housing action: two case studies compared' In: Krampen (1984).

Harris, H. (1977) *The relationship between illuminance, energy use and quality in the visual environment: user evaluations.* Cardiff. Unpublished internal report, Welsh School of Architecture.

Harris, H. (1984) Recorded personal Interview by Pol. Cardiff.

Harris, H. & Lipman, A. (1979a) 'Social symbolism of space usage'. (P. 49)In: Canter (1979).

Harris, H. & Lipman, A. (1979b) 'Social symbolism of space usage' In: comunicació a Surrey-II (1979).

Harris, H. & Lipman, A. (1980a) 'Architecture and knowledge: control or understanding' *Architecture and Behavior.* 1 (2): 135-147.

Harris, H. & Lipman, A. (1980b) 'Social symbolism and space usage in daily life' *Social Review.* 28: 415-428.

Harris, H. & Lipman, A. (1981) 'Open letter to Roy & Jane Dake' *Architectural Psych. Newsletter.* 11, (2): 36-37.

Hart, R. & Watts, N. (1979) 'The Development of Self through Environment: Implications for Theory in Environmental Psychology' In: Canter (Ed) (1979).

Hatfield, M. R. & Cain, J. G. D. (1979) 'Design Management and the Shipboard Environment' In: Canter (Ed) (1979).

Haumont, A. (1972) 'L'image de la campagne chez les citadins' *Métropolis.* 2, (2).

Haumont, A. (1976) 'Les espaces industriels et les pratiques urbaines de l'espace' In: Korosec (Ed.) (1976).

Haumont, N. (1968) 'Habitat et modèle culturel' *Revue Française de Sociologie.* 1968, 9.

Haumont, N. (1972) *Habitat et practiques de l'espace. Etude des relations entre l'interieur et l'exterieur du logement.* Paris. Institut de Sociologie Urbaine.

Haumont, N. (1976a) 'Rapport sur la séminaire consacré à "L'appropiation de l'habitat'. In: Korosec. (1976).

Haumont, N. (1976b) 'Les practiques d'appropriation du logement'. In: Korosec. (1976).

Haumont, N. & Raymond, H. (1966) 'L'habitat Pavillonnaire'. Paris Centre de Recherches Urbaines.

Hawley, A.H. (1950) *Ecología humana.* Madrid, Tecnos 1966.

Heath, T. F. (1974) 'Should We tell the Children about Aesthetics, or Schould We let Them Find out in the Street?' In: Canter & Lee (Eds.) (1974).

Hedge, A. & Travis, N. (1979) 'User Evaluation of Open-Plan Offices' In: Canter (Ed) (1979).

Heidmets, M. (1976a) 'On the problems of interrelationship of man phisical environment: An overview'. *Studies in Psychology IV.* Tartu. Tartu State University pp 41-8.

Heidmets, M. (1976b) 'Some socio-psychological problems of urbanization' *Perception and Relations.* Tartu., Estonia Tartu State University.

Heidmets, M. (1977) 'Spatial regulation of human interaction: some curreent problems' *Studies in Psychology, VI.* Tartu., Estonia Tartu State University pp. 72-84.

Heidmets, M. (1978) 'The city inside the man' (Linn inimeses). *Looming, n. 4.* pp. 630-9.

Heidmets, M. (1979) 'Spatial factors in human interaction: Development of concept'. *People, Environment and Space.* Tartu. Estonia Tartu State University pp. 4-28.

Heidmets, M. (1980) 'The socio-psychological problems of environmental personalization' *People, Environment and Relations.* Tallinn. Tpegi 26-49.

Heidmets, M. (1981) 'Enviromental components of the human "self" -a problem for psychology and architecture' In: T. Niit et al. (eds.) (1981) *People and Environment: Psychological Problems.* Tallinn. EOOP 120-21.

Heidmets, M. & Kruusvall, J. (1979) *Psychological problems of spatial organization of club buildings and city centres. (Interim Report of Grant 78/9).* Tallinn. Tallinn Pedagogic Institute.

Heidmets, M. & Kruusvall, J. (1980) 'Studies on the psychological problems of spatial organization of club buildings and city centres. Final report contract number 78/6'. Tallinn. Tpegi. UDK 301. 085: 153. 71006. 7. State registration number 79033507.

Heimeds, M. & Kruusvall, J. & Kilgas, R. (1979) 'Human interaction in public squares' *People, Environment and Space.* Tartu. Estonia Tartu State University 82-99.

Heidmets, M. & Kruusvall, J. & Niit, T. (1977) 'An interim report of grant 78/6' . Tallinn. Tallinn Pedagogic Institute.

Heidmets, M. & Niit, T. (1984) "Leisure groups and their environmental context: results of a comparative study" In: Krampen (1984).

Hellpach, W. (1911) *Geopsyche.* Leipzig. Engelmann. (Quoted by Kruse & Graumann. 1984. Preprint).

Hellpach, W. (1924) 'Psychologie der unvelt' In: Abderhalden (1924). (Quoted by Kruse & Graumann. 1984. Preprint).

Hellpach, W. (1939) (Quoted by Kruse & Graumann. 1984. Preprint).

Hellpach, W. (1951a) *Mensch und volk der grobstadt.* Stuttgard. Enke. (Quoted by Kruse & Graumann. 1987).

Hellpach, W. (1951b) *Sozialpsychologie.* Stuttgard. Enke. (Quoted by Kruse & Graumann. 1987).

Helphand, K. I. (1979) 'Environmental Autobiography' In: Canter (Ed) (1979).

Henderson, J., Brown, J. & Guilford, E. (1984) 'The significance of cnvironmental factors and attitude towards nuclear energy' In: Krampen (1984).

Hernández-Hernández F. (1980) 'Una revisión de los modelos de la Psicología ambiental' In: en Remesar, A. et alt. *Ecología.* Public. Facul. Bellas Artes. Universidad de Barcelona.

Hernández-Hernández F. (1982a) 'Perspectiva interdisciplinar del entorno: la Psicología ambiental desde la Sociología' In: en Remesar, A. et alt. *Lecturas de Conducta y Entorno.* Public. Facul. Bellas Artes. Universidad de Barcelona.

Hernández-Hernández F. (1982b) 'La metodología Ecológica en la Institución Escolar'. Comunic. en las Jorn. sobre Educ. &Ciber. Univ. Barcelona.

Hernández-Hernández F. (1983a) 'Los escenarios. Propuesta metodológica sobre conducta y entorno' In: en Remesar, A. et alt. *Lecturas sobre Conducta y Entorno.* Public. Facul. Bellas Artes. Universidad de Barcelona.

Hernández-Hernández F. (1983b) 'La calidad como medio educativo desde la perspectiva de la educación en áreas urbanas'. San Sebastian. Jornadas sobre Comunicación y Ciencias Educ. Univ. Zorroaga.

Hernández-Hernández, F. (1983c) 'El entorno en la educación'. III Jornadas sobre Organiz., Entorno y Educ. Univ. Barcelona.

Hernández-Hernández, F. (1985) *La perspectiva ecológica en psicología. La psicología ecológica de R. G. Baker como metodología de análisis de la cotidianidad.* BarcelonaTesi Doctoral.

Fac. Psic. Universitat Barcelona (no publ.).

Hernández-Hernández, F. & Sancho, J. M. (1982) 'Imágenes de la Ciudad y Enseñanza de lo Urbano'. Barcelona Comunicación a la VII IAPS Conference.

Hernández-Hernández, Riba & Remesar. (1985) *En torno al entorno*. Barcelona. Leartes.

Hernández-Hernández, F. & Sancho, J. M. (1983) *Enseñamza de lo urbano e Imagen de la ciudad*. I. C. E. Universidad de Barcelona.

Hernández-Hernández, F. & Sancho, J. M. (1984) 'Children's image of their city as a starting point for an urban environmental education" In: Krampen (1984).

Hernández-Ruiz B. (1983) *Percepción del Ambiente Urbano. Mapa Cognitivo de Santa Cruz de Tenerife*. Universidad de La Laguna. Tesis Doctoral.

Hernández-Ruiz B. (1984) 'Mapa Cognitivo de Santa Cruz de Tenerife: componentes evaluativas'. Paper to III Encuentro Nacional de Psicología Social.

Herranz, P. & Miren, K. (1992). 'Aspectos Metodológicos en la Valoración de la Molestia frente al Ruido Ambiental'. In Corraliza Ed. 1993.

Herrou, M. (1979) "Les Marelles: participation d'habitants à la conception de leur logement". In: Simon. (Ed.) (1979).

Hess, S. & Hernández-Ruiz, B. (1992). 'Elaboración de un Inventario de Conducta Ecológica Responsable'. In Corraliza Ed. 1993.

Hesselgren, S. (1954): Arkitecturens Uttrycksmedd, Estocolmo. Almkvist och Wiksell.

Hesselgren, S. (1959) *The language of architecture*. Lund. Studentlitteratur.

Hesselgren, S. (1970) 'Fundamental Needs: Mans'perception of man-made environment' In: Honikman (Ed.) (1970).

Hesselgren, S. (1971) *Experimental studies on architectural perception (Document D2)*. Stockholm. The Swedish Council for Building Research.

Hesselgren, S. (1973a) 'Opening of the C. I. E. study group A' In: Küller (1973).

Hesselgren, S. (1973b) 'Architectural semiotics'. In: Kuller. (1973).

Hesselgren, S. (1975) *Man's perception of man-made environment*. Lund. Studentlitteratur.

Hesselgren, S. (1977) *The psychology of aesthetic value. Why people want beautiful houses and towns. (Rapport T32)*. Stockholm. The Swedish Coucil for Building Research.

Hesselgren, S. (1984) 'Emotional loadings of environmental perceptions'. In: Krampen 1984.

Hesselgren, S. (1984) Recorded personal Interview by Pol. Berlin.

Hienaux, J. P. & Remy, J. (1979) 'Exclusion spatiale et construction d'identité. Les marginalités et leur dynamique dans les agglomérations urbaines' In: Simon (1979).

Higbee, B. L. (1975) 'Psychological classics: Publication that have made lasting and significant contribution' *American Psychologist*. 30, 182-184.

Hill, A. R. (1969) 'Visibility and privacy'. In: Canter. (1969).

Hilleret, M. (1976) 'L'appropriation des éléments de l'ecosystème' In: Korosec (Ed.) (1976).

Hillier, B. (1979) 'Organic Cities: Objective Morphology and Subjective Perception' In: Canter (Ed) (1979).

Hillier, Wrg. (1979) 'Psychology and the subjective matterof Architectural Research' In: Canter (Ed.) (1979).

Hirch, W.,& Singleton, J.F. (1965). *Research Support, Multiple Authorship and Publication Sociological Journals*. Unpublished.

Hiromichi Tomoda, H. T. (1984) 'Outdoor behaivior and consiousness of habitability -The case of a middle-rise apartment-house complex' In: Krampen (1984).

Hoege, H. (1984) 'Ist "Schön - Hässlich"eine asthetisch relevante dimension?' In: Krampen (1984).

Honikman, B. (1970) 'Introduction' In: Honikman (Ed.) (1970).

Honikman, B. (ed.) (1970) *Proceedings of the architectural psychology. Conference at Kingston Polytechnic*. Kingston. Kingston Polytechnic-RIBA.

Hoogdalem, Herbert Van. (1984) 'Comparative floorplan analysis as a mean to develop spatial organizational concepts in the briefing stage of the design process' In: Krampen (1984).

Hope, T. J. & Winchester, S. W. C. (1979) 'The Targets of Crime" In: Canter (Ed) (1979).

Hormuth, S. E. & Lalli, M. (1984) 'The role of urban environments in self-presentation' In: Krampen (1984).

Horwitz, J. (1984) 'The Environmental Psychological context of computer home-Use: 1976-1984' In: Krampen (1984).

Hostein, B. (1976) 'L'espace scolaire vécu par les enfants des travailleurs émigrés' In: Korosec (ed.) (1976).

Howell, S. C. (1979) 'Missing Paradigms in the Evaluation of Elderly Housing' In: Canter (Ed) (1979).

Huber, J. W. (1984) 'Possessions of Personal Value' In: Krampen (1984).

Huber, J. J. (1977) 'Bibliometric models for journal productivity' *Social Indicators Research*. 4, 441-473.

Hudson, A. & Selby, B. (1970) 'Behaviour, Happiness and Home Extensions' In: Honikman (Ed.) (1970).

Hudson, McGrew, PL. (1979) 'Attention structure in a group of pre-school infants' In: Canter (Ed) (1979).

Huici. & Macia, M. A. (1981) 'Constructos en la apreciación del Paisaje' In: en F. Jimenez-Burillo *Psicología y Medio Ambiente*. Madrid. CEOTMA 333-334.

Humphreys, M. A. (1974) 'Relating Wind, Rain and Temperature to Teacher's Reports of Young Children's Behaviour' In: Canter & Lee (Eds.) (1974).

Ibañez, T. & Iñiguez, L. (1984) 'El entorno construido como forma de control social: relaciones de poder en un nuevo modelo escolar'. In: Pol, Morales, Muntañola. (1984)

Imamoglu, V. (1973) 'The effect of furniture density on the subjective evaluations of spaciousness and estimation of size of rooms'. In: Kuller. (1973)

Imamoglu, V. (1979a) 'Students' Images and Evaluations of their Faculty Building'. (P.57) In: Canter (1979)

Imamoglu, V. (1979b) 'Assestement of living rooms by householders and architects'. In: Simon. (1979)

Imbert, M. (1984) 'Production, practice and symbolic appropiation of public spaces in a new town'. In: Krampen 1984

Inman, M. & Graff, Ch. (1984) 'Effects of house style and life cycle stage on family social climate'. In: Pol, Muntañola, Morales. (1984)

Inman, M.A. (1979) 'Influence of Environmental Setting on Human Behavior'. (P.58) In: Canter (1979)

Inui, M. & Miyashiro, C. (1984) 'Effects of number of storeys on the evaluation of living conditions in high-rise flats'. In: Krampen 1984

Iñilguez, L. (1983) *Ecopsicología del control social. Análisis experimental de algunos de sus efectos*. Tesis de licenciatura. Departamento de Psicología. Universidad Autónoma de Barcelona.

Ittelson (1973) 'Environment and cognition'. *La psicologia dell'ambiente*. V. Italiana.

Ives, S.M. (1984) 'Community cognition and response to beach erosion hazard in Carolina beach, N.C.'. In: Krampen 1984

Ives, S.M. & Ingalls, G.L. & Walcott, W.A. (1984) 'Lifestyle and housing preference: Monitoring local housing markets'. In: Krampen 1984

Janssen, J. (1984) 'Personality and perception on building exteriors'. In: Pol, Muntañola, Morales. (1984)

Janssen, J. (1984) 'The effect of professional education and experience on the perception of building exteriors'. In: Krampen 1984

Janssens, J. (1976) 'How people perceive and identify building exteriors. A methodological study'. Lund Unpubl, paper, Lund Intitute of Technology.

Javaloy, F., Ruiz, M. & González, L. (1986) 'Mapa Cognitivo de Barcelona'. Comunicación a I Jornadas de Psicología Ambiental. Madrid, Nov. 1986.

Jenkins, T.H., O'Brien, P. & Alexander, T. (1979) 'Diferences parmi de la conception de l'espace construit et l'appropiation de l'user'. In: Simon. (1979)

Jiménez-Burillo, F. (1981a) 'Psicología ambiental'. In: en Jiménez-Burillo, F. (Ed.) *Psicología y Medio Ambiente*. Madrid. CEOTMA. 197-208.

Jiménez-Burillo, F. (1981b) *Psicología Social*. Madrid. UNED.

Jiménez-Burillo, F. (1981c) 'Hacinamiento y Conducta'. In: F.J.B. (Ed.) *Psicología y Medio Ambiente*. Madrid. CEOTMA.

Jiménez-Burillo, F. (Ed.) (1981d) *Psicología y Medio Ambiente*. Madrid. CEOTMA.

Jiménez-Burillo, F. (1982) *Psicología y medio ambiente*. Madrid. CEOTMA-MOPV.

Jimenez-Burillo, F., Aragonés, J.F. (Eds.) (1986) *Introducción a la Psicología Ambiental*. Alianza editorial, Madrid.

Joardar, S.D. (1979) 'Defined Segregation'. In: Canter (1979)

Jodelet, D. (1982) 'Les representations socio-spatiales de la vile'. In: Derycke, D.H. (Ed.) (1982) *Conception de l'espace*. Paris. Nanterre, Université.

Jodelet, D. (1984 preprint) 'Approaches to people/Environment Relations in France'. In: Altman & Stokols. *Handbook of environmental psychology*. Forcoming by John Wiley & Sons. 1987.

Jodelet, D. (1987) 'The study of people-environment relations in France'. In D. Stokols & I. Altman *Handbook of Environmental Psychology* Vol. 2 pp 1171-1193. Wiley, N.Y.

Jodelet, D. et al. (1978) *Etude de la communication et de la participation dans une ville da moyenne importance*. Paris. Edicions de CNRS.

Joedicke. (Ed.) 'Psychologie und Baven'. Stuttgart. Krämer.

Joiner, D. (1970) 'Social ritual & architectural space'. In: Honikman. (1970)

Joiner, D. (1982) 'The three conferences reflexions: 7 IAPS, Barcelona; Design Policy, London; 20 IAAP, Edimburgh'. *Architectural psychology newsletter*. Kingston-up-thamesn. 2 & 3 vol. XII: 24-30.

Jones, P.J., Natter, W. & Schetzki, T.R. (1993). *Postmodern Contentions*. N.Y. Guilford Press.

Jonge, D. (1979) 'Differences in conceptios and attitudes between designers and users of the built environment'. In: Simon. (1979)

Jurgen, H.P. (1984) 'Wohnquartiersbeschreibung -ein instrument zum erfassen sozial-räumlicher differenzierung städtischer bevölkerung'. In: Krampen 1984

Kafry, D. (1979) 'Fire Survival Skills of Young Children'. (P.60) In: Canter (1979)

Kalekin-Fishman, D. (1984) 'The Kindergarten as a Controlled Environment'. In: Krampen 1984

Kaliaden G.J. & Mahadevan, I. (1984) 'Experience of Crowding and Housing -Environment Priorities in the High-Density Chawls of Bombay'. In: Krampen 1984

Kaminski, G. (1970) *Verhaltenstheorie und Verhaltensmodifikation*. Stuttgart. Klett.

Kaminski, G. (1974) 'Umweltpsychologie. Bericht über ein symposium'. In: Tack, W.H. (Ed) (1974) *Bericht über den 19 Kongress der Deutschen Gesellschaft für Psychologie in Salzburg*. Band 2. Gotinga, Hogrefe, 263-280.

Kaminski, G. (1975) 'Psychologie und Bauen'. In: Einführung. SBF 64 (1975). Stuttgart, Mitteilungen 28, 6-30.

Kaminski, G. (1976a) *Umweltpsychologie: Perspektiven, Probleme, Praxis*. Stuttgart, Ernst Klett.

Kaminski, G. (1976b) *Studi di psicologia ambientale*. Roma. Città Nuova.

Kaminski, G. (1976c) 'Theoretische Komponenten handlùngspsychologischer. Ansätze'. In: Thomas, A. (ed.) (1976) *Psycologie der Handlung und Bewegung*. Meisenheim. GlannHain.

Kaminski, G. (1976d) *Psicología Ambiental*. Buenos Aires. Troquel.

Kaminski, G. (1978) 'Environmental Psychology' *The German Journal of Psychology*, vol.2 n.3 225-239.

Kaminski, G. (1983a) ' The enigma of ecological psychology'. In: *Journal of environmental Psychology*, 3, 85-94.

Kaminski, G. (1983b) 'Potentielle Beiträge Handlungstheoretischer Konzeptionen zur Neuorientierung motivationspsychologischer Perspektiven im Sport'. In J.P. Jansen ande E. Hahn (eds.), *Motivationn Handlung*, Aktivierung, und Coaching im Sport, Schorndorf, Hofman.

Kaminski, G. & Fleischer, F. (1984a) 'Ökologische Psychologie: Ökopsychologische Untersuchungs - und Beratungspraxis' In: H. Hartmann, R. Haubl,(eds.). *Psychologische Begutachtung*. Urban & Schartzenberg. München: Wien - Baltimore, pp. 329-358.

Kaminski, G. (1984b) 'Die koordination psychologischer und nichtpsychologicher Aspekte in der Umweltgestaltung (Beispiel: Städtischer Freiraum)'. In: M. Krampen (ed.): Environment and Human Action Proceedings. 8th. International Conference of the IAPS, West-Krampen 1984

Kaminski, G. (1985a) 'Behavior Setting - Analyse'. In: Graumann, Kruse & Lantermann (eds.), *Umweltpsychychologie*. München, Urban & Schwarzenberg.

Kaminski, G. (1985b) 'Handlungstheorie'. In: Graumann, Kruse & Lantermann (eds.), *Umweltpsychologie (Ökopsychologie)*. München: Urban & Schwarzenberg.

Kaminski, G. (1985c) '10 Jahre Ökopsychologie'. In: D. Albert (eds.). *Ber. 344 Kong. DGFPs Wien 1984*. Gotinga, Hogrefe, 1985, pp. 837-840.

Kaminski, G. (1986a) 'Ökopsychologie und Umweltpolitik'. In: R. Günther & G. Winter (eds.), *Umweltbewusstsein und persönliches Handeln*. Weimnheim: Beltz.

Kaminski, G. (1986b) 'Paradigmengebundene Behavior Setting- Analyse'. In: Kaminski (ed.), *Ordnung und variabilität im Alltagsgeschenhen*. Gotinga, Hogrefe, pp. 154-176.

Kaminski, G. (1986c) *Ordnung und Variabilität im Alltagsgeschenhenn*. Gotinga, Hogrefe.

Kanetkar, V. (1979) 'How Much Time Should be Spent in Conveying Environment?'. (P.61) In: Canter (1979).

Kartashova, K.K. (1980) 'Social aspects of the development of dwelling architecture'. *Architecture USSR*. 10, 26-7.

Kastka, J. (1979) 'On the Influence of Environmental Conditions on the Annoyance Reaction to Outdoor Pollution'. (P.62) In: Canter (1979)

Katz, Z. (1979) 'Clinicians' Room Conceptions: A Facet Approach'. (P.63) In: Canter (1979)

Keenan, S. & Atherton, P. (1964) *The journal literature of physics*. N.Y. American Institut of Phisics.

Kennedy, D. (1984) 'Space for food for people'. In: Krampen 1984

Kenny, C. & Canter, D. (1979) 'The Evaluation of Hospital Wards for Adults'. (P.64) In: Canter (1979)

Kirchner, M. (1975) *La psicología aplicada, a Barcelona*. Tesi Doctoral, Universitat de Barcelona.

Kirillova. (1961) *Scale in architecture*. Moscow.

Kitaura, K. (1979) 'The Distance-Cognization on Vertical Movement in Structures'. In: Canter (1979)

Klein, H.J. (1984) 'Stabilität und wandel von wohnpräferenzen'. In: Krampen 1984

Klein, H.J. (1984) 'Stabilität und wandel von wohnpräferenzen'. In: Krampen 1984

Knnaper, C.K. & Cropley, A.J. (1979) 'Driving Behavior and the Social Environment'. (P66) In: Canter (1979)

Komninos, N. (1984) 'Social relations at production and urban space'. In: Pol, Morales, Muntañola. (1984)

Koop, A. (1975) 'Changer la vie, changer la ville' In: *De la vie nouvelle aux problèmes urbains*. Paris. Union Genérale d'Editions.

Korosec, P. (1973) 'The case of newly constructed zones: Freedom, constrait and the appropiation of spaces'. In: Kuller. (1973)

Korosec, P. (Ed) (1978) *L'appropiation de l'espace*. 3 IAPC. Louvain-la-Neuve. Ciaco.

Korosec, P. (1982) *The main square*. Hassleholm. Aris.

Korosec, P. (1984) 'Recorded personal interview'. Strasbourg.

Korosec, P. (1991a) 'La ville et ses restes'. In A. Germain (dir) *L'amenagement Urbain: Promesses et défis Institut Quebecois de Recherche sur la Culture*. pp.234-267

Korosec, P. (1991b) 'Le Public et ses domaines. Contribution de l'histoire des mentalités à l'étude de la sociabilité publique et privée.' Dans Espaces publics et complexité du social. Rev. *Espaces et Sociétés* n° 62-63 pp29-63

Korosec, P. (1992) Recorded personal interview, by Pol. Thessaloniki, July 1992.

Korosec, P. Levy, M., Decker, S. & Tramoni, M.L. (1976) 'Sauvegarde des sites urbains et appropiation des places publiques'. In: Korosec. (1976)

Krampen, M. (1974) 'A Possible Analogy Between (Psycho) Linguistic and Architectural Measurement - The Type-Token Ratio (TTR)'. In: Canter i Lee. (1974)

Krampen, M. (1979) 'An analogy between Linguistics and Architecture -The Metaphor'. (P.67) In: Canter (1979)

Krampen, M. (1984) 'The achitectural environment a mirored in Children Drawings -A cross comparison'. In: Pol, Morales, Muntañola. (1984)

Krampen, M. (Ed.) (1984) *Environment and Human Action 8-IAPS*. Berlin: Hochschule der Kürnste.

Kruse. (1974) *Rävmliche um welt*. Berlin. De Gruyter.

Kruse. (1975) 'Crowding. Ditche und enge aus sozialpsychologischer sicht'. *Zeitschrift für sozialpsychologie*. 6, 2-30.

Kruse, L. (1992) Recorded personal interview, by Pol. Thessaloniki, July 1992.

Kruse & Arlt. (1984) *An International and Multidisciplinary Bibliography 1970-1981*. Münc./N.Y./Lond./Par. SAUR.

Kruse, L. & Graumann, K. (1984, preprint)(1987) *Environmental psychology in Germany*. Heidelberg. In: Craik & Stokols. *Handbook of Environmental Psychology*. by John Wiley & sons.

Kruse,L. Graumann, CF., Lanbermann, Ed. *Ökologische Psychologie. Ein Handbuk in Sellüsselbegoihen*. München: Psychologie Verlags Union. 1990

Kruusvall, J. (1979a) 'Principles for creating a system of parameters for the comparative study of human activities in urban and rural environments'. *Problems of the countryside and the city*. Tallinn. Tartu State University Proceedings from an all-union seminar.

Kruusvall, J. (1979b) *Psychological problems of spatial organization of club buildings*. Tallinn. Tallinn Pedagogic Institut.

Kruusvall, J. (1980) 'Determination of life-style of families en the urban environment'. *People, environment and space*. Tallinn. Tpegi. 50-89.

Kruusvall, J., Heidmets, M. & Kilgas, R. (1979) 'Psychological problems of spatial organization of soviet club environments'. *Man, environment, space*. Tartu. Tartu State University. 100-117.

Küller, R. (1971) 'Rumsperception'. Lund Institute of Technology. Unpbu. M.Sc. Thesis.

Küller, R. (1972) '(Doctoral dissertation, Department of Theoretical and Applied Aesthetics, Lund Institute of Technology)'. Stockholm. The Swedish Council for Building Research.

Küller, R. (Ed) (1973) *Architectural Psychology. Proceedings of the Lund conference*. Lund, Studentlitteratur/Stroudburg, Penn. Dowden, Hutchinson & Ross.

Küller, R. (1973) 'Beyond semantic measurement In: Küller (Ed) (1973)

Küller, R. (1975) *Semantic environmental description*. Stockholm. Psycologiförlaget.

Küller, R. (1976a) 'The use of space - Some Phisiological and Philosophical aspectes'. In: Korosec. (1976)

Küller, R. (1976b) 'A social science theory of the man-environment interaction'. *The phisical environment and man*. Trondheim Institute for Research in Psychology & Sociology, University of Trondheim.

Küller, R. (1976c) 'The use of space. Some phisiological and philosophical aspects'. In: P.Korosec-Serfat (ed.) *Appropriation of Space. Proceedings of the 3th International Architectural Psychology Conference*. Louis Pasteur Univ. Strasbourg. 154-63.

Küller, R. (1978) 'The activation level of naval environments. An environmental psichology analysis'. Lund. Lund Intitute of Techonology. Unpublished paper.

Küller, R. (1979a) 'Differing demands om interior space in naval environment'. *Conflicting experiences of space. Proceedings of the 4th. Internat. Archit. Psychol. Conference, Catholic Univ. of Louvain*. Belgium. 645-54.

Küller, R. (1979b) 'Differing demands on interior space in naval environments'. In: Simon. (1979)

Küller, R. (1980) 'Architecture and emotions'. In: B. Mikkelides (ed.) (1980) *Architecture for people*. London. Studio Vista. 87-100.

Küller, R. (1981) *Non-visual effects of light and colour. Annonated bibliography. (Document D15)*. Stockholm. The Swedish Council for Building Research.

Küller, R. (1984a) 'Environmnet and Retirement'. In: Pol, Morales, Muntañola. (1984)

Küller, R. (1984b) 'Adress by the Chairman of IAPS'. In: Pol, Morales, Muntañola. (1984)

Küller, R. (1984c) Recorded personal Interview by Pol. Berlin.

Küller, R. (1992) Recorded personal interview, by Pol. Thessaloniki, July 1992.

Küller, R. & Janssen, J. (1979) 'Social Interest Patterns and the Evaluation of the Built Environment'. (P.68) In: Canter (1979)

Küller, R. & Laike, T. (1992) 'Metamorphosis in traffic behavior.' In proceeding XII IAPS.

Lakowski, R., Dwernychuk, L.W., Ganshorn, D. (1979) 'Scaling Techniques for a River Pollution Study'. (P.69) In: Canter (1979)

Langdon, F. (1963) 'The design of mechanised offices: a user study'. *Architects Journal*. 18: 943-947 / 21: 1081-1086.

Langdon, F. (1965) 'Study of annoyance caused by noise in automatic data processing offices'. *Building Science*. 1: 69-78.

Langdon, F. (1966) *Modern offices: A user survey*. LondonBuild.Research Stat., Minist.of Public Build. & Works.

Lantermann, E.D. (1976) 'La capacité dans l'environnement: un important concept dans l'architecture destinée aux personnes âgées'. In: Korosec. (1976)

Lau, J.J.H. (1969) 'Differences between full-sise and scale-model rooms in assessment of lighting quality'. In: Canter. (1969)

Laufer, R. & Wolfe, M. (1976) 'Les intrusions de l'intimité, l'environnement, le contexte interpersonel et les réactions aux intrusions'. In: Korosec. (1976)

Laufer, R.S., Proshanski, H.M. & Wolfe, M. (1973) 'Some analytic dimensions of privacy'. In: Kuller. (1973)

Lawrence, R.J. (1979) 'Comparative experiences of domestic space'. In: Simon. (1979)

Lawson, B. (1970) 'Open and closed ended problem solving in architectural design'. In: Honikman. (1970)

Lawson, B.R. & Walters, D. (1974) 'The effects of a New Motorway on a Established Residential Area'. In: Canter i Lee. (1974)

Lawton, M.P. (1984) 'Supportive housing: Changes in people, services and environments'. In: Krampen 1984

Le Men, J. (1966) *L'espace figuratif et les structures de la personnalité*. Paris. Puf.

Lecompte, W.A. (1979) 'Cross-Cultural Differences in the perception of Crowding'. (P.70) In: Canter (1979)

Lecuyer, H. (1975) 'Psychosociologie de l'espace -I Disposition spatiale et communication en groupe'. *Année Psychologique*. 1975, 75.

Lecuyer, H. (1976a) 'Psychosociologie de l'espace - II Rapports spatiaux interpersonnels et la notion d'espace personnel'. *Année Psychologique*. 1976, 76.

Lecuyer, H. (1976b) 'Adaptation de l'home a l'espace, adaptation de l'espace a l'home'. *Travail Humain*. 1976, 39.

Ledrut, R. (1968) *L'espace social de la ville*. Paris. Anthropos.

Ledrut, R. (1975). *Les images de la ville*. Paris. Anthropos.

Lee, T. (1954) *A study of urbain neighbourhood*. Cambridge. Tesis doctoral.

Lee, T. (1957) 'On the relation between the school journey and social and emotional adjustment in rural infant children'. *British Journal of Education Psychology*. 27: 101-114.

Lee, T. (1961) 'A test of the hypothesis that school reorganization is a cause of rural depopulation'. *Durham Research Review*. 3: 64-73.

Lee, T. (1963) 'The optium provisium and siting of social clubs'. *Durham Research Review*. 14: 53-61.

Lee, T. (1968) 'Urbain neighbourhood as a socio-spatial schema'. *Human Relations*. 21: 241-267.

Lee, L. (1969) 'Do we need a theory?'. In: Canter. (1969)

Lee, T. (1971) 'Psychology and architectural determinism'. *Architects Journal*. 4.8.71/1.9.71/22.9.71.

Lee, S.A. (1973) 'Environmental perception, preferences and designer'. In: Kuller. (1973)

Lee, T. (1976a) *Psychology an Environment*. V.E. Penguin Books. Barcelona. Ed. CEAC.

Lee, T. (1976b) *Psicología e ambiente*. Bologna. Zanichelli.

Lee, T. (1979a) 'Evaluation Study of Visitor Centres'. (P.125) In: Canter (1979)

Lee, T. (1979b) 'Is Spatial Cognition Isomorphic with Behavior?'. (P.126) In: Canter (1979)

Lee, T. (1984) Recorded personal Interview by Pol. Guilford.

Lee, S.A. (1992) Recorded personal interview, by Pol. Thessaloniki, July 1992.

Lefebvre, H. (1968) *Le droit a la vile*. Paris. Anthropos.

Lefebvre, H. (1970) *La revolution urbaine*. Paris. Gallimard.

Lefebvre, H. (1974) *La production social de l'espace*. Paris. Anthropos.

Leger, J.M. (1984) 'Du currieux a l'inutile: Les Habitants face a l'innovation architecturale'. In: Pol, Morales, Muntañola. (1984)

Lenartowicz, K. (1979) 'Topology - A Tool of an Architectural Space Theory'. (P.71) In: Canter (1979)

Lenartowicz, K. (1984) 'Psycological thinking in Architectural theory till 1960'. In: Pol, Morales, Muntañola. (1984)

Lepik, J. (1981) 'The psychological aspect of maintenance of recreational woods'. In: T.Niit, et al. (eds.) (1981) *People and environment: Psychological problems*. Tallinn. EOOP. 219-22.

Leroy, C. (1976) 'La formation historique de l'image de soi & des mécanismes de défense par rapport à l'environnement et autrui;son rôle dans les relations au territoire actuel'. In: Koresec. (1976)

Levy-Leboyer, Cl. (1976) 'La psychologie de l'environnement. Recherches actuelles aux EXATS-Unis'. *Revue de Psychologie Appliquée*. 1976- 4.

Levy-Leboyer, Cl. (1980a) *Psychologie et environnement*. Paris. PUF.

Levy-Leboyer, Cl. (1980b) *Psicologia dell'ambiente*. Bari. Laterza.

Levy-Leboyer, C. (1984a) 'Psychologie, environment et vandalisme'. In: Pol, Muntañola, Morales. (1984)

Levy-Leboyer, Cl. (1984b) Recorded personal Interview by Pol. Paris.

Levy-Leboyer, Cl & Bernard, Y. (1987) 'Psychologie de l'environnement'. *Psychologie Française*. (1987). Tome 32, 1-2.

Lewin, K. (1931) 'Ubergang von der aristotelischen zur galileschen denkweise in biologie und psychologie'. *Erkenntnis*. 1930/31, 1: 421-460.

Lewin, K. (1935).' The Psychological Situation of Reward and Punishment'. In: *Dynamic Theory of Personality*. pp. 114-170. NY: Mc. Graw-Hill.

Lewin, K. (1963) *Principles of topological psychology*. New York. McGraw-Hill.

Lewis, G. (1970) 'Investigation into some design criteria for underwater habitats'. In: Honikman. (1970)

Liljefors, A. (1973) 'Light planning with minimum energy consumption. The quality of lighting'. In: Kuller. (1973)

Lindberg, E., Hartig, T., Garvill, J. & Gärling, T (1992).'Residential-location preferences across the life span'. *Journal of Environmental Psychology*, 12, 187-198.

Lipman, A. (1970) 'Ideology and professional commitment: The architectural notion of community'. In: Honikman. (1970)

Lipman, A. (1982) 'The marketing of meaning: aesthetics incorporated'. In: comunicació a la 7 IAPS, Barcelona(1982) *Recull d'abstracts*. no publicat.

Lipman, A. (1984a) 'Personal Comunication'. Cardiff.

Lipman, A. (1984b) 'G. Orwell 1984'. Guilford, Surrey. Unpublished.

Lipman, A. & Harris, H. (1979) 'Environmental psychology, a sterile research enterprise?'. In: Simon. (1979)

Lipman, A. & Harris, H. (1980) 'Environmental psychology: a sterile research enterprise'. *Architectural Psychology Newsletter*. X, 2: 18-28.

Lipman, A. & Harris, H. (1983a) 'Design and politics of building use, or peeing into wind'. *Designing for building utilisation*. Portsmouth.

Lipman, A. & Harris, H. (1983b) 'Social process, space usage: reflections on socialisation in homes for children'. *Architectural Psychology Newspaper*. 12 (4): 10-27.

Lipman, A. & Russell-Lacy, S. (1974) 'Some social-psychological correlates of new town residential location'. In: Canter i Lee. (1974)

Lipman, A. & Slater, R. (1976) 'Occupaation spatiale et attitrée des sièges dans huit maisons de retraite pour vieillards'. In: Korosec. (1976)

Liverley, W. & Browley, D. (1973) *Person perception in children and adolescence*. London. Wiley.

Llorens, T. (1972) 'El lenguage de la arquitectura en el marco urbano'. CAU14, 42-49.

Llorens, T. (Ed.) (1973) *Hacia una psicología de la arquitectura*. Barcelona. La Gaya Ciència. Col-legi Ofi. d'Arquit. de Catalunya.

Llorens, T (1974a) 'Información y Semiosis'. *Teorema*. vol.IV, n.1, 55-89.

Llorens, T. (1974b) *Información y Semiosis II*

Llorens, T. (1974c) 'L'Arquitectura i l'Urbanisme Valencians'. *Arguments*. n.1, 59-96.

Lofberg, H.A., Lofstedt, B., Nilsson, I. & Wyon, D.P. (1973) 'The effect of heat and light on the mental performance of school children -introduction to a climate chamber experiment'. In: Küller. (1973)

López-Barrios, I. (1986).'El Ruido'. In: Jimenez-Burillo & Aragonés (Eds. 1986).

López-Barrios, I. (1992) 'Nuevas Perspectivas para la Evaluación de la Dimensión Sonora de los Espacios Urbanos'. In Corraliza (Ed. 1993.)

López Piñero, J.M. (1972) *El análisis estadístico y sociométrico de la literatura científica*. València. Centro de Document. e Informát. Facultad Medicina.

Lugassy, F. (1970) *La relation habitat, forêt. Significations et fonctions des boisés*. Paris. Publication de la Recherche Urbaine.

Lugassy, F. (1976) 'La spatialisation de l'identité étayée sur l'image du corps et sur le logement'. In: Korosec. (1976)

Lugassy, F. (1979) 'L'arme ideologique pour le contrôle de l'espace et l'orientation sociétale'. In: Simon. (1979)

Lukesch, H. & Schneewind, K.A. (1978) 'Themen un Probleme der Familiaren Sozializations Forschung'. In: Scheneewind & Lukesch (Eds.), *Familiäre Sozialisation*. Stuttgart: Klett-Cotta, 9-23.

Lynch. (1960a) *The image of the city*. Cambrige, Mass; Barna, MIT Press/v.e. Nueva Vision; Gustavo Gili.

Lynch. (1960b) *L'immagine della città*. Padova. Marsilio.

Maaloe, E. (1973) 'The aesthetic joy and repetition of the ever unpredictable'. In: Kuller. (1973)

MacDonald, R.G. (1979) 'A Study Inside the English Working Class Home'. (P.72) In: Canter (1979)

MacDonald, R. (1980) 'Comments 2 (on Lipman & Harris). '*Architectural Psychology Newsletter*. 10 (2): 31-32.

Macià, M.A. (1979) *Factores de personalidad preferencias en la elección de paisajes*. Tesis Doctoral. Univ. Autónoma de Madrid.

Macià, M.A. (1980) 'Paisaje y Personalidad'. *Estudios de Psicología*. 1, 30-38.

Madge, Ch. (1951) 'Private and public spaces'. *Human relations*. 3: 187-99.26.

Mainardi Peron, E., Baroni, M.R., Job, R. & Samalso, P. (1984) 'Cognitive Strategies In Remembering And Describing Unfamiliar Places'. In: Krampen 1984

Mainardi Peron, E.. Baroni, M. R.,e Zucco, G. (1988) 'The effects of the salience and typicality of objects in natural settings upon their recollection. In: H. Van Hoogdalem, N. L. Prak, T .J. M. Van der Voordt e H. B. R. Van Wegen (Eds.), *Looking back to the future.IAPS 10/1988*. Vol. II, Delft, Delft University Press. 563-572.

Mainardi Peron, E., (1989) Ti ricordi dove? In: M. R. Baroni, C. Cornoldi, R. De Beni, V. D'Urso, D. Palomba, E. Mainardi Peron e L. Stegagno. (1989) *Emozioni in Celluloide. Come si Ricorda un Film*. Cortina Editore, Milano (1989). 117-148.

Mainardi Peron, E. (1990) *Alcuni appunti di psicologia ambientale*. Padova. pp. 50.

Mainardi Peron, E.,& Baroni, M. R. (1991) 'Conoscenza, valutazione e preferenza: la teoria dello schema come chiave interpretativa della relazione individuo-ambiente'. In: C. Ferrari. *Psicologia e Turismo*. Iniziative Culturali/ETS, Sassari. 201-214.

Mainardi Peron, E., Baroni, M. R., e Job, R. (1990) 'Schema versus set-size effects in memory for places'. *Cahiers de Psychologie Cognitive/European Bulletin of Cognitive Psychology*. 10, 231-249.

Mainardi Peron, E., Baroni, M. R., Job, R.. e Salmaso, P. (1985) 'Cognitive factors and communicative strategies in recalling unfamiliar places'. *Journal of Environmental Psychology*. 5, 325-333.

Mainardi Peron, E., Baroni, M. R., Job, R., & Salmaso, P. (1990) 'Effects of familiarity in recalling interiors and external places'. *Journal of Environmental Pychology*. 10, 255-271.

Mainardi Peron, E., Baroni, M. R. e Falchero, S. (1991) 'Describing sport grounds: An investigation of 'functional' and 'acquaintance' familiarity'. *Perceptual and Motor Skills*. 73, 583-590.

Mallet. (1978) 'Discurs en el XXI Congrés Internacional de Psicologia'. Quoted by Montoro (1982)

Mark, F. (1979) 'Participatory'. In: Simon. (1979)

Markus, T. (1967a) 'The role of building performance measurements and appraisal in design methods'. *Architects Journal*. 1567-1573.

Markus, T. (1967b) 'The function of windows - a reappraisal'. *Building Science*. 2: 97-121.

Markus, T. (1970) 'Foreword'. In: Canter (ed.) (1969)

Martinou, S. (1979) 'Conflictual experience of urban space between women and men: reality and tendencies'. In: Simon. (1979)

Martyniuk, O., Flynn, J.E., Spencer, T.J. & Hendrick, C. (1973) 'Effect of environmental lighting on impression and behavior'. In: Küller. (1973)

Masides, M. (1981) 'Una proposta d'equipament per espais de joc'.

Maslow & Mintz. (1956) 'The effects of esthetic surroundings: I. initial effects of three conditions upon perceiving "energy" and "well-being" in faces'. *Journal of Psychology*. 41: 247-254.

Mazerat, B. (1976) 'Appropiation et classes sociales'. In: Korosec. (1976)

Mazis, A; Karaletsou, C. Tsoukala, K. (Ed) (1992). *Socio-Environmental Metamorphoses: Builtscape, Landscape, Ethnoscape, Euroscape*. Proceedings 12-IAPS Conference. Thessaloniki, Greece

Mehmet, E. & Velioglu, S. (1973) 'The therapeutic environment. An attempt at studying the emotional content of architectural space'. In: Kuller. (1973)

Meiss, P. (1984) 'The home: a matrix of past and present'. In: Pol, Morales, Muntañola. (1984)

Menzies, M.R., Goodey, B. & Donnelly, D.M. (1973) 'The place of meaning in perception-related studies in planning'. In: Küller. (1973)

Menzies, M. (1979) 'Rejection by half=acceptance: the fate of man in environmental desing'. *Architectural Psychology Newsletter*. 9 (3): 22-26.

Mercer, J.C. (1974) 'Towards Standing Room Only'. In: Canter i Lee. (1974)

Merton. (1969) 'The Mathew effect in science'. *Science*. 159: 56-63.

Merton. (1977) *La sociología de la ciencia*. Madrid. Alianza Editorial.

Michotte. (1938) 'Lecture to the XI Internacional Conference of Psichology'. In: Quoted by Montoro (1982)

Michotte. (1959) 'Lecture to the XV Internacional Conference of Psicology'. In: Quoted by Montoro (1982)

Mikellides, B. (1973) 'The role of psichology in architectural education'. In: Küller. (1973)

Mikellides, B. (1979) 'Conflicting experiences of colour space'. In: Simon. (1979)

Mikellides, B. (1980) *Architecture for people*. London. Studio Vista.

Mikellides, B. (1984a) 'Psicology as an integral part of architectural Education'. In: Pol, Morales, Muntañola. (1984)

Mikellides, B. (1984b) 'A question of interest'. In: Krampen (Ed) (1984)

Miller, R.L. (1979) 'Middle Class Residents in Gheto Housing: Attitudes and behavior'. (P.75) In: Canter (1979)

Mimura, M. (1984a) 'The Semiological Study of Urban Environment'. In: Pol, Morales, Muntañola. (1984)

Mimura, M. (1984b) 'The interactive structure of the urban environment for man'. In: Krampen 1984

Mintz, N.L. (1956) 'Effects of aesthetic surroundings'. *Journal of Psychology*. 41: 459-66.77.

Mitropoulos, E. (1973) 'Space notation'. In: Kuller. (1973)

Mitropoulos, E. (1976) 'Cheminements et barrières par rapport aux relations humaines dans des espaces semi-privés/publics'. In: Korosec. (1976)

Mitropoulos, E. (1979) 'Comunications and control'. In: Simon. (1979)

Mitscherlich, A. (1963) *Psychanàlyse et urbanisme*. Paris. Gallimard.

Mitscherlich, A. (1965) La inhospitalidad de nuestras ciudades. Madrid. Alianza Editorial.

Moar, I. (1979) 'The Nature and Acquisition of Cognitive Maps'. (P.76) In: Canter (1979)

Moles, A. (1976) 'Aspects psychologiques de l'appropiation de l'espace'. In: Korosec. (1976)

Moles, A. (1984) 'Recorded personal Interview by Pol'. Paris.

Moles, A. & Rohmer, E.. (1964) *Psychologie de l'espace*. Paris. Casterman.

Moles, A. & Rohmer, E. (1970) *Psicología del Espacio*. Aguilera.

Mollo, S. (1984) 'Souvenirs d'ecole'. In: Pol, Morales, Muntañola. (1984)

Montoro, L. (1982) *Los congresos internacionales de psicología. 1889-1960*. València. Tesi Doctoral, Facultat de Psicologia.

Moore, G.T. (1974) 'The Development of Environmental Knowing: An Overview of an Interactional-Constructivist Theory and some Data on Within-Individual Development Variations'. In: Canter i Lee. (1974)

Moore, R.C. (1974) 'Patterns of Activity in Time and Space: The Ecology of a Neighbourhood Play ground'. In: Canter i Lee. (1974)

MOPU-Diputacion de Barcelona(1983) *Actas de las Primeras Jornadas de Educación Ambiental*. Sitges (Barcelona).

Morales, M. (1981) 'Els espais de joc. Valoració d'una proposta d'equipament de parcs infantils'. Barcelona. Sol.liQuoted by Anònima de Disseny Associats.

Morales, M. (1982) 'Utilización del espacio y práctica Pedagógica'. *Cuadernos de Pedagogía*. Revista mensual de Educación. 86 Año VII pp.13-15.

Morales, M. (1983) *El nen i l'entorn. Activitats i orientacions per la primera infancia*. Barcelona. Oikos-Tau.

Morales, M. (1984a) 'Taller Educació Ambiental'. In: Pol, Morales, Muntañola. (1984)

Morales, M. (1984b) *El Niño y el Medio Ambiente. Actividades y Orientaciones para la primera infancia*. Barcelona. Oikos-Tau.

Morales, M. (1984c) 'L'enfant et l'environnement, une relation necessaire'. In: Krampen 1984

Morales, M. & Pol, E. (1980) 'Apunts per a un estudi de l'evolució de l'espai dedicat a l'infant, en el marc escolar a Catalunya'. *L'Entorn escolar: problemàtica psicològica, educativa i de disseny*. Sèrie Seminari-7 ICE. Universitat de Barcelona 5-19.

Morales, M. & Pol, E. (1982)'El espacio escolar. Un problema multidisciplinar'. *Cuadernos de Pedagogía*. Revista Mensual de Pedagogía. 86 Año VII pp.4-6.

Monreal, P., Martínez, M. & Iñiguez, L. (1992).'Evaluación de la Calidad de los Centros de Atención a la Tercera Edad'. In Corraliza Ed. 1993.

Moreno, E. & Ferrer,A. (1991). 'Gentrificación y Degradación de un Barrio'. In: De Castro (Ed.1991).

Moreno, E., Garrido, S., Sandoval, G., Saura, C. & Pol, E. (1988) 'Impacto Social frente a un Servicio Contaminante.'In proceeding II Jornadas de Psicología Ambiental.

Moreno, E. & Pol, E. (1990). 'A Service that Produces Pollution: Its Impact on the Environment and the Resulting Social Pressure'. In proceeding XI IAPS.

Moscovici et al. (1983) *La sensibilité des français au problèmes d'environnement.* Paris. Ministère de l'Environnement.

Moscovici, et al. (1984) *Psychologie Sociale.* Paris, P.U.F.

Moser, G. & Levy-Leboyer, C. (1979) 'Genesis and Evolution of Environmental Research in France: The case of Noise and Air Pollution'. (P.77) In: Canter (1979)

Mougenot, C. (1984) 'Histoire d'une brique dans le ventre'. In: Krampen 1984

Muchow & Muchow. (1935) Vital space in urban children. Quoted by Kruse & Graumann (1984, preprint).

Mueller, W. (1979a) 'Translation of User Requirements into House Designs: An M. D. Analysis'. (P.78) In: Canter (1979)

Mueller, W. (1979b) 'Usage of indoor and outdoor spaces of the house: architects' and dwellers' percepcions'. In: Simon. (1979)

Mueller, W. (1985) 'A utilization-focused model of design evaluation'. In: Pol, Morales, Muntañola. (1985)

Mugerauer, R. (1984) 'Openings in the city: Openings for the city'. In: Krampen 1984

Muhlich, E. (1976) 'L'appopiation'. In: Korosec. (1976)

Mukerjee. (1940) *Social Ecology.* citado por Kruse & Graumann (1984, preprint).

Mullins, C. (1968) 'The Distribution of social and cultural properties in informal communication. Network among Biological Scientists'. *American Sociological Review.* 33, 1968, 786-797.

Mulvihill, R. (1979) 'The application of Social-Scientific Knowledge and Insight to the Design of the Housing Environment'. (P.79) In: Canter (1979)

Munné, F. (1985) '¿Dinámica de grupos o actividad del grupo?', (Pre-print) Publ in *Boletín de Psicología.* 1985, 9 pp. 29-47.

Muntañola, J. (1973a) 'The child's Conceptions of Place To Live' In: *Environmental Research And Practice.* Dowden, Hutchinson & Ross. Proceedings Preiser, W.

Muntañola, J. (1973b) 'Towards an Epistemological Analysis of Architectural Design Activity'. *Behavior and Meaning.* London. Broadbent and Llorens eds, Wiley and Sons.

Muntañola, J. (1974) *La Arquitectura como Lugar.* Barcelona. Gustavo Gili.

Muntañola, J. (1975) *Materiales para un análisis crítico de la Enseñanza de la Arquitectura.* Escuela de Arquitectura de Barcelona.

Muntañola, J. (1978a) 'Towards an Environmental Childhood' In: *The Child in the World of Tomorrow.* Prentice Hall.

Muntañola, J. (1978b) *Topos y Logos.* Barcelona: Paidós.

Muntañola, J. (1978c) *Psicología y Arquitectura.* Barcelona. Escuela de Arquitectura.

Muntañola, J. (1978d) *El Barri del Pi.* Barcelona. Escuela de Arquitectura.

Muntañola, J. (1979a) 'Strategies for the Invetion of Architectural Objects'. (P.80) In: Canter (1979)

Muntañola, J. (1979b) *Topogenesis uno: Ensayo sobre el Cuerpo y la Arquitectura.* Barcelona. Oikos-Tau.

Muntañola, J. (1979c) *Topogénesis dos: Ensayo sobre al Naturaleza Social del Lugar.* Barcelona. Oikos-Tau.

Muntañola, J. (1979d) *Quatre Lectures d'Un Saber fer Arquitectura.* Barcelona. Escuela de Arquitectura.

Muntañola, J. (1979e) 'Strategies for the Invention of Architectural Objects'. Guildford Presented in the Intern. Congres of Envir. Psychol.

Muntañola, J. (1980a) *Topogénesis tres: Ensayo sobre la Significación en la Arquitectura.* Barcelona. Oikos-Tau.

Muntañola, J. (1980b) *Arquitectura de los Setenta.* Barcelona. Oikos-Tau.

Muntañola, J. (1980c) *Didáctica Medioambiental: Fundamentos y Posibilidades.* Barcelona. Oikos-Tau.

Muntañola, J. (1981) *Actividades Didácticas del Medio Ambiente 8-12 años.* Barcelona. Oikos-Tau Colección didáctica medioambiental.

Muntañola, J. (1982) 'Architect's Experience of Places'. Edimburg 20th. Intern. Conference of Applied Psychology.

Muntañola, J. (1983a) *Actividades Didácticas del Medio Ambiente para los 7-8- años de edad.* Oikos-Tau.

Muntañola, J. (1983b) *Comprendre l'arquitectura/ Conocer la arquitectura.* Barcelona. Teide.

Muntañola, J. (1983c) *El Niño y la Arquitectura.* Barcelona. Oikos-Tau. Colección del Medio Ambiente.

Muntañola, J. (1983d) *Adolescencia y Arquitectura.* Barcelona. Actividades didáct. medio amb. 12-17 años.Oikos-Tau.

Muntañola, J. (1984a) 'La concepció de la ciutat en el nen'. In: Pol, Morales, Muntañola. (1984)

Muntañola, J. (1984b) 'Disseny Arquitectonic i Ciències Socials. Malentesos del passat i alternatives pel futur'. In: Pol, Morales, Muntañola. (1984)

Muntañola, J. (1984c) 'Architectural Design as a way of Acting and as a way of thinking'. Berlin. For the 8th. Conference IAPS.

Muntañola, J. (1985) *Els nens i la seva ciutat. Barcelona - New York.* (versió Catalana i Castellana). Barcelona. Centre del Medi Urbà. Programa MAB-II UNESCO - Ajuntament de Barcelona. Informe tècnic nº 4.

Muntañola, J. & Capel, H. (1978) *Aprender de la Ciudad.* Barcelona. Escuela Superior de Arquitectura.

Muntañola, J. & Hart, R. (1981) 'Learning from the City'. Internat. Exchange Progr. between New York-Barcelona.

Muntañola, J., Morales, M. & Pol, E. (1979) *Llista de Jocs Espacials.* Barcelona. Col.legi Oficial d'Arquitectes de Catalunya.

Muntañola, J., Morales, M. & Pol, E. (1980) 'L'apropiació de l'Espai. L'infant i el Mestre, dues estrategies'. *Guix.* Barcelona. Elements d'acció educativan. 33-34 pp. 47-50.

Muntañola, J., Morales, M. & Pol, E. (1981) 'L'escola en la imatge que el nen té de la ciutat'. *Imatge de l'escola, interacció ambiental, vers una nova normativa.* Sèrie Seminari n.10 ICE Universitat de Barcelona 39-51.

Muntañola, J., Pol, E. & Morales, M. (1981) 'L'escola en la imatge que el nen té de al ciutat'. In: en Pol, E. & Morales, M. (Eds.) (1981) *Imatge de l'escola/ Interacció ambiental/ Vers una nova normativa.* ICE Universitat de Barcelona.

Murphy, J.J. (1973) 'Lotkas Law in the Humanities'. In: Quoted by Montoro(1982) *Jasis.*

Murray, R. (1974) 'The Influence of Crowding on Children's Behaviour'. In: Canter & Lee (1974)

Murray, R., Boal, F.W. & Poole, M.A. (1979) 'Ethnicity and Cognitive Space in Belfast'. (P.81) In: Canter (1979)

McDowell, K. (1984) 'Perceived satisfaction, control and quality in the home environment'. In: Pol, Muntañola, Morales. (1984)

McDowell, K. & Kishchuk, N. (1984) 'School teacher's assessment of their schools'. In: Pol, Morales, Muntañola. (1984)

McIntyre, D.A. & GRIFFITHS, I.D. (1974) "The termal Environment: Buildings and People". In: Canter & Lee. (1974)

McKennan, G., Parry, C.M. & Tilic, M. (1979) 'Bringing Lighting from the Ceiling to the Desk'. (P.74) In: Surrey. (1976)

McKenzie, W. (1979) 'Toward Better Desicions in Moving or Choosing a House'. (P.73) In: Canter (1979)

McLachlan, G.M. (1984) 'Clients' perception of a social model recovery facility'. In: Krampen 1984

Naka, Y. (1984) 'Research on complicated passenger flow in railway station'. In: Krampen 1984

Nakamura, Y. & Kitamura, S. (1979) 'The analysis of associative structure of image map'. (P.82) In: Canter (1979)

Nandi, S. & Dasgupta, S.K. (1984) 'A study on the body-buffer zone of female prison inmates'. In: Krampen 1984

Navarro, V. (1975) 'Nota acerca de los métodos de ajuste de la ley de productividad de Lotka'. *Medicina Española.*

Neill, S. & Denham, E. (1979) 'The influence pre-school design on children and staff'. (P.83) In: Canter (1979)

Niit, T. (1978) 'Crowding: its models, determinants and consequences for the social activities of people' In: *Problems of contemporary ecology. II.* Tartu, Estonia. Tartu State University.

Niit, T. (1979) 'Work environment, work content, and human social activity'. (P.83) In: Canter (1979)

Niit, T. (1980a) 'General trends in the development of theories about the interrelationship of man and his surroundings'. *People, Environment and Space.* Tallinn. Tpegi. 5-25.

Niit, T. (1980b) Final report of grant 78/9In: *Psychological problems of spatial organization of cinema buildings.* Tallinn. Tallinn Pedagogic Institute.

Niit, T. (1981a) 'Residential density and social "pathology": Which pathology to seek?'. In: T. Niit, et al. (eds.) (1981) *People and environment: Psychological problems.* Tallinn. EOOP. 127.

Niit, T. (1981b) 'Density and crowding: theories and hypotheses'. In: *People in social and physical environment.* Tallinn. Tpegi.

Niit, T. (1992) 'Recorded personal interview, by Pol'. Thessaloniki, July 1992

Niit, T. & Heidmets, M. (1984) 'Cinemas as recreational settings: Psychological problems'. In: Krampen 1984

Niit, T., Heidments, M. & Kruusvall, J. (Eds.) (1981) *Man and environment: Psychological aspects.* Tallinn. Estonian Branch of Soviet Psychological Society.

Niit, T., Kruusvall, J. & Heidmets, M. (1981) 'Environment psychology in the Soviet Union'. *Journal of Environmental Psychology.* 1 (2): 157-178.

Niit, T. & Lehtsaar. (1981) *Privacny, crowding and the activity of man.*

Niit, T., Raudsepp, M., Liik, K. (Ed) 1991 *Environment and social development.* Proceedings of the east-west colloquium in Environmental Psychology. Tallinn, Pedagogical Institut

Noble, A. (1979) 'Some social parameters in the design of hospital wards'. (P.85) In: Canter (1979)

Norberg-Schulz, D. (1967) *Intensions in architectural design.* Olso. Universitetsförlaget.

Noschis, K. (1975) 'L'appropiation dans un laboratoire d'expérimentation: le rapport entre représentation et appropiation'. In: Korosec. (1975)

Noschis, K. (1979a) 'The environment while experienced: A Methodology'. (P.86) In: Canter (1979)

Noschis, K. (1979b) 'Identity in the urban environment: appropiation as a potential sources for conflicts '. In: Simon. (1979)

Noto, P. & Panyella, M. (1986) *Introducció a la biologia Social.* Barcelona, Edicions 62.

Null, R. (1979) 'Student Perceptions of Social and Academic Climates of Dormitory Suites'. (P.87) In: Canter (1979)

Oddono. (1979) *Psicologia dell'ambiente: fabbrica e territorio.* Torino. Giappichelli.

Ojala, H. (1980) 'Personal space of man and factors influencing it' Graduation thesis. Tartu. Tartu State Univ. Depart. Psychology. Unpublished.

Olivegren, H. (1970) 'A better social-psychological climate in our housing areas'. In: Honikman. (1970)

Olivier, M. (1972) *Psychanalyse de la maison.* Paris. Le Sevil.

Orlow, P.B. (1981) 'The provision of the space and stages of development of family structure in new cities'. In: T.Niit, et al. (Eds.) (1981) *People and environment: Psychological problems.* Tallinn. EOOP. 135-8.

Osgood, C.E., Suci, G.J. & Tannenbaum, P.H. (1957) *The measurement of meaning..* Urbana, Ill. University of Illinois Press.

Pagés, Fourcade & Cafson. (1974) 'La psychologie ecologique: aplication validatrice ou analise de mecanismes'. *Psychologie et Espace.* Paris. Institut de l'environnement.

Pagés, R. (1984) 'Recorded personal Interview, by Pol'. Paris.

Palmade, J. (1971) 'Le psychologie et l'urbanisme'In: *Le psychologie dans la société.* Paris. ESF.

Palmade, J. (1977) *Système symbolique idèologique de l'habiter.* Paris. Centre Scientific et Technique du Bâtiment.

Palmade, J. et al.. (1970) *La dialectique du logement et de son environnement; etude explorative. Contribution a une psychologie de l'espace urbain*. Paris. Copedith.

Pamir, H. (1979) 'Diferences in environmental control, construal style and the use of space'. In: Simon. (1979)

Pamir, Imamoglu, Teymur (Eds) (1990) Culture, Space, History. Proceedings 11-IAPS Conference Ankara, Turkey; METU Press.

Papadopoulus, A.T. (1984) 'Formes architecturales et groupes sociaux: y-a-t-il une relation?'. In: Krampen 1984

Papageorgiu-Sefertzi, R. (1984a) 'Some methodological issues on the investigation of the socio-psysical space in schools '. In: Pol, Morales, Muntañola. (1984)

Papageorgiu-Sefertzi, R. (1984b) 'Social and ideological implications of the architectural form observations in greek school buildings'. In: Krampen 1984

Park., Burgess. & McKenzie. (1925) *The city*. Chicago. Chicago University Press.

Pauls, J. (1979) 'Documenting the "Stair Event" using film and video methods'. (P.88) In: Canter (1979)

Payne, I. (1969) 'Pupillary responses to architectural stimuli'. In: Canter. (1969)

Pearlman, W. (1979a) 'Theorical aspects of environmental psychology'. (P.89) In: Canter (1979)

Pearlman, W. (1979b) 'On the semiology of space'. In: Simon. (1979)

Peiró,J.M. (1980) 'Colegios invisibles en psicología'. *Análisis y modificación de conducta*. 11-12, 25-50.

Peiró, J.M. (1981) 'Colegios invisibles en psicología'. In: Carpintero y Peiró (1981).

Peiro, J.M. & Carpintero. (1981) 'Revistas en modificación de conducta: un estudio de la red de comunicación en la especialidad'. In: Carpintero & Peiró (eds.) (1981)

Peled, A. (1979a) 'The experience of role and the division of space'. In: Simon. (1979)

Peled, A. (1979b) 'Symbolic integration- Notes on the spatiality of places'. (P.90) In: Canter (1979)

Peled, A. & Vos, M. (1976) 'Une étude sur l'appropiation de l'espace dans un lycée par un exemple de 34 élèves'. In: Korosec. (1976)

Pelletier, F. (1975) 'Quartier et communication sociale, perspective pour une Anthropologie urbaine'. *Espaces et sociétés*. 1975, 15.

Pennartz, P. (1984a) 'Homeliness and beauty in the urban environment'. In: Pol, Morales, Muntañola. (1984)

Pennartz, P. (1984b) 'Qualitative research: an aid to designers. A case-study of a home for psyco-geriatric patiens'. In: Krampen 1984

Penyarroja, J. & Penyarroja, T. (1983) 'La vivienda rural como alternativa'. CIMAL. 22, 1983.

Perico, A. (1984) 'La arquitectura popular como "patern lenguage 'avant la lettre"'. In: Pol, Morales, Muntañola. (1984)

Perussia, F. (1979) *Il disegno in psicologia ambientale: un contributo di ricerca nell'età evolutiva*. Milano. Unicopoli.

Perussia, F. (1983) 'A critical approach to environmental psychology in Italy'. *Journal of Environment Psychology*. 3 (3): 263-279.

Pezanou, A. (1984) 'Tipologie architecturale du pavillonaire 1925-75. Le cas de la proche banlieue parisienne'. In: Krampen 1984

Philip, D. (1979) 'Aesthetic judgement of alternative building facades'. (P.91) In: Canter (1979)

Piche, D. (1979) 'The geographical understanding of children: Some theorical issues'. (P.92) In: Canter (1979)

Pidwell, D.M. (1979) 'A cross-cultural study of city environment perception'. (P.93) In: Canter (1979)

Pineau, C. (1979) 'Differential Psychology of home comfort'. In: Canter (1979)

Pinillos, J.L. (1977) *Psicopatología de la vida urbana*. Madrid, Espasa Calpe

Pinillos, J.L. (1990) 'Personal Comunication'.

Platz, A. (1965) 'Lotka's law and research visibility' *P. Refers*. 16, 566-568.

Poche, B. (1979) 'La consultance architecturale, pédagogie de l'espace ou processus de légitimation sociale?'. In: Simon. (1979)

Pol, E. (1979) *Home-Entorn. Psicologia de l'Environament*. Tesi de Llicenciatura. Universitat de Barcelona.

Pol, E. (1980) 'A la descoberta de l'Entorn. Pindoles psicològiques'. In: *Guix, 33-34*, pp. 51-53. Barcelona.

Pol, E. (1981) *Psicología del Medio Ambiente*. Barcelona. Oikos-Tau.

Pol, E. (1982) 'Programes institucionals de psicologia de l'environament/Institutional Programs on Environmental Psychology'. In: Pol, Muntañola i Morales (eds.) (1984).

Pol, E. (1984b) 'Aproximation a la psychologie de l'environnement en europe'. In: Krampen (Ed.) 1986.

Pol, E. (1984c) 'Home-environment: Aspectes qualitatius'. In: Pol, Muntañola, Morales. (1984)

Pol, E. (1984d) 'Home-Environament. Aspectes qualitatius'. In: Pol, Muntañola, Morales (eds.) (1984)

Pol, E. (1992a).'Procesos Psicológicos en el Uso y Organización del Espacio'. In J.A. Corraliza (Ed) 1993.

Pol, E. (1992a) *Perfils professionals del psicòleg ambiental*. Monografies Psico-socio-ambientals. Màster en Intervenció Ambiental. Department of Social Psychology. University of Barcelona.

Pol, E. (1993) 'La Apropiación del Espacio' .In Fernández, *El Debate sobre el Espacio y la Familia*. Barcelona. Contextum.

Pol et al. (1984) 'Approximation to european environmental psychology'. Berlin. Paper in 8th.IAPC.

Pol, E.,& Dominguez, M. (1986). 'Satisfaccion ambiental, salud mental y calidad de vida'. In Proceeding I Jornadas de Psicología Ambiental.Madrid 1986.

Pol, E. & Dominguez, M. (1987) *Representació Social de la Qualitat de Vida a Barcelona*. Unpublished repport for Environmental and Health Department, Barcelona City Council.

Pol, E., García Borés, J.M., Esteve J.M., Llueca, J. (1992). 'Centro de Reforma de Menores: La Dimensión Ambiental'.*Intervención Psicosocial*. Vol. 2. pp. 47-58.

Pol, E., García-Borés, J. Esteve, J.M., Llueca J. (1992). 'Patterns and Environmental Orientations for the Desing of Youth Custody Centres'. Proceedings in XII IAPS.

Pol, E. & Morales, M. (1979) 'Psicologia de l'Environament: una necessitat en la transformació ecològica de l'ensenyament'. *Ecologia i Educació*. Sèrie Seminari-4-ICE. Universitat de Barcelona67-74.

Pol, E. & Morales, M. (1980) 'Evolución del espacio dedicado a los niños en el marco escolar. Aportaciones concretas del proceso en los últimos 100 años en Cataluña'In: *Le jeune enfant, citoyen à part entière?*. Quebec. Actes del XVI Congrés Mundial de la OME.

Pol, E. & Morales, M. (1980) 'Reflexions sobre l'entorn escolar'. *Guix*. Elements d'acció educativan.33-34 pp.4-8.

Pol, E. & Morales, M. (Eds.) (1980) *L"entorn escolar: problemàtica psicològica, educativa i de disseny*. Sèrie Seminari-7-ICE. Universitat de Barcelona.

Pol, E. & Morales, M. (Eds.) (1981) *Imatge de l'escola, interacció ambiental, vers una nova normativa*. Sèrie Seminari n.10 ICE. Universitat de Barcelona.

Pol, E. & Morales, M. (1983) 'Calidad ambiental en la escuela'. In: Jornadas sobre la Cal.Amb.en la Esc. (1983). Madrid. Ministerio de Educación.

Pol, E. & Morales, M. (1984) 'Estudi sobre les relacions sinonòfiques en els espais d' esbarjo de les escoles'. In: Pol, Morales, Muntañola. (1984)

Pol, E., Morales, M. & Muntañola, J. (1983) *Descubrim...L'Entorn Urbà. Aula Visual*. Video didáctico. Generalitat de Catalunya.

Pol, E., Morales. & Muntañola. (Eds.) (1984) *Vers un millor entorn escolar / Towards a better school environment / Hacia un mejor entorno escolar*. Public. de l'ICE.Edic. i public. Univ. Barcelona.

Pol, E. & Moreno, E. (1992). 'Gentrification and Degradation of a Neighborhood. Social and Environmental Factors'. Proceedings in XII IAPS.

Pol, E., Muntañola., Morales. (Eds.) (1984) *Home-Environament. Aspectes qualitatius / Man-environment. Qualitative Aspects. / Hombre-Entorno. Aspectos cualitativos. 7 IAPS.*

Barcelona. Edicions Universitat de Barcelona.

Pol, E., Jiménez-Burillo. & Sánchez-Robles. (Un-published) 'Environmental psychology in Spain'.

Pol, E. & Pich, J.(Eds.) (1989). Proceedings of *II Jornadas de Psicología Ambiental*. P. Mallorca.

Pol, E., Valera, S. & Freixa J. (1988). 'The Barcelona District's Look'. In proceedings X IAPS.

Pomeroy, W. (1979) 'Attitudes of individuals and groups in assessing the design of buildings'. In: Simon. (1979)

Poodry, D. & Lee, S. (1979) 'Obsevations on neighborhood selfdefinition through inhabitation'. In: Simon. (1979)

Powell, J. A. & Yerrell, P. (1984) 'The systemic assessment of utilisation of energy resources in primary schools'. In: Pol, Morales, Muntañola. (1984).

Power, R. (1979) 'Hypotheses in perception: Some cross-environmental data'. (P.95) In: Canter (1979)

Poyner, B. (1979) 'House purchasing behavior of council tenants'. (P.96) In: Canter (1979)

Prak, N.L. (1979) 'Different meanings in architecture'. (P.97) In: Canter (1979)

Preiser, W.F.E., Harrington, D.L. & Trujillo, M.A. (1979) 'Architectural/Aesthetic Guidelines: A case study of the stronghurst neighbourhood in Alburquerque, New Mexico'. (P.98) In: Canter (1979)

Preiser, W.F.E. (1984) 'A combied tactile/electronic guidance system for visually impaired persons in indoor and outdoor spaces'. In: Krampen 1984

Preiser, W.F.E. (1984) 'Senior centers: Literature evaluation, walkthrough post-occupancy evaluations and a generic program for the city of alburquerque'. In: Krampen 1984

Pressey. (1921) 'The influence of colour upon mental and motor efficiency'. *American Journal of Psychology*. 32: 326-354.

Pressman, N.E.P. (1979) 'Images, plans and realities in the development of new settlesments'. In: Simon. (1979)

Preusler, B. 'Ein architektenleben im wandel sozialer figurationen'. In: Krampen 1984

Price & Beaver. (1966) 'Collaboration in an invisible college'. *American Psychology*. 21: 1011-1018.

Price, D.J.S. (1951) 'Quantitative measures of the development of science'. *Arch. Inst. Hist. Sci.*. 14, 85-93.

Price, D.J.S. (1961) *Science since Babilon*. Yale University Press.

Price, D.J.S. (1963) *Little science, big science*. N.Y. Columbia University Press.

Price, D.J.S. (1971). *Little Science, Big Science*. NY. Columbia University Press.

Priestley, T. (1984) 'The enviromenment behavior perspective and assessment of landscape aesthetics-powerline siting and analysis in North America'. In: Krampen 1984

Prieto, F. (1985) *Impacto y presencia de la obra de John Broadus Watson en la psicología actual (1966-1982). Un estudio objetivo*. València. Disertation. Facultat de Psicologia.

Proceedings of the East-West colloquium in Environmental Psychology. See Niit, T., Raudsepp, M., Liik, K. (Ed) 1991

Proceedings of *I Encuentro Iberoamericano de Psicología Social* (1980). Dep. Social Psychology. University of Barcelona.

Proceedings of *II Encuentro Iberoamericano de Psicología Social* (1981). Dep. Soc. Psych. Univ. Complutense, Madrid.

Proceedings of *III Encuentro Nacional de Psicología Social (1983)*. Dep. Soc. Psych. Univ. La Laguna.

Proceedings of IV Encuentro P.S. - *I Congreso Nacional de Psicología Social* (1985). Dep. Soc. Psych. Unv. Granada.

Proceedings IAPS. Conference of Dalandhui, U.K. (1969). See Canter (Ed) 1969d.

Proceedings IAPS. Conference of Kingston ,U.K. (1970). 1th IAPC. See Honikman (Ed) 1970.

Proceedings IAPS. Conference of Lund, Sweden, (1973). 2th IAPC. See Küller (Ed) 1973.

Proceedings IAPS. Conference of Surrey-I, U.K. (1974). ICEP. See Canter & Lee (Ed) 1974.

Proceedings IAPS. Conference of Strasbourg, France (1976). 3th IAPC. See Korosec (Ed) 1978.

Proceedings IAPS. Conference of Louvain-la-Neuve, Belgium (1979). 4th IAPC. See Simon (Ed) 1979.

Proceedings IAPS. Conference of Surrey-II, U.K. (1979). ICEP. Unpublished. See Canter 1979 (Ed)

Proceedings IAPS. Conference of Barcelona, Spain (1982). 7th IAPS. See Pol, Muntañola & Morales (Ed) 1984

Proceedings IAPS. Conference of Berlin, Germany (1984). 8th IAPS. See Krampen (Ed) 1984.

Proceedings IAPS. Conference of Haifa, Israel (1986). 9th IAPS. See Churchman (Ed) 1986.

Proceedings IAPS. Conference of Delft, Netherlands (1988). 10th IAPS. See Van Hoogdalem et al. (Ed) 1988

Proceedings IAPS. Conference of Ankara, Turkey (1990). 11th IAPS. Pamir, Imamoglu, Teymur (Eds) (1990)

Proceedings IAPS. Conference of Thessaloniki, Greece (1992). XII-IAPS. See Mazis et al. (Ed) 1992

Proceedings School & Env. *Jornadas sobre Ambientes Escolares* (1973). ICE: Universidad de Barcelona. INCIE, UNESCO.

Proceedings Schol & Env. Jornadas (1979) *Ecología y Educación*. ICE: Universidad de Barcelona.

Proceedings School & Env. I Jornades *Entorn Escolar. Problemàtica Psicològica, Educativa i de Disseny*. (1979). See Pol & Morales (Ed) (1980).

Proceedings School & Env. II Jornades sobre l'Entorn Escolar 1980. See Pol & Morales (Eds) (1981) *Imatge de l'escola, Interacció Ambiental, vers una nova normativa*. ICE Universitat de Barcelona-BCD.

Proceedings School & Env. *Jornades sobre Edificació Escolar (1981)*. Generalitat de Catalunya.

Proceedings School & Env. III Jornades sobre l'Entorn Escolar 1982. Pol, Morales, Muntañola (Eds) (1984)eedings School & Env. Jornadas sobre *Calidad Ambiental en la Escuela (1983)*. Madrid. Ministerio de Educación.

Proceedings School & Env. *Jornadas sobre Organización, Entorno y Educación (1983)*. Dept. de Pedagogía, Universidad de Barcelona.

Proceedings School & Env. *Jornadas sobre Educación Ambiental (1983)*. Diputació de Barcelona y M.O.P.U., Sitges.

Proceedings School & Env. *Jornades d' Educació Ambiental* (1983). ICE Universitat de Barcelona.

Proceedings School & Env. IV Jornades sobre l'Entorn Escolar (1984) *De l'Arquitecte al Mestre, del Dissenyador a l'usuari*. ICE Univ. Barcelona.BCD. Facul. de Psicologia.

Proceedings of I Jornadas *de Psicología Ambiental (1986)*. See Aragonés & Corraliza (1986).

Proceedings of II Jornadas de Psicologia Ambiental (1989) Palma de Mallorca. See Pol & Pich (Ed) (1989)

Proceedings of III Jornadas de Psicologia Ambiental (1991) Sevilla. See De Castro (Ed) (1991).

Proceedings of Jornadas de Psicologia Ambiental. (1992) Orellana. See Corraliza (Ed) (1993).

Proshanski, H.M. (1976) 'Appropiation et non appropiation (Misappopiation) de l'espace'. In: Korosec. (1976)

Proshansky, Ittelson & Rivlin. (Eds.) (1970) *Psicología ambiental. El hombre y su entorno físico*. México. Trillas.

Ragon, M. (1975) *L'homme et les villes*. Paris. Albin Michel.

Rambaud, P. (1969) *Société rurale et urbanisation*. Paris. Seuil.

Ramírez, J. (1982) 'La participación pública en el urbanismo'. Consejo Regional Murciano.

Rapoport, A. (1979) 'An approach to the study of conflicts in space'. In: Simon (1979)

Rappaport, A.G. (1981) 'Categories of perceptual experience'. In: T.Niit et alt. (Eds.) (1981) *People and Environment: Psychological Problems*. Tallinn. EOOP. 85-86.

Rapport d'Activite (1991) Laboratoire de Psychologie de l'Environnement. Université René Descartes-Paris V. CNRS (URA 1270) Juin 1991

Raudsepp, M. (1981) *Image of Environment (Cognitive Maps) and the Factors Influencing its Development*. Tartu State Univ. Graduation Thesis (unpublished).

Raudsepp, M. & Semerov, A. (1991). 'Environmental Means for Regulating some Intergroup Relations'. In Niit, Raudsepp & Liik. 1991. Tallinn. pp.127-134.

Raymond, H. (1976) 'Quelques aspeccts practiques et théoriques sur l'appropriation de l'espace'. In: Korosec (1976)

Raymond, H. & Haumont, N. (1966) *L'habitat pavillonaire*. Paris. Inst. Sociologie Urbaine.

Rees, K. & Canter, D. (1979) 'Comparing Married Couples' Satisfaction with Their Housing Environment'. In: Canter (Ed) (1979)

Remesar. (Ed.) (1981) *Ecología*. Barcelona. Facultat de Belles Arts (Mimeo)

Remesar, A. & Hernández, F. (1982) *Lecturas de Psicología y Entorno*. Facultad de Bellas Artes. Univ. de Barcelona.

Remesar, A., Hernández, F. & Marce, F. (1980) *Ecologia*. Facultat Belles Arts. Univ. de Barcelona (Mineo)

Riba, C. (1982a) 'Conocimiento del entorno y conducta en el entorno'. In: Riba, C. et alt. (1982) Publicacions de la Facultat de Belles Arts.

Riba, C. (1982b) *Color y entorno natural*. Publicacions de la Facultat de Belles Arts.

Riba, C., Hernández, F. & Remesar, A. (1982a) 'Cognitive Maps and Conductual Maps'. In: Pol, Muntañola, Morales (Eds.) (1984) *Man-Environment. Qualitative Aspects*. Proceedings of 7th. IAPS Confer. Edic.Univ.Barcelona.

Riba, C., Hernández, F. & Remesar, A. (1982b) 'Interaccions Ambientals i agrupacions espacials al pati d'una Guarderia: Anàlisi d'un registre fotogràfic.' Barcelona. Paper in 7th. IAPS Conference.

Riba, C., Hernández, F. & Remesar, A. (1984) 'Interaccions ambientals i Agrupacions espacials al pati d'una Guarderia: Anàlisi d'un registre fotogràfic'. In: Pol, Morales, Muntañola (1984)

Riba, C., Hernández, F. & Remesar, A. (1984)' Mapes Cognitius i Mapes de Conducta: Aportacions als diferents nivells del problema'. In: Pol, Muntañola, Morales (1984)

Ribot. (1889) 'Discurs en el 1er. Congrés Internacional de Psicología'. Quoted by Montoro (1982)

Richards, B.A. (1976) 'Rapport sur le séminaire consacré à L'appropriation des espaces institutionnels''. In: Korosec (1976)

Richet. (1889) 'Discurs en el 1er. Congrés Internacional de Psicología'. Quoted by Montoro (1982)

Ridruejo, P. (1981) 'Semántica del Clima Social'. República Dominicana. Actas del XVIII Congreso Interamericano de Psicol.

Ridruejo, P. (1983) 'Potenciales Implicativos del clima Social: su Sintàctica'. In: J.R.Torregosa et al. (Eds.) (1983) *Perspectivas y contextos en Psicología Social*. Barcelona. Hispano Europea.

Riley, R.B. (1984) 'The Concept of Garden: Some Theoretical Questions About Its Place in Person/Enviro. Interaction'. In: Krampen (1984)

Rips, H. & Leipik, J. (1981) 'Nature trail-environment for exploratory recreation'. In: T.Niit et alt. (Eds.) (1981) *People and Environment: Psychological Problems*. Tallinn. EOOP. 223-6.

Rivlin, L.G. (1976) 'Quelques problèmes des résidences institutionnelles'. In: Korosec (1976)

Robinson, J.W. et alt. (1984) 'Architectural parameters of normalization: an exploratory study'. In: Krampen (1984)

Rohrmann, B. (1984) 'Grundsätzliche Probleme bei der Nutzung sozialwissenschaftlicher Daten für umweltpolitische ...'. In: Krampen (1984)

Romhild, T. (1984) 'Die Genesis Der Künstlichen Beleuchtung - Versuch Einer Entwicklungstheorie Symbolbildungpro'. In: Krampen (1984)

Rossi, A.(1966) *La Arquitectura de la Ciudad*. Barcelona. Gustavo Gili.

Rovira, A. (1992). 'El Psicólogo Ambiental en el Ambito Empresarial'. In Corraliza Ed. 1993.

Rubio, V. (1987)' Evaluación del Impacto Ambiental'. In: R. Fernández-Ballesteros (Coord.) (1987), pp. 184-202.

Ruggeri, R. & Giuliani, M. V. (1992) 'Building renewal: a case study'. *5th International*

Research Conference on Housing. Montreal.

Sachenco, M.R. (1980) 'The clubs'. In: (1980) *Architecture of the Public Buildings.* Moscow. Stroiizdat.

Sachenco, M.R. (1981) 'Six kinds of architectural space-time'. In: T.Niit et alt. (Eds.) (1981) *People and Environment: Psychological Problems.* Tallinn. EOOP. 47-52.

Säisä, J., Svensson-Gärling, A., Gärling, T, & Lindberg, E. (1986) 'Intraurban cognitive distance: The relations between judgments of straight-line distances travel distances and travel times'. *Geographical Analysis, 18,* 167-174.

Salmaso, P., Baroni, R., Job, R. & Mainardi Peron, E. (1981) 'Aims, attention and natural settings. An investigation into memory for places'. *Italian Journal of Psychology.*

Salmaso, P. & Baroni, R. & Job, R. & Mainardi Peron, E.(1983) 'Schematic information, attention and memory for places'. *Journal of Environmental Psychology: Learning, Memory, and Cognition.*

Salmaso, P. & Legrenzi, P. (1984) 'Categorization of Environments: An Examination of Prototypes and Memory for Places'. In: Krampen (1984)

Samie, A. (1976) 'L'appropriation de l'espace dans les gares de banlieue parisienne'. In: Korosec (1976)

Sánchez-Robles, J.C. (1976) *La conceptualización Social de la Vivienda.* Tesis Doctoral. Escuela de Arquit. Univ. Valencia.

Sánchez Robles, J.C. & Ros, J.L. (1981) *El Almudin de la Ciudad de Valencia.* Valencia. E.T.S. Arquitectura.

Sancho, J.M. & Hernández, F. (1981a) *Interacción ambiental en el parvulario.* I.C.E. Universidad de Barcelona.

Sancho, J.M. & Hernández, F. (1981b) *Interacción ambiental en el aula.*

Sancho, J.M. & Hernández, F. (1981c) 'Análisis de la Interacción ambiental en el aula. Una investigación en el parvulario'. *Infancia y Aprendizaje.* 16, (4)

Sandrisser, B. (1984) 'Aesthetics and the Elderly'. In: Krampen (1984)

Sandstrom, C. (1970) 'What do we perceive in perceiving?'. In: Honikman (1970)

Sandstrom, S. (1973) 'A sociocultural theory on aesthetic visual estimation and use'. In: Kuller (1973)

Sangrador-García, J.L. (1981) 'Medio Construido y Conducta Social'. In: F. Jimenez-Burillo (Ed.) (1982) *Psicología y Medio Ambiente.* Madrid. CEOTMA. 423-450.

Sangrador-García, J.L. (1986) 'El medio Físico construido y la interacción social'. In: Jiménez-Burillo & Aragonés (Eds.) (1986).

Sanoff, H. (1984) 'Participatory Design of Community Arts Facilities'. In: Krampen (1984)

Sanoff, H., Adams, G. & Walker, Ch. (1984) 'Shaping school Environments for Adolescents'. In: Pol, Morales, Muntañola (1984)

Sanoff, J. & Sanoff, H. (1984) 'Designing Children's Environments Through Telecommunications Technology'. In: Krampen (1984)

Sansot. (1972) *La poètique de la ville.* Paris, Klincksieck.

Sansot, P. (1976) 'Notes sur le concept d'appropriation'. In: Korosec (1976)

Sansot, P. (1984) 'Nous vieillirons bien ensemble'. In: Krampen (1984)

Sauer, L. (1984) 'Joining Old and New: Open Space in the Making of a Neighborhood'. In: Krampen (1984)

Saup, W. (1984) 'Nouvel Environnement de logement dans le troisième âge: strain stress, copin'. In: Krampen (1984)

Saura, A. (1984). *L'Edifici Escolar.* Tesis de Licenciatura. Univ. Autónoma de Barcelona.

Saura, M. (1976) 'Le Musée-Théâtre Dali'. *NEUF.* Bruxelles. 61.

Saura, M. (1980) 'Prólogo al libro de J. Muntañola'. *La Arquitectura de los Setenta.*

Saura, M. (1981) 'La Palestra de Iradier'. *On.* Barcelona. 23.

Schmidt, A.J. (1979) 'Participatory design with a scale model- a primer'. In: Simon (1979)

Schmidt, J.A. (1984a) 'Stadtgestaltung Nach Prinzipien Des Lebendigen'. In: Krampen (1984)

Schmidt, J.A. (1984b) 'Urban Design In Accordance With Natural Principles'. In: Krampen (1984)

Schneider, G. & Graumann, C.F. (1984) 'Urban Identity And Identification'. In: Krampen (1984)

Schorr (1975) 'Lotkas Law in the Humanities'. *Jasis*. Quoted by Montoro (1982)

Schreiber, K. (1984) 'Von Der 'Alien' Zur 'Neuen' Politik - Die Raketen-Diskussion Als Motor Des Soziao-Kulturellen'. In: Krampen (1984)

Schubert, H.J. (1984) 'Zu gründen und folgen der selbest-steuerung von beziehungen zur umwelt....'. In: Krampen (1984)

Schultz-Gambard, J. (1973) 'Social determinants of crowding'. In: W.A. Le conte & M.R. Gürkaybak (Eds.), *Human Consequences of Crowding*. N.Y.: Plenum.

Schultz-Gambard, J. (1979)' Social Determinants of Crowding'. In: W.A. LeComte & M.R. Gürkaynak (Eds.), *Human Consequences of Crowding*. New York: Plenum.

Schultz-Gambard, J. (1983) 'Crowding: Dichte und Engeals Gegenstand angewandter sozialpsychologisher Forschung'. In_ J. Haisch (Ed.), *Angewandte Sozialpsychologie*, pp. 171-193. Bern, Switzerland: Huber.

Scott, J.K. (1984) 'Community Design Research'. In: Krampen (1984)

Seager, C.P. (1973) 'Problems arising through transfer of psychiatric care from large mental hospitals...'. In: Küller (1973)

Seassaro, L. (1980) 'Perché ed in che architectura ed urbanistica possono riferirsi alla psicologia'. *Bolletino della Ricerca*. 1, 49-64.

Secchiaroli, G. (1979) 'Progrettazione dell'ambiente per l'uomo e psicologia ambientale: una nuova prospettiva?'. *Psicologia Contemporanea*. 33, 30-35.

Secchiaroli, G. & Bonnes-Dobrowolny, M. (1980) 'La psicologia ambientale tra indagine teorica e ricerca applicata'. *Psicologia Italiana*. 2, 11-6.

Segaud, M. (Ed.) (1992) **'Le propre de la ville: practiques et symbolismes'** Editions de l'Espace Européen, Paris

Seidel, A.D. (1984) 'Myths in Local Goverment Toward Urban Growth and the Physical Environment'. In: Krampen (1984)

Seiwert, M. & Krampen, M. (1984) 'Development and Validation of Scales on Environmental experiences'. In: Krampen (1984)

Selby, B. (1970) 'Environments and Meaning - A Linguistic Model'. In: Honikman (1970)

Semple, R.Mcl. & Phillips, S.D. (1979) 'Individual Differences in Environmental Dispositions within a Faculty of Environmental Studies'. In: Canter (Ed) (1979)

Seredyuk. (1979) *Perception of Architectural Environment*. Lvov. Visa Skola.

Serrano, J. (1992) '¿Hacia una Psicología Social Postmoderna?'. In: 3th. Encuentro Luso-Español de Psicología Social. Lisboa: 15-16 December.

Servais, E. & Lienard, G. (1976) 'Espace habité et ethos de classe'. In: Korosec (1976)

Setala, M-L. (1984) 'The Culture of Children's Plays and Games from the Perspective of Environmental and Ecolo.Psych'. In: Krampen (1984)

Shafer, E.L. & Richards, T.A. (1974) 'A Comparison of Viewer Reactions to Outdoor Scenes and Photographs of those Scenes'. In: Canter & Lee (1974)

Sharon, E. (1984) 'Evaluation of an Architects-in-Schools Program'. In: Krampen (1984)

Shemyakin, F.N. (1962) 'Orientation in space'. In: B.G. Anan'yev & col. (Eds.) *Psychological Science in the U.S.S.R.* Washington D.C. Office of Technical Services.

Shemyakin, F.N. (1971) 'Psychological aspects of perception of a city'. In: P.S.Mejlaha (Ed) *Artistic Perception*. Mocow. Nauka. 309-24.

Shimizu, H. (1979) 'The Structure of Image in Central Districts'. In: Canter (Ed) (1979)

Shinnick, T.G. & Brebner, J.M.T. (1979) 'Nurses' Subjective Evaluation of Hospital Ward Design'. In: Canter (Ed) (1979)

Siguán, M. (1958) *Del Campo al Suburbio*.

Siguán, M. (1981) *Història de la Psicologia a Catalunya*. Barcelona. Edicions 62.

Sime, J.D. (1979) 'The Use of Building Exits in a Large-Scale Fire'. In: Canter (Ed) (1979)

Sime, J.D. (1984) 'Creating Places or Designing Spaces: Place Affiliation and Architectural Design'. In: Krampen (1984)

Sime, J.D. (1992) Recorded personal interview, by Pol. Thessaloniki, July 1992.

Simmel. (1903) 'The Metropolis and Mental Life'. *Social Science III.Selections and Selected*

Readings. Chicago. University of Chicago Press (1984)

Simon, C. & Simon, J. (1984) 'Housing and Community Design Responses to the Canadian Climate'. In: Krampen (1984)

Simon, J.G. (1979) 'Preface'. In: Simon, J.G. (Ed) (1979)

Simon, J.G. (Ed) (1979) *L'Experience conflictuel du l'espace*. 4 IAPC. Louvain-la-Neuve. CIACO.

Simon, J.G. (1984) 'Le desire de construire et son image publicitaire'. In: Krampen (1984)

Simon, J.G. (1981) 'Recorded Personal Interview, by Pol'. Barcelona.

Simon, J.G. (1984) 'Recorded Personal Interview, by Pol'. Cluny.

Simon, J.G. & Delvaux, L. (1979) 'Analyse de la charge sémantique de l'habitation parlée'. In: Simon (1979)

Simon, J. et alt. (1984) 'Residential Design Integrating Social-Cultural Aspects into the Basic Program'. In: Krampen (1984)

Sivik, L. (1973) 'General and applied research on colour perception. A review of current Swedish projects'. In: Kuller (1973)

Sivik, L. (1974a) 'Color meaning and perceptual color dimensions: A Study of color samples'. *Goteborg Psychological Reports*. 1.

Sivik, L. (1974b) 'Color meaning and perceptual color dimensions:A Study of exterior colors'. *Goteborg Psychological Reports*. 11.

Sivik, L. (1974c) 'Measuring the meaning of colors:Problems of semantic bipolarity'. *Goteborg Psychological Reports*. 13.

Sivik, L. (1974d) 'Measuring the meaning of colors:Reliavility and Stability'. *Goteborg Psychological Reports*. 12.

Sivik, L. (1974e) 'Studies in color meaning'. *Goteborg Psychological Reports*. Goteborg. University of Goteborg14.

Sivik, L. (1975) 'Studies of color meaning'. *Man-Environment Systems*. 5, 155-60.

Skouteri-Didaskalou, E. (1984) 'Mothers' in, Fairies 'Out: on Some Contradicting Socio-Sexual Conceptualizations of Space'. In: Krampen (1984)

Smith, P.F. (1979) 'Architecture: the inavoidable art'. In: Simon (1979)

Smith, P.F. (1984) 'A Taxonomy of Architectural Aesthetics'. In: Krampen (1984)

Smith, P.K. (1974) 'Aspects of the Playgroup Environment'. In: Canter & Lee (1974)

Smith, P.K. (1979) 'Experimental Studies of How the Pre-School Environment Affects Behaviour'. In: Canter (Ed) (1979)

Sokolov, S.I. (1971) 'Psychological aspects of the perception of the city'. In: B.S.Meilakh (Ed) *Art Peception*. Mocow. Nauka.

Soler. (1984) *El factor humano el conducción de vehículos automóviles*. Valencia. Tesis Doctoral. Facultad de Psicología.

Solzhenkin, V.V. (1980) 'Distances in interpersonal interaction'. *Social Interaction in the light of Reflection Theory*. Frunze. Ilim. 61-69.

Solozhenkin, V.V.et al. (1981) 'Factors of social regulation of behavior and parameters of socio-physical environment..'. In: T.Niit, et al. (Eds) (1981) *People and Environment:Psychological Problems*. Tallin. EOOP. 115-19.

Sombart. (1907)' Der Bergriff der Stadt und das Wesen der Städtebeildung'. *Archiv für Sozialwinssenschaft und Sozialpolitik*, 25, 1-9.

Sommer, R. (1974) *Espacio y Comportamiento Individual*. Madrid. ILEAL.

Sorte, G.J. (1971) 'Perception av landskap'. Vollebeck. Norwegian University of Agriculture. Unpub. Thesis.

Sorte, G.J. (1973) 'Significance of components in environmental settings'. In: R.Kuller (Ed.) *Architectural Psychology . Proceedings of the Lund Conference*. Stroudsburg. P.A.Dowden, Hutchinson & Ross, 198-210.

Sorte, G.J. (1975) 'Methods for presenting planned environment'. *Man-Environment Systems*. 5, 148-54.

Sorte, G.J. (1981) *Visually discernible qualities of components in the build environment*. Stockholm. The Swedish Council for Building Research.

SPIEE (1983) *Interacció Ambiental en els Espais d'Esbarjo de l'Escola.* I.C.E. Univ. Barcelona.

Stafford, P.B., Henderson, N. & Frush, J. (1984) 'Nursering Home Ethnography'. In: Krampen (1984)

Stea, D. & Taphanel, S. (1974) 'Theory and Experiment on the Relation Between Environ. Modelling (Toy Play)and Environ.Cognition'. In: Canter & Lee (1974)

Stern, W. (1903). 'Angewandte Psychologie'. In: *Beiträge zur Psychologie der Aussage.* Vol.1, pp.1-45. Leipzig, FRG: Barth.

Stern, W. (1930) 'Studien zur persohwissen schaft'. In: Teil, I. (Ed.) *Personalistik als wissenschatf.* Leipzig. Barth.

Stern, W. (1938) *General psychology from the personalistic standpoint.* New York: Macmillan.

Stilitz, I.B. (1969) 'Pedestrian Cogestion'. In: Canter (1969)

Stockdale, J.E. & Hale, D.A. (1979) 'Subjective Crowding: Conceptual Dimensions and Preferences'. In: Canter (Ed) (1979)

Stokols, D. (1977) *Perpectives on Environment and Behaviour.* New York-London. Plenum Press.

Stokols, D. (1978) 'Environmental Psychology'. *Annual Review of Psychology.* 29: 253-295.

Stokols, D. (1985) 'Recorded personal interview'. California. Univ. Irvine.

Stringer, P. (1969) 'Architecture, Psychology, the game's the same'. In: Canter (1969)

Stringer, P. (1970) 'Spatial Ability in Relation to Design Problem Solving'. In: Honikman (1970)

Stringer, P. (1974) 'Individual Differences in Repertory Grid Measures for a Cross-Section of Female Population'. In: Canter & Lee (1974)

Suárez, E., Martínez, J.& Hernández-Ruíz, B.(1992). 'Definición de un Modelo Explicativo de la Conductas Proambientales'In: Corraliza (Ed.) 1993

Summerer, S. (1984) 'Umweltethik Als Handlungsimpuls'. In: Krampen (1984)

Sureda, J. (1984) 'La Educación ecológico-ambiental y la Educación integral'. *Educació i Cultura.* 4.

Sureda, J. & Picornell, C. (1980) 'Educació i espai. Pedagogia Ambiental i Educació Ambiental'. *Educació i Cultura.* Palma. 1.

Sureda, J., Rios, P. & Picornell, C. (1983) 'La incorporació de l'Educació Ambiental en l'ensenyament bàsic i obligatori'. Sitges. Primeras Jornadas sobre Educación Ambiental.

Sureda, J., Rios, P. & Picornell, C. (1983) 'La ciutat: medi i objecte d'Educació Ambiental. En el cas de Palma'. Sitges. MOPU & Diputació de Barcelona.

Sureda, M. (1990,). *Pedagogía Ambiental.* Anthropos.

Syme, G.J. et alt. (1979) 'Identification of Early Leavers from a Remote Mining Community'. In: Canter (Ed) (1979)

Symes, M. (1984) 'Designes Involvemont in Design'. In: Krampen (1984).

Symes, M. & Worthington, J. (1979) 'The conflict over standards in the use of workspace'. In: Simon (1979)

Szecsenyi, J. (1984) 'Wahrnehmung Und Bewertung Von Wohnhäusen - Eine Vergleichende Studie Zur Erfassung Relevanter..'. In: Krampen (1984)

Tadeusz, G. (1984) 'Functioning of Selected Values, Needs and Decisions in the Situation of Ecological Pollution'. In: Krampen (1984)

Tagg, S.K. (1973) 'The use of multidimensional scaling type techn.in the structuring of the archit.psych. of places'. In: Küller (1973)

Tagg, S.K. (1979) 'Some Theoretical Facets of the Multivariate Methods of Environment Psychology'. In: Canter (Ed) (1979)

Takahashi, T. (1984) 'Seven Wonders of the Japanese House'. In: Krampen (1984)

Tappuni, R.R. (1979) 'Social Interaction and the Spatial Concept of the House in Baghdad'. In: Canter (Ed) (1979)

Teixidor, M.J. (1982) *València, la construcció d'una ciutat.* Valencia. Diputación Provincial.

Teklenburg, J. & Timmermans, H. & Van Wagenberg, A. (1991) ' Space syntax: standardised integration measures and some simulations'. Eindhoven University of Technology.

Teklenburg, J. & Timmermans, H. (1992) 'The distribution of use of public space in urban areas'. *Eindhoven University of Technology.*

Terradas, J.A. (1983) 'Desarrollo histórico y concepción actual de la educación ambiental'. Sitges. MOPU & Diputación de Barcelona.

Teymur, N. (1982) 'Economic signification of Physical Surroundings'. In: Pol, Muntañola, Morales (Ed.) 1984.

Teymur. (1982) *Environmental Discourse.* London. Blackwell Press.

Teymur. (1984) 'Economic signification of physical surroundings'. In: Pol, Muntañola, Morales (1984)

Thorne, R. (1984) 'The Environmental 'Psychology' of Theachers and Movie-Palaces 1902 to 1930'. In: Krampen (1984)

Thorne, R. & Turnbull, J.A.B. et alt. (1979) 'High-Rise Living for Low-Income Elderly: Satisfaction and Design'. In: Berlin (1979)

Thronberg, J.M. (1976) 'L'identité du corps et l'appropriation des lieux'. In: Korosec (1976)

Thurwald. (1904)'Stadt und Land im Lebensprozeß der Rasse'. *Archiv für Rassen: und Gesellschaftsbiologie, 1*, 718-735.

Thurstone. (1947) *Multiple-factor analysis.* Chicago. University of Chicago Press.

Timmermans, H. (1992) 'Retail environments and spatial shopping behavior'. *University of Tecnology, Eindhoven.*

Tognoli, J. & Hornberger, F. (1984) 'Images of Household Roles and Children's Television Programs'. In: Krampen (1984)

Tolman, E. (1948) 'Cognitive Maps in rats and men'. *Psychological Review.* 55: 198-208.

Tomas, J.L. (1983) *Una Aplicación del diferencial semántico en el campo del arte.* Tesis de Licenciatura. Univ. de Valencia.

Tong, D. & Wishart, J. (1984) 'Meeting Psychological Criteria in the Design of Fire Alarm Systems'. In: Krampen (1984)

Tortosa, F. (1980) 'Evolución de la Psicología en España en el siglo XX'. *Revista de Historia de la Psicología.* Valencia. 1, (3-4): 353-391.

Tortosa, F. (1985) 'Recorded personal Interview, by Pol'. Valencia.

Touraine, A. (1969) *La société post-industrielle.* Paris. Denoel.

Trieschmann, G.V. (1979) 'Dynamic Design Language'. In: Canter (Ed) (1979)

Turnbull, J.A.B. (1984) 'The Hability of Communal Spaces in High Rise Buildings for the Elderly'. In: Krampen (1984)

Turner, J.A. (1979) 'Successful design for public authorities'. In: Simon (1979)

Tzamir, Y. & Churchman, A. (1984) 'The Use of the Realm Knowledge by Architecture Students - an Ethical Perpective'. In: Krampen (1984)

Uexkül. (1909) *Streifzuge durch die Umwelten von Tieren und Menschen.* Hamburg. Rowohlt.

Universal Decimal Classification (1969) 'British Standards Institution'. English full.

Uppenbrink, M. (1984) 'Das umweltbundesamt im Umweltschutz der letzten zehn Jahre'. In: Krampen (1984)

Urbina, J. & Ortega, P. (1984) 'La organizaciómambiental de los centros de desarrollo infantil y sus efectos sobre las intercciones'. In: Pol, Morales, Muntañola (1984)

Uzzell, D. (1979) 'Participation and Representation: A Psychological Interpretation'. In: Canter (Ed) (1979)

Valach, L. & Kalbermatten, U. (1984) 'A paper-directed action, environment and the protection of the environment'. In: Krampen (1984)

Valera, S. (1991) 'Satisfacción residencial en el centro histórico de Barcelona' In R. de Castro 1991

Valera, S. (1992) 'El simbolismo del espacio urbano. Consideraciones teóricas' in Corraliza (Ed.) 1993

Valera, S., Freixas, Mollevi,J. & Pol, E. (1990).'The Image of the Districts os Barcelona II'. In proceedings XI IAPS.

Valera, S. & Iñiguez, L. (1991) 'Evaluación ambiental' in R. de Castro 1991

Valera, S. & Pol, E. (1992). 'The Imatge of the Districts of Barcelona. A Theoretical Approach.'In proceedings XII IAPS.

Valsiner, J. & Heidmets, M. (1976) 'Density and spatial behavior: Some problems of environmental Psychology'. *Eesti Loodus*. 172, 33-37, 89-95.

Van Hoogdalem, H. (1976) 'Quelques outils conceptuels pour l'analyse de systèmes homme-environnement'. In: Korosec (1976)

Van Hoogdalem, H; Prak, N.L.; Van der Voordt T.J.M. & Van Wegen H.B.R (Eds.) (1988), *Looking back to the future.IAPS 10/1988*. Vol. II, Delft, Delft University Press

Van der Wurff, A., Stringer, P.F. & Timmer, F. (1984) 'Feelings of insafety in residential environments: integrating divergent research'. In: Krampen (1984)

Vanderweiden, J. (1984) 'Unplanned planning. Social processes cripling physical planning'. In: Krampen (1984)

Ventalló, J. (1980) *Les escoles populars ahir i avui*. Barcelona, Nova Terra.

Vendenik, F. & Hart, R. (1984) 'Individual differences in how children code and decode enviro. meanings through different media'. In: Krampen (1984)

Veshniski, V.G. (1979) *New city districts-through the eyes of the inhabitants*. Mosproektovec.

Veyssiere, M. & Levy-Leboyer, C. (1979) 'An Epidemiological Noise Study'. In: Canter (Ed) (1979)

Villela Petit, M. (1976) 'Espace approprié - espace appropriant'. In: Korosec (1976)

Vincienne. (1972) *Du village a la Ville. Les système de mobilité des agriculteurs*. Paris. Mouton.

Voelker, W.J. (1984) 'Identification of alternative techniques for improving office work station illumination'. In: Krampen (1984)

Voltelen, M. (1973) 'On various ways to insight'. In: Kuller (1973)

Voss, K.F. (1976) 'Une étude psychologique sur les caractéristiques d'un modèle d'habitation intégré pour des étud'. In: Korosec (1976)

Voyé, L. (1976) 'Production et appropriation de l'espace: régles et modalités différentielles'. In: Korosec (1976)

Voye, L. (1979) 'Concepteurs, usagers et légitimité du bâti. Dynamyque des discordances'. In: Simon (1979)

Wade, F.M., Chapman, A.J. & Foot, H.C. (1979) 'Child Pedestrian Accidents: The Influence of the Physical Environment'. In: Canter (Ed) (1979)

Wallinder, J. (1973) 'Borderline against the future. Place of the theory of architectural psych.in tomorrow's planning'. In: Küller (1973)

Walters, D. (1969) 'Annoyance due to railway noise in residential areas'. In: Canter (1969)

Warfield, J.P. (1984) 'Cultural Saturation: A methodological lavoratory in environmental design'. In: Pol, Muntañola, Morales.

Watts, J. & Smith, M. (1979) 'User Participation in the Early Stages of Design'. In: Canter (Ed) (1979)

Wedin, C.S., Avant, Ll.L. & Wolins, L. (1973) 'Communication of the residential spaces by architectural graphics'. In: Kuller (1973)

Weinberg, A.M. (1961) 'The impact of the large-scale science on the United State'. *Science*. 134, 161-164.

Weinberg, A.M. (1967) *Reflexions on Big science*. Cambridge, MassMIT. Press.

Weinstock. (1971) 'Citation index'. In: *Enciclopedia of library and infortonal science*. New York. M.Dekker.

Wells, B.W.P. (1964) *Office design and the office worker*. Liverpool. Ph.D. Thesis (Unpublished)

Wells, B.W.P. (1965a) 'Subjetive responses to the lighting installa. in modern office building and their design inplic.'. *Buildings Science*. 1: 57-68.

Wells, B.W.P. (1965b) 'The psycho-social influence of building environment'. *Building Science*. 1: 153-165.

Wells. (1965) 'Towards a definition of environmental studies- A psychologist's contribution'. In: *Architects Journal*.

Weresch, K. (1984) 'Farben in der Architektur in Ihrem Bedeutungswandel als Anzeiger Langfristiger ....'. In: Krampen (1984)

Wineman, J. (1979) 'Resident Response to the Design of Waterfront Communities'. In: Canter (Ed) (1979)

Wrona, S. (1979) 'Some Aspects of the Decision-Making Process in Design Participation'. In: Canter (Ed) (1979)

Wyon, D.P. (1970) 'Studies of children under imposed noise and heat stress'. *Ergonomics*. 13, 598-612.

Wyon, D.P. (1973) 'The role of the environment in buildings today: thermal aspects (factors affecting the choice...'. *Build International*. 6, 39-54.

Wyon, D.P. (1974a) 'Noise in dwellings'. *Build International*. 7, 1-15.

Wyon, D.P. (1974b) 'The effects of moderate heat stress on typewriting performance'. *Ergonomics*. 17, 309-18.

Wyon, D.P., Andersen, I. & Lundqvist, G.R. (1972) 'Spontaneous magnitude estimation of thermal disconfort during changes in the ambient temperature'. *Journal of Hygiene*. Cambridge. 70, 203-21.

Wyon, D.P., Andersen, I. & Lundqvist, G.R. (1979) 'The effects of moderate heat stress on mental performance'. *Scandinavian Journal of Work, Environmental and Health*. 5, 352-61.

Wyon, D.P., Fanger, P.O. et al. (1975) 'The mental performance of subjects clothed for confort at two different air temperatures'. *Ergonomics*. 18, 359-74.

Wyon, D.P., Lofberg, H.A. & Lofstedt, B. (1975) 'Environmental resarch at the climate laboratory of the National Swedish Inst. of Build. Research'. *Man-Environment Systems*. 5, 107-200.

Wyon, D.P. & Nilsson, I. (1980) 'Human experience of windowless environments in factories, offices shops and colleges in Sweden'. Oslo, Norway. Paper presented at the 8th. CIB Congress.

Xhingnesse, L.V. & Osgood, Ch.E. (1967) 'Bibliographical citation characteristics of the psychological journals network in 1950 and 1960'. *American Psychologists*. Quoted by Peiró i Carpintero (1981) 22, 778-791.

Yamada, Y. (1984) 'Factors influencing the 'noisiness' of sound (an investigation of the personality of sub...'. In: Krampen (1984)

Yanitsky, O. (1991).'Environmental Movements and the Emergence of Civil Society in the Soviet Union"'In Niit, Raudsepp & Liik. 1991. Tallinn. (Estonia)

Young, D. (1979) 'Less of the lip, man'. *Architectural Psychology Newsletter*. 9, (3): 46.

Young, D.W. (1979) 'Semiotics in Place'. In: Canter (Ed) (1979)

Young, D. (1980) 'The iconoclast who isn't'. *Architectural Psychology Newsletter*. 9, (4): 24.

Ziman. (1968) *Public knowledge. The social dimension of science*. Cambridge. University of Cambridge Press.

Zimring, C., Weitzer, W. & Knight, C.R. (1979) 'Control and Behaviour: An Institution for the Developmentally Disabled'. In: Canter (Ed) (1979)

Zrudlo, L.R. (1979) 'Differing aspirations in user participation programs'. In: Simon (1979)

Zwirner, W.G. (1984) 'On site visits, over lunch or elsewhere: the behavioural action spaces of agents for environ...'. In: Krampen (1984).